ADVANCES IN
CARBENE CHEMISTRY

Volume 2 • 1998

ADVANCES IN CARBENE CHEMISTRY

Editor: UDO H. BRINKER
Institut für Organische Chemie
Universität Wien
Wien, Austria

VOLUME 2 • 1998

JAI PRESS INC.

Stamford, Connecticut   *London, England*

CONTENTS

LIST OF CONTRIBUTORS

Roland Bonneau

URA 348 du CNR
Université de Bordeaux 1
Talence Cedex, France

Udo H. Brinker

Institut für Organische Chemie
Universität Wien
Wien, Austria

Peter Chen

Laboratorium für Organische Chemie
ETH-Zentrum
Zürich, Switzerland

Maitland Jones, Jr.

Department of Chemistry
Princeton University
Princeton, New Jersey

Boleslaw Kozankiewicz

Institute of Physics
Polish Academy of Sciences
Warsaw, Poland

Michael T.H. Liu

Department of Chemistry
University of Prince Edward Island
Charlottetown, P.E.I., Canada

Eva Migirdicyan

Laboratoire de Photophysique
    Moléculaire du CNRS
Université de Paris-Sud
Orsay Cedex, France

Matthew S. Platz

Department of Chemistry
The Ohio State University
Columbus, Ohio

Murray G. Rosenberg

Department of Chemistry
State University of New York at Binghamton
Binghamton, New York

Hideo Tomioka

Chemistry Department for Materials
Mie University
Mie, Japan

John P. Toscano

Department of Chemistry
Johns Hopkins University
Baltimore, Maryland

John Warkentin

Department of Chemistry
McMaster University
Hamilton, Ontario, Canada

PREFACE

Beginning as chemical curiosities, carbenes are now solidly established as reactive intermediates with fascinating and productive research areas of their own. Five decades of divalent carbon chemistry have provided us with a vast repertoire of new, unusual, and surprising reactions. Some of those reactions, once classified as exotic, have become standard methods in organic synthesis. These highly reactive carbene species have been harnessed and put to work to achieve difficult synthetic tasks other reactive intermediates cannot easily perform.

The fruitful relationship between experiment and theory has pushed carbene chemistry further toward the direction of reaction control; that is, regio- and stereoselectivity in intra- and intermolecular addition and insertion reactions. The interplay between experiment and modern spectroscopy has led to the characterization of many carbenes that are crucial to both an understanding and a further development of this field.

Monographs devoted entirely to carbenes appeared in the late sixties, such as the classic works of Hine and of Kirmse. Second-generation surveys published in the seventies consisted of the series of three volumes edited by Jones and Moss. At the end of the eighties, the status of such knowledge was collected into the two imposing volumes of Houben-Weyl (*Methoden der Organischen Chemie*; Regitz, editor).

Our understanding of carbene chemistry has advanced dramatically, especially in the last decade, and new developments continue to emerge. Some of the recent exciting findings have been collected in the first volume of *Advances in Carbene Chemistry*. With the second volume, the series will continue to provide a periodic coverage of carbene chemistry in its broadest sense—leading into the next century.

Udo H. Brinker
Series Editor

1,2-HYDROGEN MIGRATION IN CARBENES:

LASER-FLASH PHOTOLYSIS AND BEYOND

Roland Bonneau and Michael T.H. Liu

ABSTRACT

In this chapter, we consider the rearrangement of singlet carbenes by 1,2-H migration to generate an alkene. After a short historical introduction, we discuss the kinetic aspects of the problem: The methods used for measuring the rate constant of the

Advances in Carbene Chemistry
Volume 2, pages 1–28
Copyright © 1998 by JAI Press Inc.
All rights of reproduction in any form reserved.
ISBN: 1-55938-837-4

rearrangement are presented, and tables give some of the results obtained, including some very recent ones. These results are then discussed in terms of the effect of substituents on the frequency factor and the activation energy of the reaction.

In the second section, we consider the mechanistic problems arising when carbenes are generated from diazirines, as appears to be the case for most of the studies cited in the second section. Several reaction pathways have been proposed as possible sources of the rearranged alkene formed by the 1,2-H migration in the carbene: a rearrangement in the excited state of the diazirine, the decomposition of a carbene-olefin complex, and the decomposition of a diazo intermediate generated in parallel to the carbene. Recent results indicate that the two former processes coexist and that their efficiency to produce the rearranged alkene is strongly dependent on the system investigated. The diazo intermediate is also formed with a yield depending strongly on the nature of the diazirine, but the nature of the products resulting from its decomposition is not established.

The conclusions summarize recent progress made in the field and lists some of the unsolved questions.

I. INTRODUCTION

The 1,2-shift of a hydrogen atom to the carbene center, which generates an alkene, is one of the most common rearrangement reactions of singlet carbenes such as alkylhalocarbenes, alkylmethoxycarbenes, or some dialkylcarbenes.

$$Y\text{-}CH_2\text{-}\ddot{C}\text{-}X \rightarrow Y\text{-}CH{=}CHX$$

Y = H, alkyl, phenyl, phenoxy, . . .   X = Cl, Br, F, OMe, alkyl, . . .

Prior to 1988 there was no experimental determination of absolute rate constants and activation energies for the intramolecular rearrangement. Energy barriers to this rearrangement, ranging from 0 to 27 kcal/mol, had been deduced from theoretical calculations.[1-6] The original zero value,[1,2] obtained by the semiempirical MINDO method, was considered unreliable due to the known deficiency of MINDO in favoring cyclic structures. *Ab initio* methods gave large values,[4,5] but more sophisticated calculations, involving polarized functions and correlation effects, reduced the value of the energy barrier to 2.1 kcal/mol.[6]

Experimentally, work on 1-aryl-2-diazopropanes[7] and examination of the vibrational structure of the vinylidene photoelectron spectrum[8] have produced evidence of a definite barrier for 1,2-H migration, even though no value has been assigned. The products derived from the thermolysis of 4-diazirinopentanoic acid gave an estimate of 1.1 ± 1 kcal/mol for the height of the barrier to 1,2-H shift in a dialkylcarbene.[9]

During the past decade, much progress has been made and several papers have reviewed this subject.[10-12] This chapter summarizes the major results and offers details on some recent developments on the 1,2 hydrogen migration in carbenes.

II. KINETIC ASPECTS OF THE PROCESS

A. Methods for Measurement

1. *Competition "Cyclopropanation/Rearrangement" and Product Analysis*

In 1985, Liu and coworkers[13,14] studied the decomposition of 3-chloro-3-ben-zyldiazirines in the presence of tetramethylethylene (TME). They discovered that the intramolecular hydrogen transfer to yield (Z)- and (E)-β-chlorostyrenes competed with cyclopropanation, and that the ratio between the products of intermolecular and intramolecular reactions should therefore reflect the ratio of the rate for cyclopropanation (k_r [TME]) and rearrangement (k_i). This result gave, with some reasonable assumption on the values of k_r, the first estimate for the value of k_i.

By examining the effect of temperature on the competition between inter- and intramolecular reactions of benzylhalocarbenes, Liu and Subramanian[15] estimated the barrier to be 6.4 and 4.7 kcal/mol in Ph-CH$_2$-C-Cl and Ph-CH$_2$-C-Br, respectively. Although these values may have errors of ±2 kcal/mol, they provided a valuable beginning for evaluating the factors that control this intramolecular rearrangement.

2. *Pyridinium Ylide Probe Technique*

The availability of powerful commercial lasers, transient digitizers, and other electronic components offers the organic photochemist attractive opportunities to engage in laser flash experiments as a means of understanding carbene reactivity. Every species that lacks a strongly absorbing chromophore, i.e., one with a molar absorption coefficient larger than a few 10^2 M^{-1}cm^{-1} in a spectral region accessible to UV spectroscopy, is invisible to laser experiments with UV-Vis detection. This difficulty was overcome by the pyridinium ylide probe technique developed by Jackson et al.[16] who studied the reaction of phenylchlorocarbene and *t*-butylchlorocarbene with pyridine. This new approach permitted the measurement of the first absolute rate constant of the rearrangement of an alkylchlorocarbene.

The pyridinium ylide probe technique may be used in two different ways, relying on the measurement of either the kinetics or the yield of formation of the pyridinium ylide.

   a. Direct (or Kinetic) Method. In the kinetic method, the analysis of the growth of the ylide absorption gives a pseudo-first-order rate constant, $k_{obsd} = k_1 + k_y$ [pyridine] where k_1 and k_y are, respectively, the rate constants for the decay of the carbene in the absence of pyridine and for the reaction of the carbene with pyridine. The value of k_1 is determined by plotting k_{obsd} *vs* [pyridine] and extrapolating to [pyridine] = 0. In most cases, k_1 can be identified with the rate constant of the 1,2-H shift, k_i. This method requires a flash photolysis set-up with a time resolution that is at least three times shorter than the lifetime of the carbene. For instance, laser

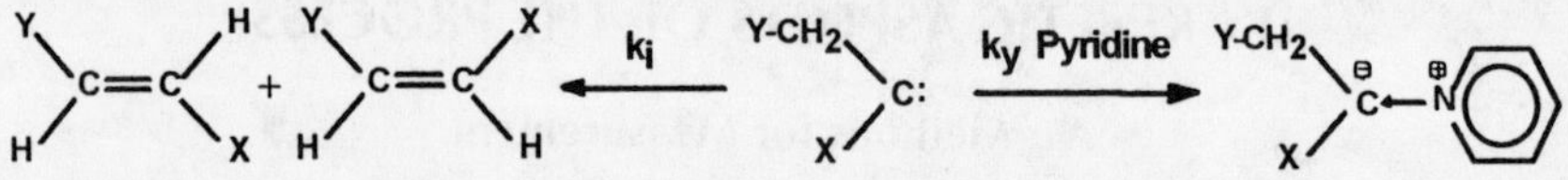

Scheme 1.

flash photolysis of 3-chloro-3-benzyldiazirine in isooctane[17] produces benzylchlo-rocarbene, which reacts with pyridine to form an ylide that is observable by its strong absorption at 380 nm or undergoes 1,2-H shift to form (Z)- and (E)-β-chlo-rostyrenes (Scheme 1). At 25 °C, $k_i = 6 \times 10^7$ s^{-1} ($\tau_0 = 17$ ns), and $k_y = 8 \times 10^9$ M^{-1}s^{-1}, close to diffusion controlled, so that the value of k_{obsd} to be measured increases from 6.4×10^7 s^{-1} ($\tau = 15.6$ ns) to 10^8 s^{-1} ($\tau = 10$ ns) when [pyridine] increases from 0.5 to 5 mM. This requires a set-up with a time resolution of ≤ 3 ns and accuracy in the measurements, because the change of the measured value, τ, is only 50% on the full range of investigation. A higher limit for the concentration of pyridine would indeed extend the range of variation of τ (2.2 ns at [pyridine] = 50 mM), but would require a time resolution of ≤ 0.7 ns. A lower limit for the values of [pyridine] would not lend much to the dynamic of the phenomenon but would raise new problems because the yield of formation of the ylide (and thus the size of the signal) would become very small.

b. Stern–Volmer Type. In this method, the amplitude of the ylide absorption is measured as a function of [pyridine] with everything else (excitation and analysis characteristics, temperature, solvent, and other variables) being constant. The amplitude of the absorbance due to the ylide, A, is then proportional to the yield of conversion of the carbene into ylide, Φ. With $\mathcal{P}$ as the proportionality factor and $\Phi = k_y$ [pyridine] / ($k_i + k_y$ [pyridine]), the relation $A = \mathcal{P}\Phi$ can be rearranged to $A^{-1} = \mathcal{P}^{-1} + k_i/(\mathcal{P}k_y)$ [pyridine]$^{-1}$. The ratio of slope/intercept of the straight line obtained by plotting A^{-1} vs [pyridine]$^{-1}$ therefore gives the ratio k_i/k_y from which k_i may be estimated with a reasonable assumption on the value of k_y.

The pyridinium ylide is usually quite stable and lasts in the solution for millisec-onds. Consequently, this method may be used with relatively simple flash pho-tolysis systems, with microsecond time resolution, to obtain an estimate of the rate of 1,2-H migration in carbenes with lifetimes as short as a nanosecond. The main disadvantage in this process is that the value of k_i depends on the value of k_y, which, measured by the direct method (Section II.A.2.a) for several carbenes, appears to be strongly dependent on the substituants attached on the carbene center as shown in Table 1. Values of k_y between 10^9 and 10^{10} M^{-1}s^{-1} are common for benzyl- and alkyl-chlorocarbenes, and dialkyl- and arylalkyl-carbenes, but the range of vari-ation was extended down to 10^8 M^{-1}s^{-1} with a series of akylchloro and fluorocar-benes;[18] values ranging from 7×10^4 to 3.5×10^7 M^{-1}s^{-1} have been reported for a series of phenyloxacarbenes.[19] The changes in the values of k_y from 1.4×10^{10} to

Table 1.

Carbene	$k_{yl}\ (M^{-1}s^{-1})$	Ref.
CH_3-C-Cl	8.8×10^9	a
CH_3-C-OCH$_2$CF$_3$	1.6×10^7	b
CH_3-C-OCH$_3$	6.6×10^5	b
$(CH_3)_3$C-C-Cl	4.2×10^9	c
	2.4×10^9	d
c-Propyl-C-Cl	4.7×10^8	d
	7.4×10^8	e
c-Propyl-C-OCH$_2$CF$_3$	1.0×10^6	b
c-Propyl-C-OCH$_3$	$< 10^5$	b
Ph-CH$_2$-C-Cl	7.5×10^9	d
$(CH_3)_3$CCH$_2$-C-F	2.1×10^9	f
Ph-C-Cl	3.4×10^8	g
	7.6×10^8	h
	1.5×10^9	b
Ph-C-F	1.5×10^9	b
Ph-C-Br	1.2×10^9	b
Ph-C-OCOCH$_3$	3.5×10^7	i
Ph-C-OCH$_2$CF$_3$	1.8×10^6	i
Ph-C-OCH$_3$	7×10^4	i
homocuban-9-ylidene	$> 10^9$	j
adamantanylidene	1.5×10^6!	k
	5.6×10^9	l

Key:
a) Liu and Bonneau, *J. Am. Chem. Soc.* **1989**, *111*, 6873
b) Moss et al., *J. Chem. Soc. Chem. Commun.* **1994**, 1479
c) Moss et al., *J. Am. Chem. Soc.* **1990**, *112*, 5642
d) Liu, Bonneau, *J. Am. Chem. Soc.* **1989**, *111*, 5973
e) Moss et al., *J. Am. Chem. Soc.* **1989**, *111*, 6875
f) Moss et al., *J. Am. Chem. Soc.* **1992**, *114*, 959
g) Platz et al., *J. Am.Chem. Soc.* **1988**, *110*, 5595
h) Platz et al., *J. Org. Chem.* **1991**, *56*, 1694
i) Moss et al., *J. Am. Chem. Soc.* **1994**, *116*, 1583
j) Platz et al., *J. Am. Chem. Soc.* **1990**, *112*, 7794
k) Platz et al., *J. Am. Chem. Soc.* **1991**, *113*, 2782
l) Bonneau, R.; Hellrung, B.; Liu, M.T.H.; Wirz, J. (submitted for publication in *J. Photochem. Photobiol. (A: Chem.))*. A similar value is given by Platz et al. in a manuscript submitted to the same Journal.

$1.2 \times 10^5\ M^{-1}s^{-1}$ for a series of 16 carbenes have been rationalized by Moss,[20] who used the frontier orbital theory.

M.S. Platz and coworkers used the Stern–Volmer type method extensively. It has several other interesting applications, such as the evaluation of the rate constant for reaction of an "invisible" carbene with a reactant[16,21] (or its determination if k_1 or k_y is known) and the evaluation of the quantum yield of formation of a carbene from the excited state of its (diazirine) precursor.[22]

3. Time-resolved Photo-acoustic Calorimetry

A large amount of energy is released into the solution during the 1,2-H shift rearrangement. The alkene product is much more stable than the carbene, by ≈ 50 to 60 kcal/mol. Time-resolved photo-acoustic-calorimetry (PAC)[24–26] monitors this heat release with time resolution of a few nanoseconds, thereby providing a direct measurement of the kinetics of the rearrangement without any requirement for the UV absorption properties of the carbene. In these experiments, the carbene is produced by the photolysis of a precursor (usually a diazirine) by a nanosecond laser pulse. Consequently, there will be at least two waves of heat release. The first one, corresponding to the difference in energy between the carbene and the excited state of the precursor, is quite fast because the lifetime of this excited state is less than 1 ns. The second one, due to the rearrangement of the carbene, must be substantially slower to be analyzed independently of the first one. Lifetimes as short as 10 ns (i.e., $k_i = 10^8\,\mathrm{s}^{-1}$) have been reported in studies that used this method, which seems to be close to the limits of the technique in the present "state of the art."

4. Laser-flash Spectroscopy

We stated that "alkylchlorocarbenes . . . are invisible to laser experiments with UV-Vis detection," but this is not always true! In addition to the absorption band in the visible spectrum,[27] characteristic of singlet carbenes but too low in intensity to be used in laser-flash photolysis, benzylhalocarbenes present an absorption in the 250 to 320 nm range, with an absorption coefficient around $1000\,\mathrm{M}^{-1}\mathrm{s}^{-1}$ at 300 nm,[28] similar to the absorption of phenylhalocarbenes. This seems rather surprising because the phenyl ring and the carbene center are (in principle) isolated by a saturated carbon, but the assignment of this absorption to the carbene has been confirmed by i) the fact that all the kinetic measurements made by analyzing the decay of this absorption fit perfectly those obtained by analysis of the growth of the pyridinium ylide and ii) measurements combining IR and UV spectroscopies in rigid matrices.[29] A transient absorption around 230 nm, due to cyclopropylchlorocarbene, has also been observed by laser-flash photolysis.[30–31]

B. Results

1. Lifetimes

During the past few years, the lifetime of many alkylhalocarbenes has been measured at room temperature in fluid solution, using one or several of these methods. Those assumed to be limited by the rate of 1,2-H migration are given in Table 2.

Neglecting the reactions of the carbene with the precursor, the solvent, and the carbene itself (dimerization), which is certainly justified for carbenes with $\tau < 100$ ns, the rates of hydrogen migration are given by the reciprocal of the lifetime values.

Table 2.

Carbene	Lifetime (at room temperature)
CH_3-C-Cl	330 ns
CH_3-C-Br	190 ns
CH_3-C-C_6H_5	> 166 ns
t-Butyl-CH_2-C-Cl	70 ns
t-Butyl-CH_2-C-F	38 ns
Cl-CH_2-C-Cl	50 ns
CF_3-C_6H_4-CH_2-C-Cl	45 ns
Cl-C_6H_4-CH_2-C-Cl	23 ns
C_6H_5-CH_2-C-Cl	18 ns
C_6H_5-CH_2-C-Br	≈18 ns
Me-C_6H_4-CH_2-C-Cl	≈8 ns
MeO-C_6H_4-CH_2-C-Cl	≈6 ns
C_6H_5-O-CH_2-C-Cl	9 ns
CH_3-CH_2-C-Cl	≈10 ns
CH_3-CH_2-C-C_6H_5	2.8 ns
C_2H_5-CH_2-C-Cl	≈3.6 ns[a]
$(CH_3)_2$CH-C-Cl	≤2 ns[a,b]
C_6H_5-CH(CH_3)-C-Cl	1 ns[a]

Notes: [a]These values are too short to be measured by any of the methods presented above. They are obtained by extrapolation at room temperature of measurements made between –40 °C and –100 °C, in fluid isooctane solution.

[b]Measurement of the lifetime of this carbene was nearly impossible because not only its lifetime is very short, even at –100 °C, but also the signal (pyridinium ylide absorption) is very small and hardly emerges from the noise.

It turns out that most of these carbenes do not rearrange with blinding speed and, therefore, an activation energy must slow down the process as predicted by the theoretical studies cited in the introduction. This phenomenon was confirmed by studies of the temperature dependence of the lifetime of several of these carbenes.

2. Kinetic Parameters

Let us assume that 1,2-H migration is the only (or the main) process controlling the carbene lifetime, τ, and that its rate constant, k_i, is correctly described by an Arrhenius type equation: $k_i = 1/\tau = A_i \exp(-E_i/RT)$. The kinetic parameters, activation energy E_i and preexponential factor A_i, measured for some carbenes are given in Table 3.

The results given in Tables 2 and 3 require some discussion.

a. *Activation Energy and By-stander Assistance.* Comparing the lifetimes of $C_6H_5CH_2$-C-Cl and C_6H_5-CH(CH_3)-C-Cl or those of CH_3-C-Cl, C_2H_5-C-Cl,

Table 3.

Carbene	$log_{10}(A_i)$	E_i (kcal/mole)	Method[i]	T(°C) range
CH_3-C-Cl	9.7	4.9[a]	PAC	+11 to +61
$C_6H_5CH_2$-C-Cl	11.1	4.8[b]	direct LFP	–5 to +30
CF_3-C_6H_4-CH_2-C-Cl	10.9	4.9[c]	direct LFP	+2 to +47
Cl-C_6H_4-CH_2-C-Cl	10.9	4.5[c]	direct LFP	–3 to +40
C_6H_5-O-CH_2-C-Cl	10.1	2.8[d]	kin P.Y.	–10 to +24
$(CH_3)C$-CH_2-C-F	8.9	3.3[e]	kin P.Y.	+20 to +61
cyclobutyl-C-F	8.5	3.8[e]	kin P.Y.	+20 to +61
CH_3-C-Cl	10.7	6.8[f]	kin. P.Y	–70 to +80
n-C_3H_7-C-Cl	10.5	3.6[g]	kin. P.Y.	–30 to –100
i-C_3H_7-C-Cl	(10.3)	$\approx$2[g]	kin. P.Y.	–50 to –100
C_6H_5-CH_2-C-Cl	11.9	5.8[h]	kin P.Y.	–80 to +30
C_6H_5-$CH(CH_3)$-C-Cl	10.7	2.7[g]	kin P.Y.	–40 to –100

Notes: [a]Ref. 25b
[b]Ref. 17
[c]Ref. 54
[d]Chateauneuf, J.E. and Liu, M.T.H. *J. Org. Chem.* **1991**, *56*, 5942
[e]Ref. 18
[f]From the linear Arrhenius plot obtained when using the values from Fig. 1 in *J. Am. Chem. Soc.* **1993**, *115*, 10424 (Dix, E.J.; Herman, M.S.; Goodman, J.L.) and recalculation of k_i with $k_{lim} = 6.2 \times 10^5$ s^{-1}
[g]Ref. 32
[h]Ref. 36
[i]PAC: Photoacoustic calorimetry; LFP: Laser-flash photolysis kin. P.Y.: kinetic pyridinium ylide.

n-C_3H_7-C-Cl, and i-C_3H_7-C-Cl, it is clear that methyl substitution on the carbon atom in α-position of the carbene center does largely increase the rate of the 1,2-H migration. Considering the lifetimes of CH_3-C-Cl and $C_6H_5CH_2$-C-Cl, it appears that the substitution of one H in the methyl group of CH_3-C-Cl by a phenyl has a similar effect. This has been discussed by Nickon[10] as "by-stander assistance." The increase of the rate of rearrangement upon substitution by a methyl has been used as an argument to support the theoretical prediction that the 1,2-H migration rearrangement proceeds by migration of a hydride ion, H$^-$.

The first experimental measurements failed to clarify the reason or reasons for this increase in the rate: Was it due to a lowering of the activation energy or to an increased frequency factor? It was only known that the lifetimes of C_2H_5-C-Cl, n-C_3H_7-C-Cl, i-C_3H_7-C-Cl and $C_6H_5CH(CH_3)$-C-Cl were $\leq$ 10 ns, too short to be measured, and the first measurements of the kinetic parameters for CH_3-C-Cl ($\log(A)$ = 9.7; E_a = 4.9 kcal/mol) and $C_6H_5CH_2$-C-Cl ($\log(A)$ = 11.3; E_a = 4.8 kcal/mol) indicated a change in the frequency factor, contrary to expectations.

Recent results have demonstrated that this assumption was erroneous. The temperature dependence of the lifetime of C_6H_5-$CH(CH_3)$-C-Cl, monitored by its UV absorption at 300 nm, and of n-C_3H_7-C-Cl, by the kinetic pyridinium ylide technique, has been recently[32] measured by laser-flash photolysis in fluid isooctane at low temperature, from –40 to –100 °C. The resulting E_a values, respectively 2.7

and 3.6 kcal/mol for C_6H_5-CH(CH$_3$)-C-Cl and n-C$_3$H$_7$-C-Cl, are clearly lower than for the unsubstituted homologous, C_6H_5-CH$_2$-C-Cl and CH$_3$-C-Cl, respectively 4.8 and 4.9 kcal/mol (or rather, as shown below, 5.8 and 6.8 kcal/mol), whereas the frequency factors are similar.

Theoretical chemistry seems unable to predict, quantitatively at least, the changes in the value of E_a induced by structural modifications. For instance, the values of E_a calculated for the reaction CH$_3$-C-X $\rightarrow$ CH$_2$=CHX are, respectively, 0.6, 4.7, 11.5, 19, or 27 kcal/mol when X = H, Me, Cl, F, or OMe.[33] The experimental value for CH$_3$-C-Cl is about twice as low (4.9 or 6.8) than predicted (11.5) and, in the series of alkyl-chloro and fluoro-carbenes studied by Moss,[18] the rates for rearrangement of fluorocarbenes are only a few times slower than those for the corresponding chlorocarbenes. The difference in the activation energies is $\approx$ 1 or 2 kcal/mol, instead of the 7.5 kcal/mol predicted by calculations.

b. Frequency Factor.

b. Frequency Factor. These frequency factors are surprisingly low. Reported values are in the range of 10^{11} to 10^{10} s^{-1}, two or three orders of magnitude lower than expected, considering that only one degree of freedom is frozen in the transition state. First measurements gave an especially low A factor, $10^{9.7}$ s^{-1}, for the long lived CH$_3$-C-Cl ($\tau \geq$ 330 ns at room temperature). More recent experiments with an extended temperature scale indicated that several processes contribute to the decay of this carbene.[34] Similar results have been obtained for C_6H_5-CH$_2$-C-Cl.[36] In both cases, the Arrhenius plots, $\log(1/\tau)$ vs $1/T$, are not linear, but show a pronounced curvature in the low T region. In fact, for the lowest temperatures, τ tends to a limit of around 1.6 µs in the case of CH$_3$-C-Cl in heptane. For C_6H_5-CH$_2$-C-Cl, the limit is around 110 ns in isooctane and 55 ns in decalin but, in chloroform, the Arrhenius plot shows no curvature between +20 and −60 °C. The reciprocal of this limit corresponds to the rate constant of a process that is either independent of (or very slightly dependent on) the temperature.

The Arrhenius plots of $\log(k_i) = \log(1/\tau - k_{lim})$ vs $1/T$ are indeed linear over the whole temperature range studied and give E_a values larger than those determined in previous experiments by 1 or 2 kcal/mol. Frequency factors increased by about one order of magnitude. The new set of A values now appears consistent, ranging from $10^{10.5}$ to $10^{11.9}$, with some logic in the changes due to substitution in the alkyl and benzyl series: A decrease of the A factor, by 33% between CH$_3$-C-Cl and R-CH$_2$-C-Cl and by 50% between C_6H_5-CH$_2$-C-Cl and C_6H_5-CHR-C-Cl, is indeed expected because of the number of H atoms that may undergo migration. The experimental A factors are still much smaller than expected, and thus, $\Delta S^{\ddagger}$ is much larger than predicted by theory.

Regarding the process responsible for the limitation of the carbene lifetime at low temperature, in the case of CH$_3$-C-Cl, this process was assigned to a 1,2-H shift by quantum mechanical tunneling.[34] This explanation has been "supported" by theoretical calculations,[35] and yet it seems strange: The deuterium kinetic isotope effect (in fact, the ratio of the lifetime of CD$_3$-C-Cl and CH$_3$-C-Cl), which is 1.8 at

70 °C, becomes around unity (1.0 ± 0.1) below 0 °C, i.e., in the region where the process assigned to tunneling becomes predominant. It is generally accepted that the deuterium kinetic isotope effect should be very important for a reaction controlled by tunneling!

In the case of $C_6H_5\text{-}CH_2\text{-}C\text{-}Cl$, this process was assumed to be a reaction with the solvent (a 1,2-H shift by an exchange mechanism between the carbene and the solvent and/or an insertion of the carbene in a C–H bond of the solvent), because the value of $k_{\lim}$ was found to be strongly dependent on the solvent.[36] Here again, the deuterium kinetic isotope effect decreases from 2.6 at 30 °C to about 1.0 ± 0.1 for temperatures lower than −35 °C. If the reaction responsible for the limitation of the carbene lifetime at low temperature is indeed a reaction with the solvent, it is easily understood that, in the low temperature region where this process is predominant, the lifetime of both benzyl- and methyl-chlorocarbenes is unsensitive to deuteriation.

The hypothesis of an H-exchange reaction with the solvent has been recently confirmed by isotopic analysis of the styrenes obtained by rearrangement of benzylchlorocarbene in octane-d_{18} at low temperature, showing incorporation of a D atom in the rearrangement product.[37] Also, photolysis of benzylchlorodiazirine in isooctane at −80 °C resulted in the detection of 5 to 10% of products of insertion of the carbene in the solvent. (See Note 1 Added in Proof)

III. MECHANISTIC PROBLEMS

A. Diazirines as Source for Carbenes

In most (if not all) of the experimental studies performed to get the above results, the carbene was generated by photolysis of a diazirine. As long as interest is limited to the kinetic study of the carbene by UV spectroscopy, either by direct observation of the carbene or by the kinetic pyridinium ylide techniques, diazirines are a perfectly convenient source of carbene for laser flash photolysis measurements. Several types of lasers (e.g., N_2, excimers, frequency-tripled Nd, frequency-doubled dye) can provide excitation in their absorption band in the 300 to 370 nm range. Because the value of molar absorption coefficient is low, usually around 100 or less, high concentrations of diazirine (from a few mM to several 100 mM, depending on the experimental set-up) must be used to obtain an appreciable absorption of the laser excitation. This presents some advantages, but also one disadvantage: the reaction of the carbene with the parent diazirine, to give an azine, must be taken into account, especially when the carbene lifetime is long and the diazirine concentration is large.

If interest is also centered on the analysis of the products of rearrangement of the carbene or on the study, based on products analysis, of the competition between rearrangement and intermolecular reactions, the use of diazirines as a source of carbene gives rise to several problems. The following discussion recapitulates the nature of these problems and explicates some recent progress achieved in this field.

1. *Differences between Photolysis and Thermolysis*

In the introduction of the chapter, *Absolute Kinetics of Intramolecular Alkylcarbene Reactions*, in Vol. 1 of this series, Moss identified the two major unsolved questions: "(1) the possible intervention of carbene/olefin complexes and (2) the extent of carbene-mimetic reactions that occurs via the excited states of the carbene precursors instead of in the carbenes themselves."

As early as in the 1960s, several studies revealed that decomposition of alkyldiazirines with heat and light yields different mixtures of alkenes.[38–40] Thus, there must be at least two pathways that lead to alkene rearrangement products. One route is obviously the relaxed carbene; but what is the other? A vibrationally hot carbene, a diazo compound, or an excited state of the precursor? What is the possible role of the diazirine excited state?

2. *Theoretical and Experimental Studies of the Diazirine Excited State*

Platz and co-workers have presented the most convincing data for excited state chemistry of diazirine photolysis.[41] Pulsed laser excitation of 3,3-dimethyl-diazirine (DMD) in pentane produces a highly structured fluorescence spectrum, with a mirror-image relationship between the emission and the fluorescence excitation spectra. A similar observation was made with other dialkyldiazirines, such as cyclohexyldiazirine, adamantyldiazirine, and alkylmethyldiazirines where alkyl = ethyl, *i*-propyl, and *t*-butyl. A study of the temperature dependence of the fluorescence intensity gave an activation energy ≈ 1 kcal/mol for DMD (lower for the other dialkyldiazirines investigated) for the non-radiative deactivation processes of the fluorescent state. If the frequency factor of the non-radiative process or processes is around 10^{12} s^{-1}, the fluorescence lifetime at room temperature should be shorter than 0.03 ns. On the other hand, the $S_0 \leftrightarrow S_1$ transition of diazirines is poorly allowed (absorption coefficient ≈ 100, oscillator strength $f \approx 0.002$), so that $k_f \approx \nu_0 \times f \approx 10^6$ s^{-1}. Since the fluorescence quantum yield, $\phi_f = k_f \tau$, is ≈ 1.5 × 10^{-4} for adamantyldiazirine in isooctane,[42] the lifetime of the excited singlet of this diazirine must be around 0.15 ns. (See Note 2 Added in Proof)

Although the excited state populated by excitation in the first absorption band, $^1n\pi^*$, has a very short lifetime, it is not dissociative; several processes may contribute to its non-radiative deactivation, each one possibly yielding a specific product. At least three of these processes must be considered as in Scheme 2: i) the formation of carbene, ii) the formation of a diazo, and iii) an extrusion of nitrogen concerted with a 1,2-H shift, the rearrangement in the excited state (RIES) mechanism.

Theoretical studies[43–45] provide a slightly different picture for the reactions of the excited diazirine. According to Jug,[44] the 3,3-dimethyldiazirine molecule excited in its S_1 ($n\pi^*$) state must overcome a small energy barrier (≈ 9 kcal/mol) to reach the S_2 ($n\sigma^*$) state, dissociative, which correlates directly with $(CH_3)_2$ $^1C{:}^*$

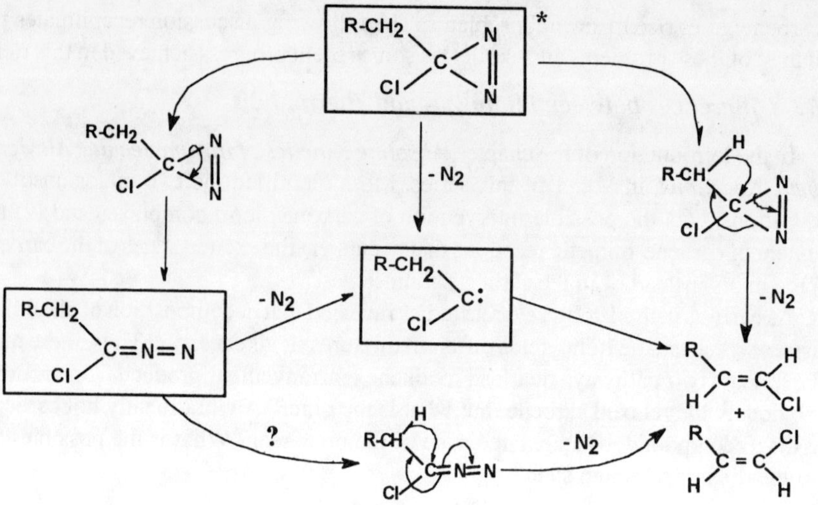

Scheme 2.

+ N_2 so that the carbene should be produced in its excited state. But when the C-N distance is ≈ 2.2 Å, the surface connecting the diazirine S_2 state and $(CH_3)_2$ $^1C{:}^*$ + N_2 comes very close to the one connecting the diazirine ground state to $(CH_3)_2$ $^1C{:}^*$ + N_2, so that some ground-state carbene would be produced by a radiationless transition from the excited-state surface. The percentage of ground-state carbene, determined by the efficiency of this process, could be sensitive to any modification of the molecular structure of the diazirine, such as isotopic substitution. This study did not consider the formation of a diazo compound.

In a recent study of the unsubstituted diazirine by *ab initio* methods,[45] a similar energy barrier ($E_{a1} \approx 12$ kcal/mol) was found between the relaxed S_1 and the crossing point to the S_2 state that is predicted to yield ground state $^1CH_2 + N_2$, *via* a conical intersection with the ground-state surface, in a diradicaloid $^1D_{\sigma\sigma}$ geometry. More importantly, the S_1 state is found to have another way of deactivation, with a much smaller activation energy ($E_{a2} \approx 0.5$ kcal/mol), to go to another conical intersection corresponding to a diradicaloid $^1D_{\sigma\pi}$ geometry that would give diazomethane. On the way between S_1 and $^1D_{\sigma\pi}$, a small fraction of the molecule could go to the $^1D_{\sigma\sigma}$ and thus give the ground state $^1CH_2 + N_2$ with a low yield.

According to this study: i) photolysis of the diazirine should not give the excited carbene, in defiance to Jug's predictions; ii) photolysis of the diazirine should give diazomethane almost exclusively, the observed carbene resulting from the decomposition of diazomethane. The discussion below shows that, at least for the 3,3-substituted diazirine, the second of these conclusions cannot be true. If the description of the system, shown in Scheme 3, is correct, either the difference between E_{a1} and E_{a2} is smaller than calculated, or the partitioning between the

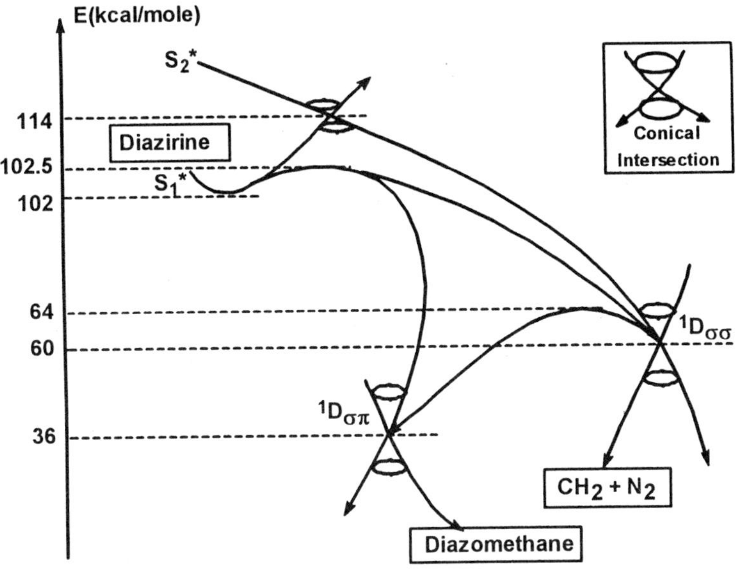

Scheme 3.

relaxation of S_1 to $^1D_{\sigma\sigma}$ and $^1D_{\sigma\pi}$ is much more equilibrated and even, in some cases at least, in favor of $^1D_{\sigma\sigma}$ (i.e., in favor of the formation of the singlet carbene).

None of these theoretical studies considered the RIES mechanism (extrusion of nitrogen concerted with a 1,2-H shift). It remains that the rearrangement products may originate from:

- a 1,2-H shift in the carbene formed directly from the excited diazirine;
- a 1,2-H shift in the carbene formed by thermal decomposition of the diazo compound;
- a rearrangement in the excited diazirine, concerted with extrusion of nitrogen;
- a rearrangement in the diazo compound, either concerted with extrusion of N_2 or acid catalyzed.

Part II of this chapter has demonstrated that much has been done during the past decade to study the first of these processes; little or nothing is known regarding the three other possible pathways. The formation of diazo compounds upon photolysis of diazirines in rigid matrices at low temperature has been reported,[46,47] but the quantum yield of formation or the thermal stability of these compounds has not been investigated. Platz has proposed the rearrangements, concerted with extrusion of nitrogen of the excited state of the diazirine and of the diazo, but whereas the former mechanism has been supported by some experimental measurements,[48] the latter suggestion seems only speculative for the moment.

B. RIES and COC as Sources for the 1,2-Hydrogen Migration Product

1. Arguments Favoring the RIES Mechanism

The RIES mechanism process has been proposed to rationalize the variations of the quantum yield of production of carbene in the photolysis of a series of alkylchlorodiazirines.[46] This quantum yield was estimated from the limiting value of the pyridinium ylide absorption when [pyridine] becomes very large, assuming that i) every carbene is then converted to ylide and ii) the same value for the absorption coefficient is maintained for the considered series of ylides. Upon photolysis of the corresponding diazirines, the quantum yields of formation of cyclopropyl-C-Cl, $(CH_3)_3C$-C-Cl, CH_3-C-Cl, CH_3CH_2-C-Cl, and $(CH_3)_2CH$-C-Cl were found to be in the ratio of 1:1:0.52:0.28:0.024, a proportion explained by the decreasing strength of the C–H bond in α-position of the carbene center (for the last three members of the series), which would make the RIES process increasingly more efficient.

The RIES process also provides a possible explanation for the differences in the products obtained under photolysis and thermolysis of diazirines,[38–40] because the nature or the stereochemistry of the products originating from this process (relevant only in the case of photolysis) may differ from those of the products formed via the carbene. For instance, as shown in Scheme 4, the thermal and triplet sensitized decompositions of t-butyldiazirine and of 1-diazo-2,2-dimethylpropane give the same distribution of products ($\approx$ 10% 2-methyl-2-butene resulting from 1,2-H shift and $\approx$ 90% 1,1-dimethylcyclopropane resulting from 1,3 C-H insertion). Direct irradiation and singlet sensitization of the same precursors yield a quite different distribution, around 50%:50%. The difference was assigned to multiplicity.[40] But it could be argued that the triplet diazirine, although in an excited state, cannot give RIES because this concerted mechanism should produce an energetically impossible triplet product. Therefore, the triplet diazirine would produce the carbene with a high efficiency, comparable to thermal decomposition, so that the 10%:90% distribution would be characteristic for the t-butylcarbene. On the contrary, the excited singlet diazirine would give 45% of 2-methyl-2-butene by

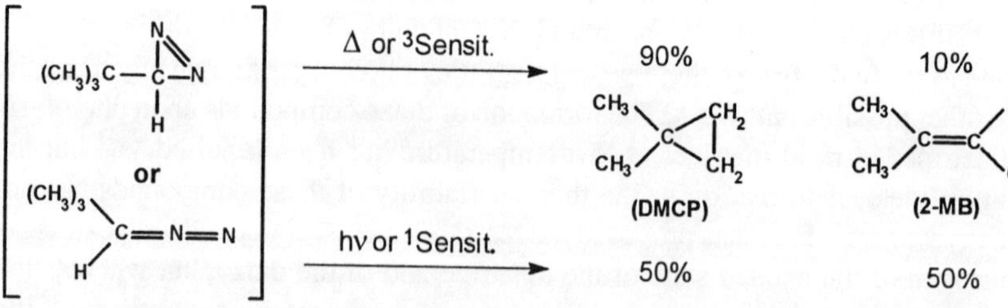

Scheme 4.

RIES and 55% of carbene, the latter giving the additional 5% of 2-methyl-2-butene and the 50% of 1,1-dimethylcyclopropane.

The RIES would be a source of olefinic rearrangement products unquenchable by efficient carbene traps such as olefins. When a diazirine is decomposed in the presence of an olefinic reactant, the simple mechanism shown in Scheme 5 predicts a linear relation between the ratio [3]/[2] and the concentration of the olefinic reactant, [olefin]. In this mechanism, the excited diazirine produces only the carbene (either directly or by both the direct and the diazo pathways), which then rearranges to give **2** (rate constant k_i) and adds to olefins (rate constant k_r) to give a cyclopropane, **3**. But, experimentally, the plots of [3]/[2] *vs* [olefin] are curved: At high concentrations of olefin reactant, the amount of **2** is larger than expected. Also, the distribution of the **2-Z** and **2-E** isomers depends on the concentration of olefinic reactant. The RIES may be the source of **2** in excess (with respect to the amount predicted by Scheme 3) and may also explain the changes in the ratio **2-Z/2-E.** As stated above, the stereochemistry of the products formed via the RIES and the carbene pathways may be different. But this phenomenon can also be explained by the formation of a carbene-olefin complex (COC), which then either collapses to cyclopropane **3** or decomposes after the rearrangement to **2** has occurred within the complex.[47–49] Turro et al.[50] had previously proposed the existence of a COC to explain the negative activation energy measured for the quenching of phenylchlorocarbene by olefins. But theoretical chemistry[51] has denied the existence of a COC as a minimum on the energy surface connecting the system carbene + olefin to the cyclopropane adduct, and this may be considered as a last argument in favor of the RIES mechanism. These calculations indicate a long-range interaction between a carbene and an olefin, yielding a well-defined orientation for the approach of the carbene, with the empty and filled orbitals of the singlet carbene pointing respectively to the filled π and empty π^* orbitals of the olefin. This is the geometry expected for a COC with a stabilization by charge transfer.

Kinetically, both the RIES and the COC mechanism account for the non-linearity of the plots of [3]/[2] *vs* [olefin]. During the last few years, there has been much controversy (at least in reviews of submitted papers) between those favoring one or the other of these interpretations.

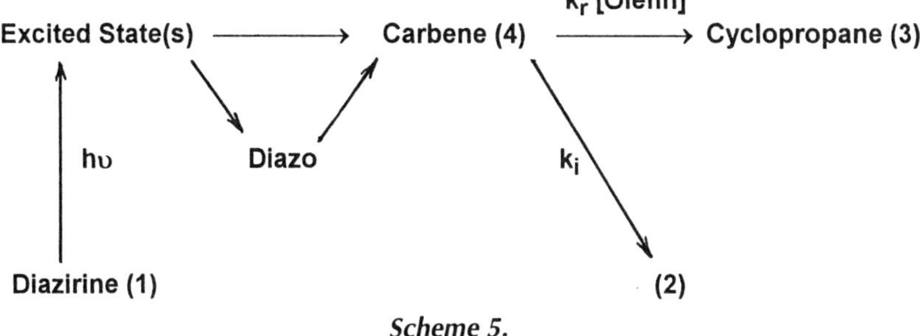

Scheme 5.

2. Arguments Favoring the COC

The existence of a COC is also supported by the following strong arguments:

1.   The curvature of the plots of [3]/[2] *vs* [olefin] may be observed under thermolysis of diazirine + olefin mixtures. Yet, under these conditions, the RIES mechanism cannot be invoked. This curvature is usually much less pronounced than that under photolysis, and sometimes it falls within the range of experimental incertitude. This is probably due to the fact that, at the high temperatures required for thermolysis, the rearrangement of the carbene by 1,2-H shift is so fast that it competes efficiently with the reaction of the carbene with the olefin, even for large values of [olefin]. A recent study of the system *n*-propylchlorocarbene + tetramethylethylene, showed the curvature to be very clear, by far beyond all possible experimental incertitudes.[52] However, see Note 3 Added in Proof.

2.   For a given diazirine (the 3-benzyl-3-chlorodiazirine), the amount of **2** in excess depends on the nature of the olefin.[49] This cannot be explained by the RIES process, the efficiency of which should depend only on the nature of the diazirine. It can be easily understood if the excess of **2** originates from a COC. This amount of **2** in excess is given by the extrapolation of the curves [3]/[2] *vs* [olefin] or, more easily, by the intercept of the plots of [2]/[3] *vs* [olefin]$^{-1}$. It amounts to around 20% or 30% with the very reactive TME, the unreactive 1-hexene, the electron deficient chloroacrylonitrile, and the electron-rich TME. However, it is around 60% with diethylfumarate and 70% with diethylmesaconate. The reason for these large values is not obvious, but might be due to a steric effect of the two ethylester substituents in *trans* position on the double bond. This steric factor would not affect the rate constant for carbene trapping by the olefin if the carbene–olefin distance is relatively long in the COC, but it might be quite decisive when the carbene–olefin distance must be strongly reduced to form the cyclopropane adduct.

3.   The negative activation energy measured for the quenching of phenylchlorocarbene by TME triggered the COC concept.[50] A negative activation energy, –4.7 kcal/mol, was also measured for the trapping of *p*-chlorobenzylchlorocarbene by TME.[54] In this case, there is an excellent agreement between the results obtained from the temperature dependence of the rate constant measured by laser-flash photolysis and that of the ratio $k_{(1,2\text{-H shift})}/k_{(\text{cyclopropanation})}$ obtained from products analysis, but the changes in the rate constant for diffusion on the relatively small temperature range (–3 to +25 °C) were not taken into account.[49] Quite recently, the measurement of the rate constant for the quenching of the α-methylbenzylchlorocarbene by TME, by LFP from 170 to 210 K, also yielded a negative activation energy, –2.7 or –1.4 kcal/mol, depending on the importance given to the effect of temperature on the rate constant for diffusion.[32] According to the Houk reaction model,[51] these negative activation energies could be explained in terms of free energies of activation strongly dominated by an unfavorable $T\Delta S^{\ddagger}$ term.

3. *Coexistence of RIES and COC*

Both sets of arguments seem rather convincing. But most of the results supporting the COC were obtained with benzylchlorocarbenes, whereas those favoring the RIES relied on alkylchloro-, or dialkyl-carbenes. Therefore, it seems quite plausible that both RIES and COC coexist, with their relative importances depending on the system being considered. Assuming that both the RIES and the COC mechanisms are efficient, Scheme 6 is yielded.

With $\alpha = k_c/(k_i^* + k_c)$ and $\beta = k_2/(k_i' + k_2)$ being the yields of formation of the carbene from the excited diazirine and of the cyclopropane from the COC, the quantum yield of formation of the 1,2-H shift product, **2**, is:

$$\Phi_{(2)} = (1 - \alpha) + \alpha k_i/(k_i + k_1[\text{TME}]) + \alpha(1 - \beta)k_1[\text{TME}]/(k_i + k_1[\text{TME}])$$

and the quantum yield of the cyclopropane, **3**, is:

$$\Phi_{(3)} = \alpha \beta k_1 [\text{TME}]/(k_i + k_1 [\text{TME}]).$$

The ratio $\Phi_{(2)}/\Phi_{(3)}$ gives, after simple mathematical calculations:

$$[2]/[3] = (1 - \beta)/\beta + (k_i^*/\beta k_C) + (k_i/\alpha\beta k_1)/[\text{TME}]$$

Under thermolysis, the RIES is irrelevant and, therefore, $k_i^* = 0$ and $\alpha = 1$. This leads to:

$$[2]/[3] = (1 - \beta)/\beta + (k_i/\beta k_1)/[\text{TME}].$$

If the COC does not exist, i.e., if $\beta = 1$, then $[2]/[3] = k_i/(k_1 [\text{TME}])$ under thermolysis, whereas $[2]/[3] = (k_i^*/k_C) + (k_i/\alpha k_1)/[\text{TME}]$ under photolysis.

In all cases, the plot of **[2]/[3]** *vs* 1/[TME] must be linear but i) the slopes of the plots obtained under thermolysis $(k_i/\beta k_1)$ and photolysis $(k_i/\alpha \beta k_1)$ must differ by a factor α; ii) the intercept, $([2]/[3])_\infty$ (value of [2]/[3] when [TME] $\rightarrow\infty$), must be larger under photolysis than under thermolysis, by an amount equal to $(k_i^*/\beta k_C) = (1 - \alpha)/(\alpha\beta)$ and iii) a non-zero value of the intercept under thermolysis demonstrates the existence of the COC mechanism.

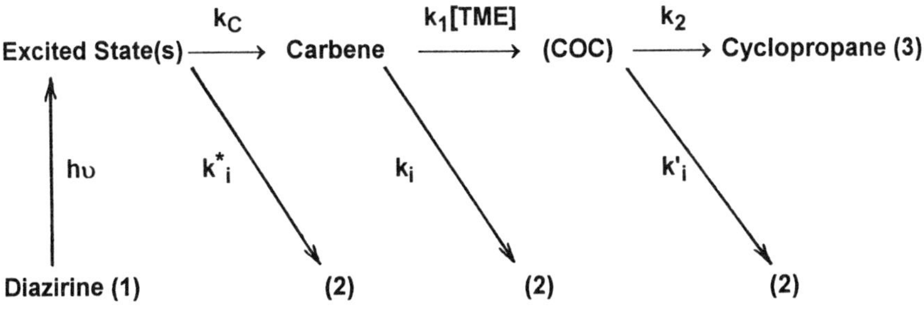

Scheme 6.

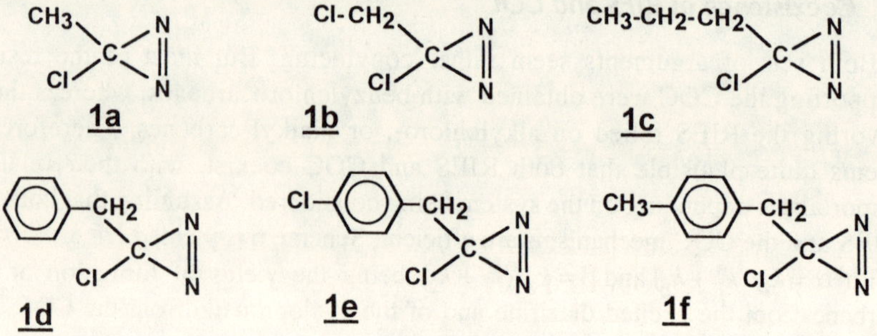

1a   **1b**   **1c**

1d   **1e**   **1f**

Scheme 7.

The amounts of the products **2** and **3** obtained under photolysis and thermolysis of the chloromethyl- and *n*-propyl-chlorodiazirines, **1b** and **1c**, have been measured at various temperatures as a function of [TME]. These data, plus similar ones previously obtained with several benzylchlorodiazirines, **1d** through **1f**, have then been analyzed according to the above equations. The results are presented in Figures 1 and 2.

In Figure 1, which shows the values of $([2]/[3])_\infty$ obtained from measurements under photolysis and thermolysis as a function of temperature, it is seen that i) the values are (nearly) independent of the temperature; ii) the values obtained under thermolysis are clearly $\neq 0$; iii) the difference between $([2]/[3])_\infty$, measured under photolysis and thermolysis, is around 0.96 and 0.50 for **1b** and **1c**, respectively, but negligible in the case of **1e**. The above equations give $\alpha \approx 0.54$, 0.72, and 1.00 and $\beta \approx 0.85$, 0.62 and 0.77 for **1b**, **1c**, and **1e**, respectively.

Similar values of α can be obtained from Figure 2 where the log of the slope of the plots [2]/[3] *vs* 1/[TME] is plotted *vs* $1/T$, for **1b**, **1c**, and **1e**. Since this slope is equal to $(k_i/\alpha\beta k_1)$ under photolysis and to $(k_i/\beta k_1)$ under thermolysis, the shift between the Arrhenius lines is equal to $\log(\alpha)$. For **1b** and **1c**, the amplitude of this shift gives $\alpha \approx 0.42$ and 0.62, respectively, whereas α must be very close to 1.00 for **1e** (and other benzylchlorodiazirines), because the data obtained under photolysis and thermolysis fall on the same line as previously noted.[49]

A quite different method confirmed these values of α—the time-resolved photoacoustic calorimetry (PAC), which monitors the amplitude and time evolution of heat depositions following photoexcitation.[24–26] Experimentally, deconvolution of the PAC waveforms obtained after photoexcitation of diazirines **1a** through **1c** in heptane indicates two heat depositions: The first one, ΔH_1, is fast ($\tau < 5$ ns) and reflects the formation of the carbene and vinyl halide **2** (*via* the RIES process) from the excited diazirine; the second one, ΔH_2, is slower, due to the formation of **2** by rearrangement of the carbene with a rate constant k_r similar to that determined by nanosecond absorption spectroscopy (Figure 3).

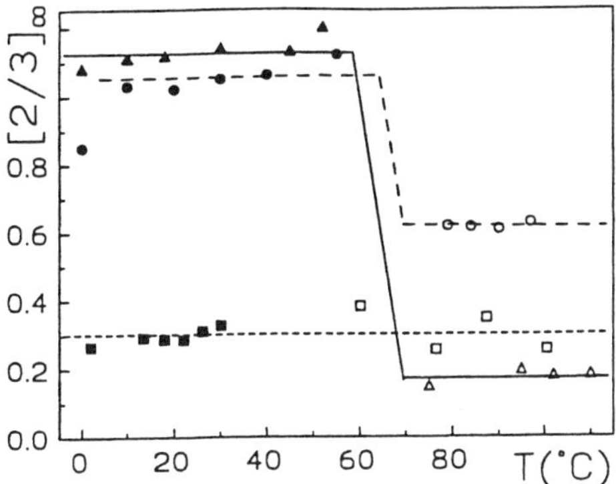

Figure 1. Values of $([2]/[3])_\infty$ *vs* temperature for thermolysis (open symbols) and photolysis (closed symbols) of diazirines **1b** (Δ, $\blacktriangle$), **1c** (o, ●) and **1f** ($\square$, ■) in the presence of TME. These values are obtained from the intercepts of the linear plots of $([2]/[3])$ *vs* $1/[\text{TME}]$.

Assuming that the overall quantum efficiency for the formation of **2** is unity, and that the excited diazirine gives the carbene **4**, and the vinyl halide **2** with efficiencies α and $(1 - \alpha)$, respectively, the value of α is given by $\alpha = \Delta H_2/\Delta H_{r(4\rightarrow2)}$, where $\Delta H_{r(4\rightarrow2)}$ is the heat of reaction for the rearrangement of the carbene **4** into the vinyl halide **2**. The values obtained for α by this method are in good agreement with the values derived from relative rate studies: 0.76 for **1a**, to be compared to 0.67 from quenching by TME[56,57] and 0.49 for **1b** and 0.94 for **1d**, to be compared to the values given above (Table 4).

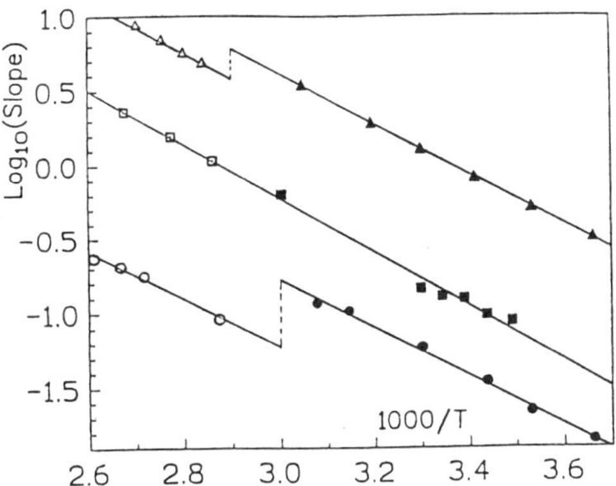

Figure 2. Arrhenius plots for the diazirines **1b** (Δ, $\blacktriangle$), **1c** (o, ●) and **1e** ($\square$, ■, upshifted by 0.3 unit) under thermolysis (open symbols) and photolysis (closed symbols). The slope used on the y-axis is the slope of the plots of $[2]/[3]$ *vs* $1/[\text{TME}]$, equal to $(k_i/\beta \, k_1)$ under thermolysis and to $(k_i/\alpha\beta \, k_1)$ under photolysis (see text).

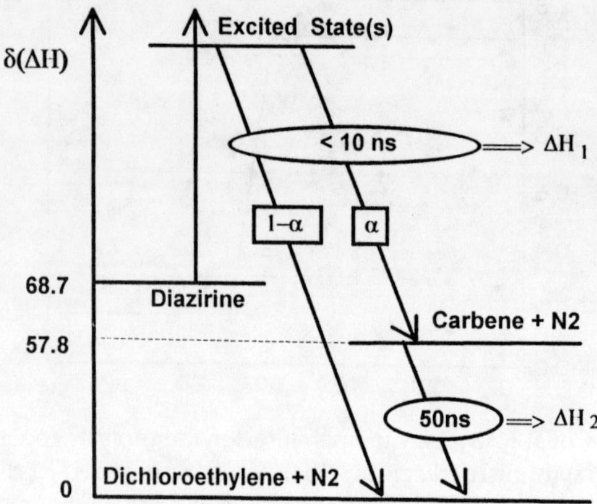

Figure 3. Energetic diagram explaining the principle of the measurements of the RIES efficiency by time resolved-photoacoustic calorimetry.

It appears that the efficiency of the RIES process, $(1 - \alpha)$, strongly depends on the nature of the diazirine, changing from ≈50% for **1b** to less than 10% for benzylchlorocarbenes. In the latter case, however, the RIES efficiency is probably not zero and could be slightly temperature dependent, as shown by the analysis of the changes of the ratio **2-Z/2-E**.[52] In most of the cases considered up to now, the carbene-TME complex collapses into the cyclopropane **3** with a high yield. But this yield is only ≈ 60% in the case of **1c** and must be even lower when **1d** faces diethyl-fumarate or -mesaconate.

C. Diazo Compounds as Source for the 1,2-Hydrogen Migration Product

Theoretical studies[45] considered above predicted that diazo compounds should be produced during the photolysis of diazirines. Indeed, diazo compounds have been detected by their infra-red (IR) absorption at a very low temperature in rigid

Table 4.

Diazirine	β (from [2/3]∞)th.	α (from PAC)	α (from [2/3]∞)	α (from slopes)	α (mean value)
1a	0.88[a]	0.76	0.67[56, b]	–	–
1b	0.85	0.49	0.54	0.42	0.49 ± 0.05
1c	0.62	–	0.72	0.62	0.67 ± 0.05
1d–f	0.77 (1f)	0.94 (1b)	≈ 1[c] (1f)	≈ 1[c] (1d–f)	0.95 ± 0.05

Notes: [a]Assuming that formation of **2** from the COC is responsible for the differences between the 2 values of α

[b]Not corrected for the formation of **2** from the COC

[c]Might be slightly temperature dependent.

matrices,[46,47] but the dialkyl- or alkylchloro-diazomethanes expected from the dialkyl- or alkylchloro-diazirines have not been observed at room temperature, probably because of their rapid thermal decomposition.

In all the previous discussion about RIES and COC processes as sources of rearrangement products that are unquenchable by alkene reactants, the role of a diazo intermediate was not considered. This is because it has been assumed that the decomposition of a short lifetime diazo intermediate would give the carbene. But Platz[58] suggested that the decomposition of a diazo ground state (perhaps catalyzed by traces of acid) could explain, better than could the COC mechanism, the excess of rearrangement products obtained under thermolysis, as well as under photolysis after correction for the RIES.

1. Evidence for the Formation of Diazo Compounds at Room Temperature

Laser-flash photolysis of isopropylchloro-, *n*-propylchloro-, and 2-adamantane-diazirines in isooctane at 25 °C produces a transient absorption in the 230 to 250 nm region ($\lambda_{max} \approx 240$ nm), which decays with a lifetime of around 0.5 second for the two former ones, and of around 1 hour for the latter.[55] These transient absorptions are assigned to the corresponding diazo compounds due to the similarity of the transient absorption spectra, shown in Figure 4, with the absorption spectrum of diazomethane ($\lambda_{max} = 220$ nm). The assignment of the 234 nm absorption band to the 2-diazoadamantane was confirmed by the reactivity of this species with acetic acid ($k \approx 10^5$ M^{-1}s^{-1}), and by comparison of the kinetics of disappearance of this UV absorption and that of the IR absorption at 2042 cm^{-1}.

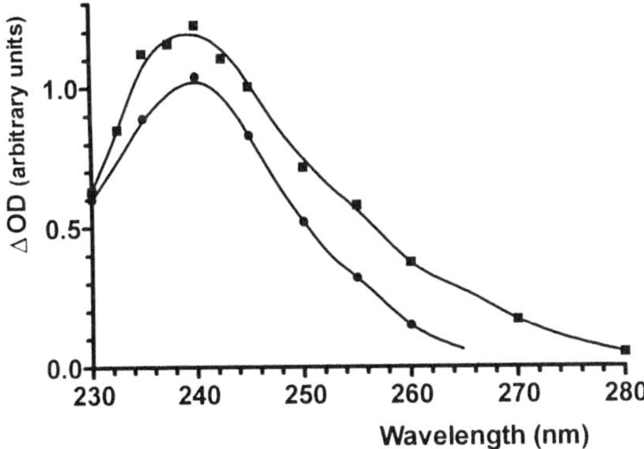

Figure 4. Transient absorption spectra recorded 1 μs after excitation by a 200 ps laser pulse (355 nm) of solutions of (■) *n*-propyl,chloro-diazirine and (●) *i*-propyl,chloro-diazirine in isooctane. These spectra are assigned to the *n*-propyl, chloro-, and *i*-propyl, chloro-diazomethanes.

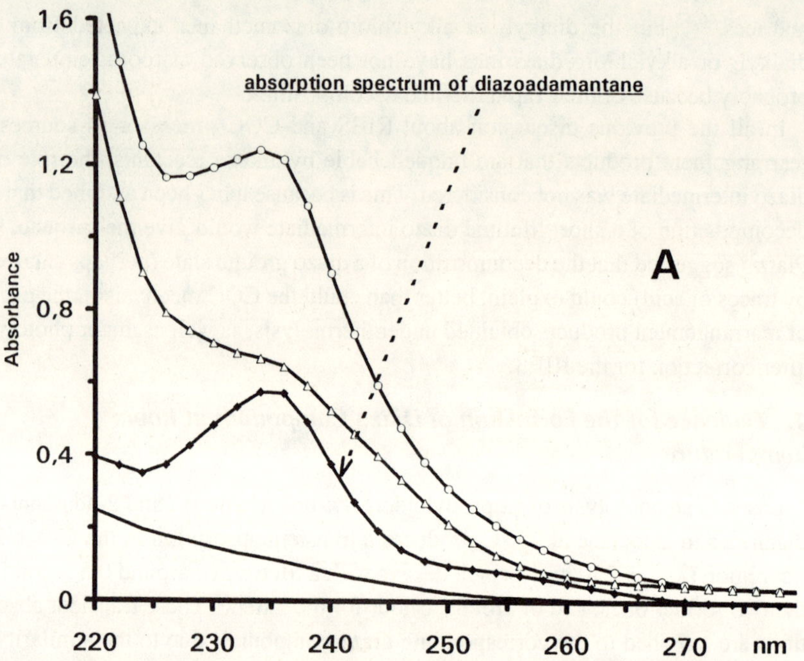

Figure 5a. Absorption spectra of diazoadamantane ($\tau \approx 2$ hs at 21 °C) obtained as the difference between the spectra recorded on a photodiode array spectrophotometer after laser irradiation during 2 sec, before (—○—) and after (—△—) addition of acetic acid (1mM). Both the diazo and the azine formed during excitation (and immediately after) absorb in this spectral region. The addition of acid completely removes the diazo without producing any other absorbing species in this spectral range. The absorption of the solution before irradiation is also shown (——).

The absorption spectra of both isopropyl- and *n*-propyl-chlorodiazomethanes, nearly identical to those recorded by laser-flash photolysis at room temperature, can easily be recorded on a conventional spectrophotometer in isooctane at –30 °C because their lifetime is then around 30 minutes (Figures 5). The temperature dependence of the lifetime from –30 to +30 °C yields an activation energy around 16 kcal/mol and a frequency factor in the range 10^{12} to 10^{13} s^{-1} for the process responsible for the disappearance of these alkylchlorodiazomethanes.

2. Quantum Yield of Formation for Diazo Compounds[55]

A quantum yield of formation around 50% and absorption coefficients ≈ 6000 M^{-1}cm^{-1} at 234 nm and ≈ 1000 M^{-1}cm^{-1} at 2042 cm^{-1} were calculated for 2-diazoadamantane. This value of the absorption coefficient at 234 nm is similar to that reported for diazomethane, $\varepsilon = 8300$ M^{-1}s^{-1} at 220 nm.[59] Assuming, by similitude, that the absorption coefficient is also in the range 5000 to 7000 M^{-1}cm^{-1}

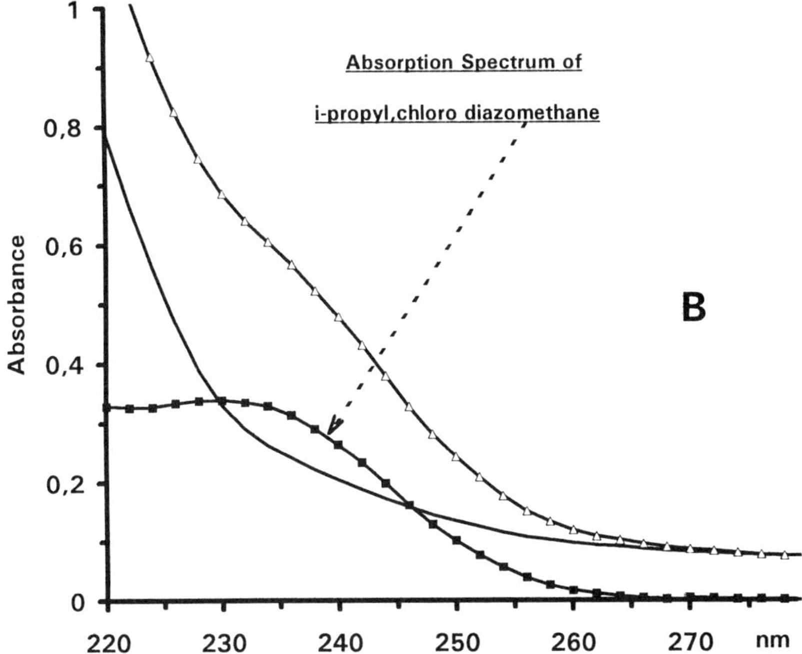

Figure 5b. Absorption spectra of *i*-propyl-chlorodiazomethane obtained as the difference between the spectra recorded, on a photodiode array spectrophotometer, immediately after laser irradiation during 2 sec (—Δ—), and one h later (———). The absorption spectrum before irradiation is nearly identical (in this spectral region) to the final spectrum (———) because only a few percent of the diazirine is decomposed and the yield of azine is extremely low (the carbene disappearing, very quickly, almost exclusively by 1,2-H migration).

for isopropyl- and *n*-propyl-chlorodiazomethanes, the quantum yield of formation of these species from the excited state of the corresponding diazirines is found to be ≈ 10 to 15% and 8 to 11%, respectively.

Therefore, even if these values of quantum yield are determined with a rather large incertitude, the formation of diazo compound is not responsible for the very low quantum yield of formation of the carbene from the *i*-propylchlorodiazirine, which is 50 and 15 times smaller than from *t*-butylchlorodiazirine and ethylchlorodiazirine, as estimated by the pyridinium ylide saturation technique.[46]

Also, the lifetime of alkylchlorodiazomethanes is such that these compounds are stable in the time range considered in the laser-flash photolysis experiments performed to study the intra- and intermolecular reactions of carbenes. By contrast, the formation of these diazo compounds should be taken into account in the PAC experiments, for instance, it should appear in the global energy balance because the heat released during the disappearance of the diazo is not seen by the PAC

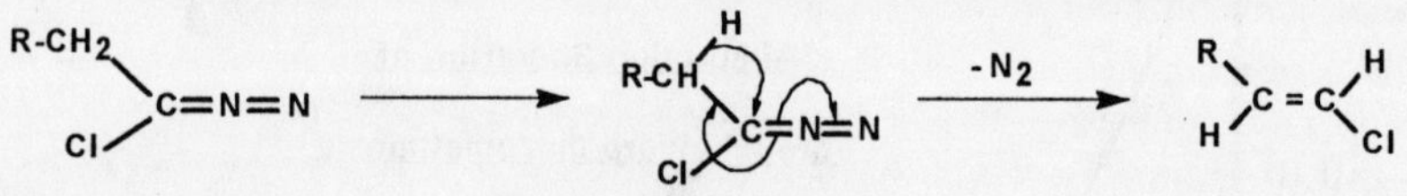

Scheme 8.

experiments. In fact, in the experiments performed on diazirines **1a**, **1b**, and **1f**, no deficit was found in the energy balance, thus, with Φ_{diazo} being the quantum yield of formation of the diazo and $\Delta H_{products}$ the enthalpy of formation of (N_2 + rearranged vinyl chloride), the product $\Phi_{diazo} (\Delta H_{diazo} - \Delta H_{products})$ must be smaller than the sum of the uncertainties on the measured total amount of heat released and on the calculated value of ($\Delta H_{diazirine} - \Delta H_{products}$). This may easily be the case if Φ_{diazo} is $\approx 10\%$. Finally, in studies conducted by products analysis after continuous irradiation, the diazo compound is entirely decomposed when this analysis is made and, therefore, we have to consider Platz's suggestion of the diazo compound as a source for rearrangement products.

3. Ratio Carbene/Rearrangement Product in the Decomposition of Diazo Compounds

Photolysis and thermolysis of diazo compounds are known to produce carbenes by elimination of nitrogen, but it has never been demonstrated that these processes are quantitative. A 1,2-H shift rearrangement more or less concerted with the extrusion of nitrogen, such as that represented in Scheme 8 and comparable to the RIES process, may be possible, especially under photolysis.

An acid-catalyzed thermal decomposition of the diazo compound, such as that shown in Scheme 9, also seems plausible and could be a non-negligible source of

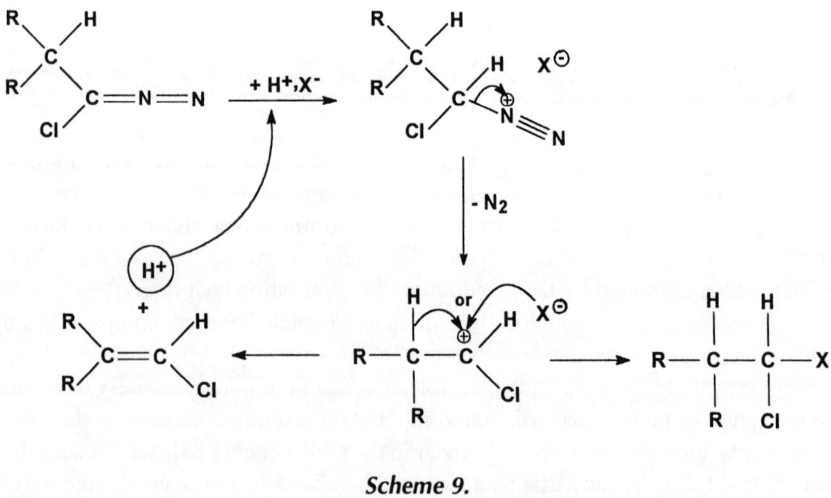

Scheme 9.

rearrangement products because the formation of small amounts of HCl has been reported during the photolysis of alkylchlorodiazirines. This leads, for instance in nonpolar solvents, to the precipitation of ammonium chloride salt as seen in the studies of ammonium ylides formed by amines with benzylchlorocarbenes.[60] Assuming that small amounts means a few percents of the diazirine concentration, i.e., a few 10^{-4} M if [diazirine] $\approx 10^{-2}$ M, and with a rate constant around 10^5 $M^{-1}s^{-1}$, as measured for the reaction between diazoadamantane and acetic acid, the rate of such a process would be comparable to the rate of the thermal decomposition of propylchlorodiazirines (≈ 10 s^{-1}, not an acid-catalyzed process as shown by the values of the kinetic parameters).

The efficiency of this attack of a diazo compound by acids to produce rearrangement products is probably rather poor because the carbocation formed after protonation and elimination of N_2 may add to the anion of the acid. IR spectroscopy shows that the decomposition of diazoadamantane upon addition of acetic acid yields mainly the 2-adamantylacetate. Most probably, the yield of formation of rearrangement products by thermal reactions of the diazo intermediate is $\leq 50\%$. Since, in the case of propylchlorodiazirines, the quantum yield of formation of the diazo compound is around 10% or less, the amount of rearrangement products formed this way should not exceed 5%.

It should be noted that for *n*-propylchlorodiazirine, the products analysis gave $\approx 38\%$ of rearrangement products under thermolysis, as well as under photolysis after correction for the RIES process. Clearly, the possible formation of rearrangement products by decomposition of a diazo intermediate is, at least in this case, a very minor (if not negligible) pathway.

IV. CONCLUSIONS

The sum of the experimental and theoretical studies on the kinetic aspects of the rearrangement of singlet carbenes by 1,2-H shift now gives a rather good picture of this process, although some points still remain to be clarified. Why is the frequency factor so low? Which are the factors controlling the stereochemistry of the rearrangement? (the chloromethylchlorocarbene gives $\approx 90\%$ of *cis*-dichloroethylene whereas benzylchlorocarbenes give $\approx 90\%$ of *trans*-chlorostyrene).[52] Does tunneling really play a non-negligible role at room temperature?

It is now recognized that, when diazirines are used as the source of carbenes, several processes contribute to the formation of the rearrangement products: A rearrangement in the excited state (of the diazirine) and the formation of a metastable diazo intermediate compete with the production of carbene. The efficiency of these (at minimum) three ways of decomposition of the diazirine strongly depends on the structure of the diazirine considered: The efficiency of RIES is low ($< 10\%$) for benzylchlorodiazirines but is around 50% for chloromethylchlorodiazirine; the quantum yield of formation of diazo is less than 10% for propylchlorodiazirines but about 50% for adamantanediazirine. Currently, no rules exist to predict the relative efficiency of these different pathways.

Theory and experiments indicate that the various decomposition processes of a diazirine in its excited state have very low, but non-zero, activation energies, ranging from 0.5 to a few kcal/mol. A structural modification may induce a small change in the difference between the E_a of two competing processes that, in turn, will give a large change in the relative efficiency of these processes. For instance, let us assume that the E_a for the RIES and carbene pathways are both equal to 1.5 kcal/mol for the chloromethylchlorodiazirine and that, when considering a benzyl-chlorodiazirine, the value of E_a for the carbene pathway decreases by 1 kcal/mol, whereas that for the RIES pathway increases by 0.5 kcal/mol. Assuming also that the frequency factors for these two processes are similar, the ratio $(k_{carbene}/k_{RIES})$ will then be equal to 1 and temperature independent for the chloromethylchlorodiazirine, in agreement with experimental results. For the benzylchlorodiazirine, the same ratio would be equal to 14 at 10 °C ($\Phi_{RIES} = 6.6\%$) but 9.5 at 60 °C ($\Phi_{RIES} = 9.5\%$), also in agreement with the experiment that gives $\Phi_{RIES} \leq 10\%$ and slightly increasing with temperature.[52] The strong changes in the yield of formation of the diazo intermediate may be explained in a similar manner. In both cases, it seems quite difficult to predict or to calculate with a 0.2 kcal/mol accuracy the changes in the values of E_a due to structural changes.

Finally, even when taking into account the formation of the rearrangement products by the RIES process and by the decomposition of a metastable diazo compound, which are unquenchable by olefinic reactants, it still remains true that the analysis of products obtained by photolysis or thermolysis of diazirines in the presence of olefins indicates an excess of rearrangement products, which can be explained only by a rearrangement occuring within a carbene-olefin complex. In most cases, this COC will collapse with high yield into the expected cyclopropane adduct, but this yield may be much lower in some cases: Values ranging from 0.9 to 0.3 have been reported and the actual value of the probability for the COC to give the cyclopropane instead of the rearrangement products depends not only on the diazirine but also on the olefin used as reactant. Steric factors, as well as electron densities at the reactive centers on both the carbene and the olefin, may be decisive, but it seems impossible to draw any firm conclusion from the small numbers of systems that have been investigated.

NOTES ADDED IN PROOF

1. For benzylchlorocarbene in isooctane, the curvature of the Arrhenius plot, $\log(1/\tau)$ *vs.* $1/T$, is reduced when the diazirine concentration is lowered (Merrer, D.C.; Moss, R.A.; Liu, M.T.H.; Banks and Ingold, K.U. *J. Org. Chem.* **1998**, in press). In addition to the reaction with the solvent considered in text, the reactions of the carbene with the diazirine and with moisture, if the solvent is not carefully dried, are reactions with a very low activation energy responsible for this curvature.

2. A detailed analysis of the fluorescence of adamantyldiazirine was recently published (Buterbaugh, J.S.; Yoscano, J.P.; Weaver, W.L.; Gord, J.R.; Hadad, C.M.; Gustafson, T.L.; Platz, M.S. *J. Am. Chem. Soc.* **1997**, *119*, 3580)

3. When benzylchlorocarbene is generated by photolysis of a phenanthridene precursor in the presence of tetramethylethylene, TME, as trapping agent, the plot of the ratio [cyclopropane adduct]/[chlorostyrene] *vs.* [TME] is linear whereas the same plot is curved when the carbene is generated by photolysis or thermolysis of a diazirine (Nigam, M.; Platz, M.S.; Toscano, J.P.; Johnson, R. *J. Am. Chem. Soc.* to be published). The curvature of these plots which was explained (see section III-B) by an hypothetical carbene-olefin complex, COC, in conjunction with the RIES mechanism in the case of photolysis of some diazirines, is thus due to some rearrangement pathway(s) specific for the diazirine precursor. This rules out one of the main arguments supporting the COC hypothesis. We thank Pr Platz for communication of these results prior to publication.

REFERENCES

1. Bodor, N.; Dewar, M.J. *J. Am. Chem. Soc.* **1972**, *94*, 9103.
2. Kyba, E.P. *J. Am. Chem. Soc.* **1977**, *99*, 8330.
3. Frenking, G.; Schmidt, J. *Tetrahedron* **1984**, *40*, 2123.
4. Altman, J.A.; Csizmadia, I.G.; Yates, K. *J. Am. Chem. Soc.* **1974**, *96*, 4196.
5. Altman, J.A.; Csizmadia, I.G.; Yates, K. *J. Am. Chem. Soc.* **1975**, *97*, 5217.
6. Nobes, R.H.; Radom, L.; Rodwell, W.R. *Chem. Phys. Lett.* **1980**, *74*, 269.
7. Su, D.T.T.; Thornton, E.R. *J. Am. Chem. Soc.* **1978**, *100*, 1872.
8. Burnett, S.M.; Stevens, A.E.; Feigerle, C.S.; Lineberger, W.C. *Chem. Phys. Lett.* **1983**, *100*, 124.
9. Stevens, I.D.R.; Liu, M.T.H.; Soundararajan, N.; Paike, N. *Tetrahedron Lett.* **1989**, *30*, 481.
10. Nickon, A. *Acc. Chem. Res.* **1993**, *26*, 84.
11. Liu, M.T.H. *Acc. Chem. Res.* **1994**, *27*, 287.
12. Moss, R.A. in: *Advances in Carbene Chemistry*, Vol.1; Brinker, U.H., Ed.; JAI Press, Stamford, CT, 1994, p. 59.
13. Liu, M.T.H. *J. Chem. Soc., Chem. Commun.* **1985**, 982.
14. Liu, M.T.H.; Soundararajan, N.; Paike, N.; Subramanian, R. *J. Org. Chem.* **1987**, *52*, 4223.
15. Liu, M.T.H.; Subramanian, R. *J. Phys. Chem.* **1986**, *90*, 75.
16. Jackson, J.E.; Soudararajan, N.; Platz, M.S.; Liu, M.T.H. *J. Am. Chem. Soc.* **1988**, *110*, 5595.
17. Liu, M.T.H.; Bonneau, R. *J. Am. Chem. Soc.* **1990**, *112*, 3915.
18. Moss, R.A.; Ho, G-H.; Liu W. *J. Am. Chem. Soc.* **1992**, *114*, 959.
19. Moss, R.A.; Xue, S.; Liu W. *J. Am. Chem. Soc.* **1994**, *116*, 1583.
20. Ge, C.-S.; Jang, E.G.; Jefferson, E.A.; Liu, W.; Moss, R.A.; Wlostowska, J.; Xue, S. *J. Chem. Soc., Chem. Commun.* **1994**, 1479.
21. Jackson, J.E.; Platz, M.S. in: *Advances in Carbene Chemistry*, Vol. 1; Brinker; U.H., Ed.; JAI Press, Stamford, CT, 1994, p. 89.
22. Modarelli, D.A.; Platz, M.S.; Sheridan, R.S.; Ammann, J.R. *J. Am. Chem. Soc.* **1993**, *115*, 10440; Platz, M.S.; White, W.R.; Modarelli, D.A. *Res. Chem. Intermed.* **1994**, *20*, 175; see also Refs. 41 and 46.
23. Bally, T.; Matzinger, S.; Truttmann, L.; Platz, M.S.; Morgan, S. *Angew. Chem. Int. Ed. Engl.* **1994**, *33*, 1964.
24. Rudzki, J.E.; Goodman, J.L.; Peters, K.S. *J. Am. Chem. Soc.* **1985**, *107*, 7849.
25. LaVilla, J.A.; Goodman, J.L. *J. Am. Chem. Soc.* **1989**, *111*, 712 and 6877.
26. Du, X.; Fan, H.; Goodman, J.L.; Kesselmayer, M.A.; Krogh-Jespersen, K.; LaVilla, J.A.; Moss, R.A.; Shen, S.; Sheridan, R.S. *J. Am. Chem. Soc.* **1990**, *112*, 1920.
27. Adamantylidene absorbs at 620 nm (see Ref. 23), dicyclopropylcarbene at 490 nm, Ammann, J.R.; Subramanian, R.; Sheridan, R.S. *J. Am. Chem. Soc.* **1992**, *114*, 7592; and $^1CH_2$ at 820 nm Green, W.H.; Handy, N.C.; Knowles, P.J.; Carter, S. *J. Chem. Phys.* **1991**, *94*, 118 and references therein.

28. Liu, M.T.H.; Bonneau, R. *J. Am. Chem. Soc.* **1990**, *112*, 3915.

29. Wierlacher, S.; Sander, W.; Liu, M.T.H. *J. Am. Chem. Soc.* **1993**, *115*, 8943.

30. Liu, M.T.H.; Bonneau, R. *J. Phys. Chem.* **1989**, *93*, 7298.

31. Ho, G-J.; Krogh-Jespersen, K.; Moss, R.A.; Shen, S.; Sheridan, R.S.; Subramanian, R. *J. Am. Chem. Soc.* **1989**, *111*, 6875.

32. Liu, M.T.H.; Bonneau, R. *J. Am. Chem. Soc.* **1996**, *118*, 8098.

33. Evenseck, J.D.; Houk, K.N. *J. Phys. Chem.* **1990**, *94*, 5518; *J. Am. Chem. Soc.* **1990**, *112*, 9148.

34. Dix, E.J.; Herman, M.S.; Goodman, J.L. *J. Am. Chem. Soc.* **1993**, *115*, 10424.

35. Storer, J. W.; Houk, K.N. *J. Am. Chem. Soc.* **1993**, *115*, 10426.

36. Liu, M.T.H.; Bonneau, R.; Wierlacher, S.; Sander, W. *J. Photochem. Photobiol. (A: Chem.)* **1994**, *84*, 133.

37. Liu, M.T.H. unpublished results.

38. Kirmse, W.; Buschoff, M. *Angew. Chem., Int. Ed. Engl.* **1965**, *4*, 692.

39. Mansoor, A.M.; Stevens, I.D.R. *Tetrahedron Lett.* **1966**, 1733.

40. Chang, K.T.; Shechter,H. *J. Am. Chem. Soc.* **1979**, *101*, 5082.

41. Modarelli, D.; Morgan, S.; Platz, M.S. *J. Am. Chem. Soc.* **1992**, *114*, 7034.

42. Bonneau, R. unpublished data.

43. Bigot, B.; Ponenc, R.; Sevin, A.; Devaquet, A. *J. Am. Chem. Soc.* **1978**, *100*, 6575.

44. Müller-Remmers, P.; Jug, K. *J. Am. Chem. Soc.* **1985**, *107*, 7275.

45. Yamamoto, N.; Bernardi, F.; Bottoni, A.; Olivucci, M.; Robb, M.A.; Wilsey, S. *J. Am. Chem. Soc.* **1994**, *116*, 2064.

46. White, W.R.; Platz, M.S. *J. Org. Chem.* **1992**, *57*, 2841.

47. Tomioka, H.; Hayashi, N.; Izawa, Y.; Liu, M.T.H. *J. Am. Chem. Soc.* **1984**, *106*, 454.

48. Liu, M.T.H. *J. Chem. Soc., Chem. Commun.* **1985**, 982.

49. Liu, M.T.H.; Soundararajan, N.; Paike, N.; Subramanian, R. *J. Org. Chem.* **1987**, *52*, 4223.

50. Turro, N.J.; Lehr, G.F.; Butcher, J.A., Jr; Moss, R.A.; Guo, W.; Munjal, R.C. *J. Am. Chem. Soc.* **1982**, *104*, 7576.

51. Houk, K.N.; Rondan, N.G.; Mareda, J. *J. Am. Chem. Soc.* **1984**, *106*, 4291.

52. Bonneau, R.; Liu, M.T.H.; Kim, K.C.; Goodman, J.L. *J. Am. Chem. Soc.* **1996**, *118*, 3829.

53. Liu, M.T.H.; Bonneau, R. *J. Phys. Chem.* **1989**, *99*, 4802.

54. Liu, M.T.H.; Bonneau, R. *J. Am. Chem. Soc.* **1992**, *114*, 3604.

55. Bonneau, R.; Liu, M.T.H. *J. Am. Chem. Soc.* **1996**, *118*, 7229.

56. LaVilla, J.A.; Goodman, J.L. *Tetrahedron Lett.* **1990**, *31*, 5109.

57. The difference between Φ_{RIES} = 0.24 obtained by PAC for **1a** and the 0.33 yield of vinyl chloride found[56] under photolysis of **1a** + TME when [TME] $\rightarrow \infty$ may be due to vinyl chloride produced, in the later case, via the COC mechanism.

58. Platz, M.S. private communication.

59. Müller, E.; Haiss; Rundel. *Chem. Ber.* **1960**, *93*, 1545.

60. Bonneau, R.; Reuter, I.; Liu, M.T.H. *J. Am. Chem. Soc.* **1994**, *116*, 3145.

CHEMISTRY OF CARBENES IN MOLECULAR REACTION VESSELS[‡]

Udo H. Brinker and Murray G. Rosenberg

[‡]Dedicated to Professor Emanuel Vogel on the occasion of his 70th birthday.

Advances in Carbene Chemistry
Volume 2, pages 29–44
Copyright © 1998 by JAI Press Inc.
All rights of reproduction in any form reserved.
ISBN: 1-55938-837-4

> *The ferocity of untamed carbenes*
> *May be quelled by the soothing confines of*
> *Accommodating host vessels;*
> *When working with sugar coated carbenes*
> *Expect remarkable new chemistry*
> *For the rewards are just as sweet!*
>
> —*Murray G. Rosenberg*

I. INTRODUCTION

Supramolecular chemistry is a rapidly growing field of research.[1] Although myriad host–guest chemistry publications have been documented in the scientific literature, nearly nothing is known about carbenes[2] within motion-restrictive host media. This chapter outlines some key techniques used to study carbene inclusion phenomena. It also highlights some exemplary studies that validate this new genre of carbene chemistry.

Recently, new techniques that afford carbenes within crystals at room temperature have been developed.[3] Nevertheless, the confinement of carbenes has been restricted, for the most part, to the use of inert-gas matrix isolation techniques.[4] The chemistry of divalent carbons entrapped within solid argon matrices is different from that in the solution or gas phase. This may be attributed to restricted motion, isolation, and low temperature effects. In constraining systems, such as cyclodextrins and zeolites, a similar change in the reactive behavior of carbenes should be expected. Indeed, any substantial quantum mechanical tunneling effects,[5] which might dominate results in a frigid argon matrix, become less pronounced in room temperature reactions of reactive intermediates inside supramolecular hosts.

The motive behind supramolecular carbene chemistry is, therefore, to modify inter- and intramolecular reactions of entrapped carbenes in order to manipulate product formation. Selectively directing carbene reactions when more than one product may be formed would, thereby, confound statistical prediction. The reactive behavior of carbenes within molecular reaction vessels is expected to be very different from that of carbenes in any other reaction medium. Host vessels may have any variety of different cavity sizes that can restrict the mobility of carbene guests and any other reaction partner that might be included.[6] One should even be able to generate, within the host domain, a carbene that has no intramolecular reaction pathways available. Situated like a model ship within a bottle, this highly reactive species will have been harnessed because no other molecules would be available with which it could react. This might introduce the possibility to spectroscopically characterize carbenes in an unfettered state. Supramolecular modification of carbene chemistry may be achieved, to some degree, by forming host–guest inclusion complexes of nitrogenous carbene precursors with cyclodextrins and zeolites. The results of these reactions should help physical organic chemists better understand the detailed mechanisms by which carbenes decay.

A. Carbenes

Carbene reactions are strictly those in which uncharged, divalent carbons are formed as reaction intermediates. These unstable molecules can undergo electronic rearrangements that either convert them to stable products or transform them into secondary carbenes of lesser energy content. The ephemeral carbene is electron deficient and, by various ways, gains an electron pair to completely fill its valence shell. Carbenes may be observed transiently under special reaction conditions. Sometimes they are paramagnetic or possess chromophores that may be detected spectroscopically. Other times, donor molecules, such as pyridine, are used to reveal their presence.[7]

Spin multiplicity is dependent upon whether or not the carbene nonbonding electrons are magnetically paired. The spin states of carbenes are an important aspect since singlet and triplet carbenes do not react in the same manner. For instance, a singlet carbene adds to carbon–carbon double bonds with retention of the substrate's geometric configuration; a triplet carbene reacts without stereospecificity under nonconcerted conditions.[8] In fact, triplet carbenes, which are geminal diradicals, behave much like other organic free radicals. They can abstract H atoms from organic solvents or can be photoreduced in protic solvents. Electron spin resonance (ESR) spectroscopy can be used to help detect triplet carbenes.[9]

The bond angle that carbon makes with the two atoms to which it is bonded is one factor that determines carbene ground and electronically excited spin states.[10] This geometric parameter might be susceptible to change when a carbene is formed in a supramolecular state. Moreover, carbenes may possess ground and excited states of unlike spin multiplicity that are significantly close in energy. When compared with that of a mobile carbene, the reacting state of an entrapped carbene might display a change in its spin multiplicity. Thus, like convincing an old dog to do new tricks, a molecular reaction vessel might foster exciting new chemistry from carbenes once dismissed as unimportant.

A carbene can react with almost any molecule that surrounds it. Numerous intramolecular rearrangements and even complex skeletal fragmentation reactions have been observed. This lack of selectivity becomes troublesome, particularly when the carbene is an intermediate used in synthetic organic chemistry.

B. Cyclodextrins

Since their discovery,[11] cyclodextrins have found a wide variety of uses in chemistry. They can protect fine chemicals from premature oxidation;[12] catalyze many classes of organic reactions, including ester cleavage[13] and the Diels–Alder reaction;[14] and assist in enantioselective chromatography.[15] They also serve as enzyme mimics[16] and are important in the study of molecular recognition in general.[17] A cyclodextrin (CD) is a cyclic oligosaccharide molecule composed of D-glucose monomers linked at the 1 and 4 positions in an α-stereoconfiguration. Its name depends on the number of pyranose monomers present: six (α-CD), seven (β-CD), eight (γ-CD), and nine (δ-CD) units (Figure 1).[18]

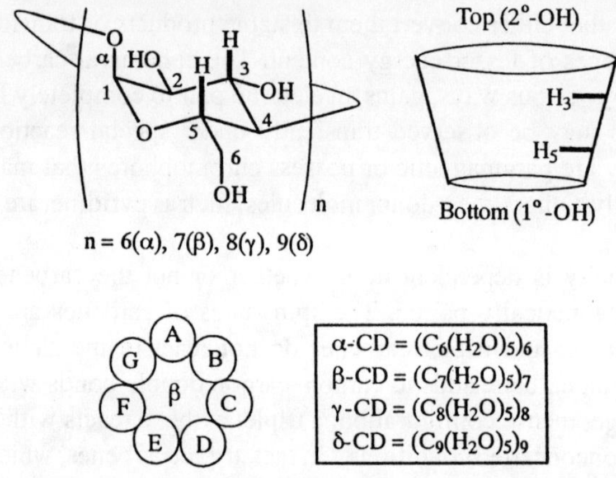

Figure 1. Cyclodextrin structure indicating positional nomenclature: Note that H3 and H5 point toward the guest and are most susceptible to NMR shift.

These torus-shaped hosts can accommodate many sorts of organic molecules that can reside within the nonpolar cavity. The strength of this noncovalent interaction is directly related to the position of the equilibrium for complex formation. This is evaluated by a binding constant[19] that may be measured in a variety of different ways, including spectrofluorometric methods based on competition experiments.[20] Direct proof of a guest@host inclusion complex may be obtained and important information such as reactant orientation and stoichiometry can be ascertained using well-developed methods (*vide infra*).

C. Zeolites

Zeolites were discovered in the middle of the eighteenth century.[21] There are many naturally occurring forms, as well as synthetic ones.[22] As with CDs, chemists have employed zeolites for a variety of tasks.[23] Most notable are their applications as catalysts for petroleum cracking, water-softening agents in laundry detergents, cat litter deodorizers, soil additives, desiccants, and molecular reaction vessels.[24]

Faujasite (FAU) zeolites are crystalline, highly polar, multicameral aluminosilicates where organic molecules may be entrapped. The two mostly used zeolites with a faujasite framework structure, X and Y, are characterized by a tetrahedral network of interconnected supercages (Figure 2).[25] Positively charged gegenions, which counter AlO_2^- moieties, are found within the FAU lattice. These cations may be exchanged,[26] thereby allowing chemists to alter the free volume of supercages by choosing ions of certain radii. According to the Lowenstein rule,[27] it is impossible to have two Al atoms connected to the same oxygen atom, but two Si atoms may do so. It is for this reason that the proportion of Si within zeolites is always higher than that of Al. Y-zeolites have a higher Si/Al ratio than do X-zeolites.

The loading factor ($<s>$) is a measure of the ratio of guest molecules to supercages. For instance, if $<s> = 1.00$ then each filled supercage is adjoined to another filled supercage; in the limit of $<s> = 0.25$, each filled supercage is surrounded by four empty ones. The uptake of guest molecules can be determined experimentally.[28] Like CDs, FAU zeolites allow a range of ultraviolet (UV) light to pass through and interact with photolabile guests, such as nitrogenous carbene precursors.

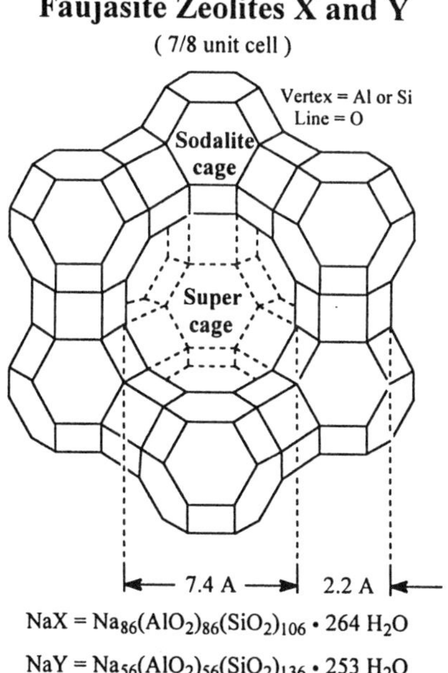

$$NaX = Na_{86}(AlO_2)_{86}(SiO_2)_{106} \cdot 264\ H_2O$$

$$NaY = Na_{56}(AlO_2)_{56}(SiO_2)_{136} \cdot 253\ H_2O$$

Figure 2. View of faujasite supercage through one of its apertures.

D. Physical Methods

Several physical methods have been employed to ascertain the existence and nature of these guest@host complexes. Generally, these inclusion compounds are solids and can routinely be analyzed by infrared (IR) and nuclear magnetic resonance (NMR) spectroscopy. Normally, subtle shifts in peak positions are indicative of the unique environment that supramolecular inclusion provides. For instance, vibrational normal mode frequencies can change for molecules within FAU zeolites.[29]

Shielding and deshielding of ^{1}H and ^{13}C resonances have also been reported.[30] Guests possessing an aromatic ring will help pronounce the effect. For these studies, it is important to use NMR solvents that do not dissociate the noncovalently bound supramolecular complexes.

The stoichiometry between the guest and CD host can also be evaluated using NMR by integrating peak areas of well-known signals. One source of error, though, could be differences in ^{1}H nuclear relaxation rates, but this error is usually small.

Intermolecular ^{1}H nuclear Overhauser effect (NOE) difference spectroscopy can also be used to characterize CD inclusion complexes.[31] Again, it is critical to use NMR solvents that do not dissociate the complexes. Also, one must make sure that the NMR solution is dilute for NOE experiments. This is usually not a problem given that the complexes are only sparingly soluble in solvents that ensure the integrity of the complex as well.

Solid state cross-polarization/magic-angle-spinning (CP/MAS) ^{13}C NMR is a powerful way to demonstrate true guest inclusion.[32] Whether inside CDs or FAU zeolites, guest signals become broadened in comparison with those of a neat sample and the contact times required for optimum CP are reduced. Peak broadening is attributed to the increased amorphous nature of the included guest within the host medium and shortened contact times indicate more efficient proton cross-relaxation due to increased guest–host proximity. For zeolites, heteronuclear ^{29}Si NMR or ^{27}Al NMR can also be used to see how the guest affects the inorganic crystal lattice.[33]

Single crystal[34a,b] and powder[34c] diffraction analyses have been used as another tool for identifying a legitimate inclusion complex. For this, a comparison between the diffraction pattern of a solid guest admixed with the host may be compared with that of a preparation of the supramolecular complex. These spectra will show different X-ray patterns.

Finally, fast atom bombardment mass spectrometry (FAB-MS)[35] can be used to study the results of a carbene interaction with a CD host. Unfortunately, the atomic bombardment process is too strong for the relatively loosely bound CD complex to stay intact and, in general, no parent ion can be observed. However, ethers derived from a formal carbene insertion reaction with the hydroxyl groups of a CD have been observed. This product is difficult to substantiate with any other physical method.

E. Theoretical Methods

Guest–host complex formation can be studied using molecular modeling.[36] This computational method has two main advantages: 1) molecular topologies of the guest and host can be investigated and modified, and 2) docking simulations can be performed giving insight about different orientations the guest may assume within the molecular reaction vessel. In addition, estimation of the equilibrium binding constant for a CD inclusion complex can be inferred. This value should be proportional to the free energy difference between the guest and host reactants and the inclusion complex formed from them.[37]

Sometimes, unfortunately, the limitations of this method are pronounced. First of all, not unlike calculations for biological enzymes, there exists the multiple minima problem. A unique supramolecular arrangement for a guest and host might not be found. Several arrangements of near equal energy may be predicted. Conversely, a unique assembly may be arrived at through computation, though two or more may coexist for the real complex. Furthermore, it is sometimes possible to make a CD complex in the laboratory that the computer predicts will not form. The flexibility of a CD host is dependent on the degree of hydrogen bonding between its own hydroxyl moieties and the water molecules that invariably surround it.[38] These special parameters need to be incorporated in the algorithms that the modeling program uses. Typically, a guest and host are more easily modeled in the absence of other molecules.

Recent advances in combining semiempirical calculations that are useful for guest and host with classical mechanics that may be used to simulate a bulk solvent environment, such as water, are now being utilized. Without such techniques, calculations can take a very long time if one desires to incorporate a large number of solvent molecules in the system. Including solvent molecules in calculations should give more accurate results for the structures of supramolecular assemblies.[39] Currently, computational methods are less reliable than physical ones, but this may soon change.

II. RESULTS AND DISCUSSION

The history of supramolecular carbene chemistry is not long. In the middle 1970s, base-induced α-eliminations of benzal halides were performed in the presence of crown ethers.[40] The halophenylcarbenes produced in this manner behaved much like those arising from 3-halo-3-phenyl-3*H*-diazirines, which are reported to yield free carbenes. Since these chlathrates are known to encompass the cations of the salt byproducts, it was reasoned that a transformation from carbenoid to carbene character was being observed.

In the early 1980s, *ortho-/para*-regioselectivity was reported for hydroxybenzaldehyde (**1**) formation by the Reimer–Tiemann reaction in the presence of CDs.[41] Although this reaction may proceed via dichlorocarbene intermediacy, a non-

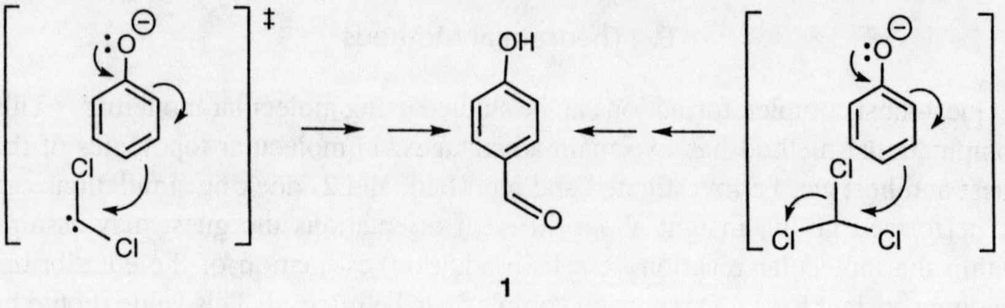

Figure 3. Tow possibilities for reaction of phenol with chloroform in alkaline medium.

carbene mechanism involving nucleophilic substitution of chloroform was not disproved (Figure 3).

Around this time, the conversion of methanol to gasoline over zeolite H-ZSM-5 catalyst was being studied. One set of experiments led to the hypothesis that carbenes were involved in this industrially important process.[42] However, a more explanatory mechanism was proposed[43] and by decade's end a rigorous study of this reaction using MAS ^{13}C NMR showed carbon monoxide to be the key intermediate.[44]

During the late 1980s, diazirines[45] and diazo compounds[46] were included within CDs.[47] The motive for these experiments was to utilize carbenes as agents to selectively modify CDs. The reactions can, however, also be explained by intermediates other than carbenes. Diazirines with α-protons, like 3-methyl-3-phenyl-3H-diazirine (**2**) (Figure 4) are susceptible to rearrangements in the excited state (RIES)

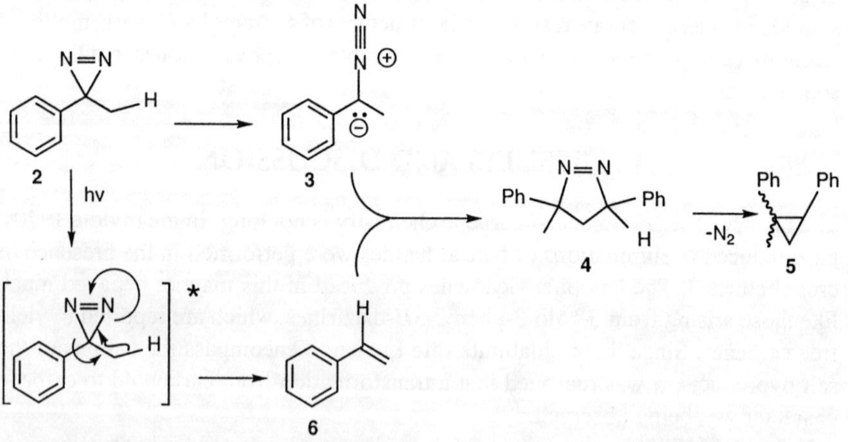

Figure 4. Formation of styrene and diphenylmethylcyclopropane *via* non-carbene pathways.

that mimic the results of carbene 1,2 H-shifts.[7] The styrene (**6**) formed also had undergone a secondary reaction giving diphenylmethylcyclopropanes (**5**). These three-membered rings, however, may be derived from homolytic breakdown of the pyrazoline **4**, formed by a dipolar addition of diazo compound **3** to styrene.

With so much uncertainty regarding true carbene intermediacy within molecular reaction vessels, a reaction demonstrating a unique process only attributable to carbenes needed to be performed.

A. Azi-adamantane

It is for this reason that the reaction of azi-adamantane (**7**) within the confines of CDs[48] and FAU zeolites[49] serves as a major achievement in carbene chemistry. This diazirine, first prepared in the early 1970s,[50] has been studied to a considerable extent. Believed to possess a singlet ground state,[51] adamantylidene (**9**) does not easily form singlet adamantene (**8**), a strained *anti*-Bredt compound (Figure 5).

Instead, it forms adamantanone azine (**12**) and other bimolecularly derived products such as 2-alkyladamantane (**13**) when reacted in solution. The dramatic increase in 2,4-didehydroadamantane (**11**) production when photolyzed in CDs and FAU zeolites is demonstrative of the powerful influence these hosts can have on reaction outcomes. This cyclopropane is only formed in trace amounts from the diazirine in organic solvents. When a physical mixture of β-CD and azi-adamantane (**7**) was photolyzed, less than 1% of 2,4-didehydroadamantane (**11**) was formed. Furthermore, photolysis of **7** in the solid state leads to no detectable **11** (Table 1).

The formation of the three-membered ring of **11** can only be easily explained as a 1,3 C-H insertion, which is not atypical of alkylcarbenes. Thus, there is little doubt that a carbene intermediate is involved.

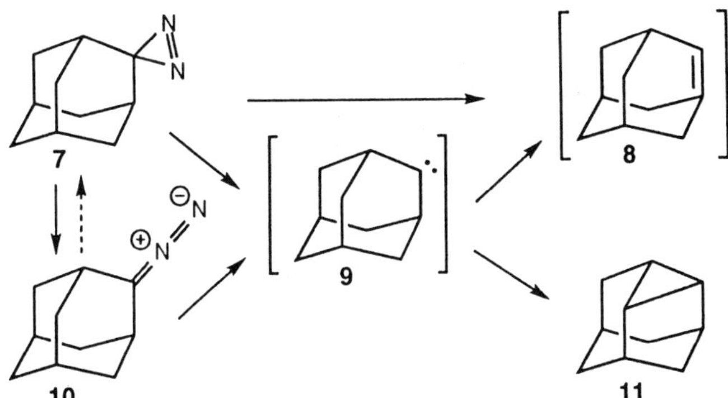

Figure 5. Formation of strained adamantanes from azi-adamantane.

Table 1. Relative Intra- and Intermolecular Product Distribution for Photolytic Reactions of Azi-adamantane

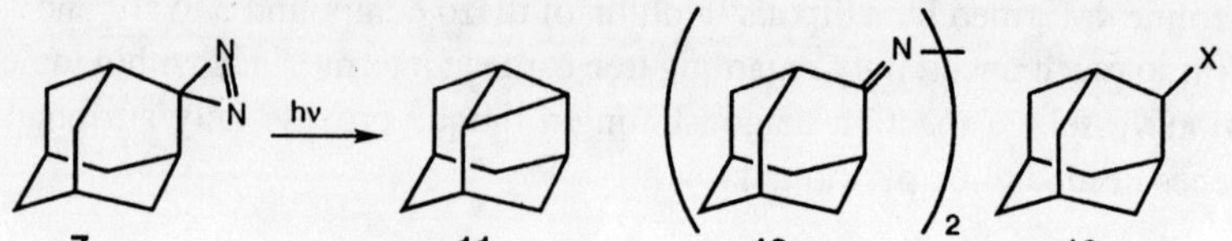

	11	12	13
n-heptane	0.2	54	10
methanol	6.5	0	93.5
β-CD	29	53	5
NaY	45	0	50

Additionally, according to FAB and tandem mass spectrometry, adamantylidene (**9**) also inserts into one of the 21 hydroxyl moieties of β-CD to make the covalently bound host–guest adduct 6-*O*-adamantylcyclodextrin (**14**) or another isomer having an (M-H)⁻ ion with a mass to charge ratio (*m/z*) of 1267 (Figure 6).

Further proof was secured by treating a sample with D_2O to exchange the number of active hydrogen atoms left and also by collisionally activating the *m/z* 1267 ion. No measurable loss of 2,4-didehydroadamantane (**11**) could be detected. A low abundance ion of *m/z* 1401.5 Daltons was also observed in the FAB mass spectrum. This was probably due to a β-CD bearing *two* adamantyl substituents. Support for this was demonstrated by the fact that only 19 hydrogen atoms were exchanged upon D_2O treatment. Thus, two insertions by two different adamantylidene inter-

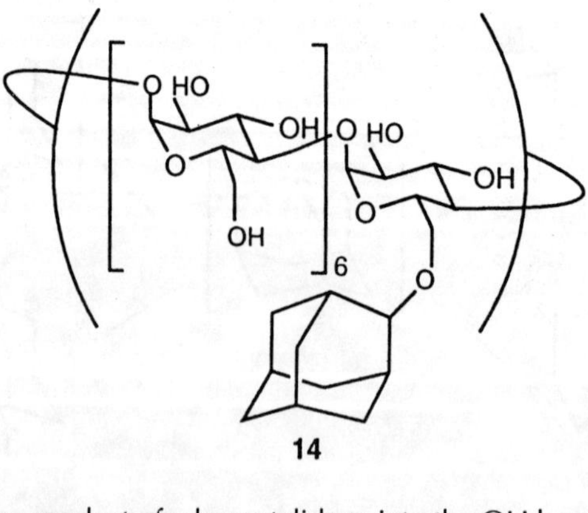

Figure 6. Insertion product of adamantylidene into the OH bond of cyclodextrin.

mediates into two separate hydroxyl sites of β-CD had taken place. The absolute amounts of O-H insertion products detected by FAB-MS are not yet known nor are the exact positions of carbene insertions, although the freely rotating primary hydroxyl groups distributed along the bottom of the CD host are more accessible than the secondary ones along the top. The guest molecules should be positioned differently inside and outside the β-CD cavity.

The effects of molecular constraint by CDs was explored further. Since aziadamantane is a fairly rigid molecule, investigation of the CD inclusion complexes of the more flexible 4-methyl-1,2-diazaspiro[2.5]oct-1-ene (**15**)[52] was pursued.[53]

This precursor of 2-methylcyclohexanylidene (**16**)[54] has an unscaffolded cyclohexane ring and probably is able to "ring flip." It was postulated that the supramolecular guest@CD complex would assume a conformation putting its methyl substituent in an axial position (Figure 7). This conformer is less stable than that with an equatorial methyl group. However, docking-like molecular mechanics simulations using the SYBYL 6.1 force field[55] showed that it is the axial conformer that fits inside the CDs with less steric repulsion interactions. This more compact conformer should be less prone to form 1-methylcyclohexene (**17**) and more able to produce 3-methylcyclohexene (**18**) and bicyclo[4.1.0]heptane (**19**). It has been shown that orbital alignment[56] plays a crucial role in carbene 1,2-H shifts.[57] With

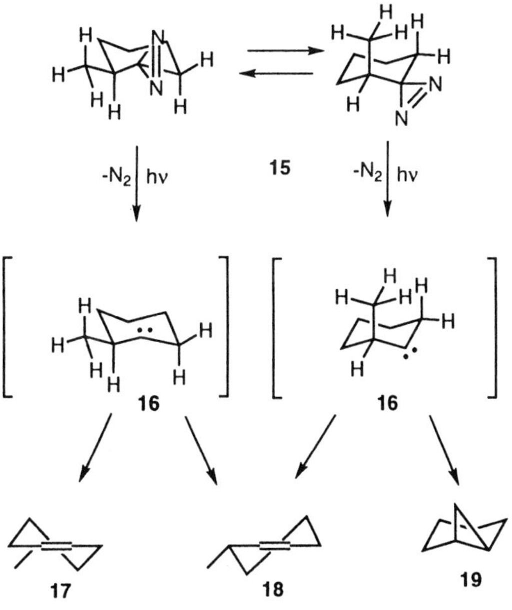

Figure 7. Conformer-dependent product distribution from 2-methylcyclohexanylidene.

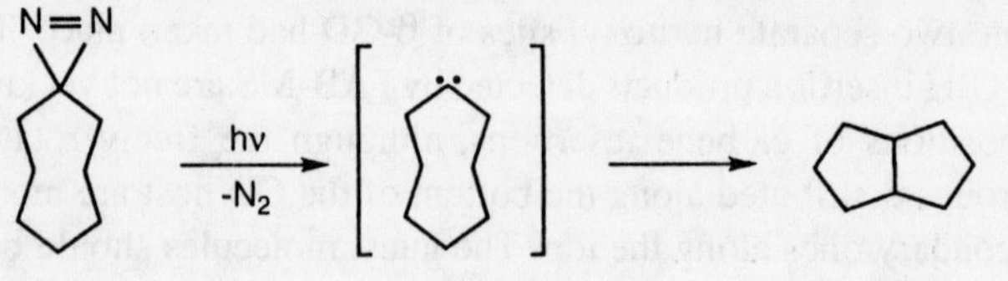

Figure 8. Long-range transannular CH insertion of cyclooctanylidene.

diazirine **15** the expected result was not observed, which leads one to conclude that CD hosts did not affect the dynamic equilibrium between the conformers of **15**.

B. Azi-cyclooctane

The effect of molecular confinement on reactant conformation can be astounding. If an activated complex is deprived of its normal decay route then new reaction alternatives may arise. This is the case with azi-cyclooctane (**20**) (Figure 8).[58]

Upon photolysis, cyclooctene (**24**) is formed, but so too are products derived from intramolecular carbene C-H insertions. Of special note is bicyclo[3.3.0]octane (**22**).

Created from a 1,5 C-H insertion in cyclooctanylidene (**21**) (Figure 9),[59] this molecule needs space to form. Since the free volume of a CD depends on the number of monomer units it comprises, it is not surprising that these factors conspire to

Table 2. Relative Allotment of Long- *versus* Short-range Carbene Rearrangements of Cyclooctanylidene from the Preceding Diazirine

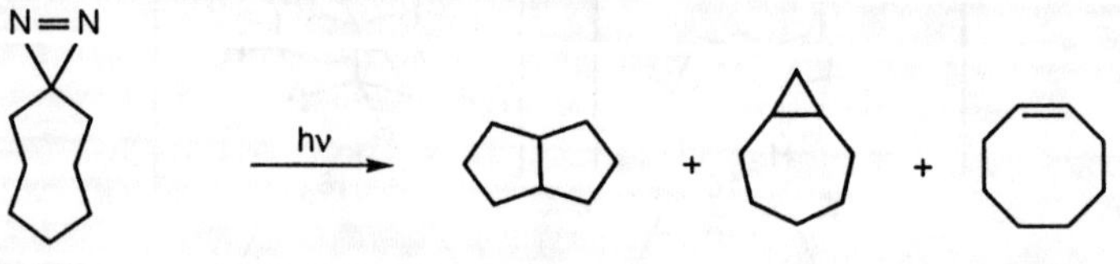

	22	23	24
hexane	54	8	35
methanol	28	9	63
α-CD	29	14	57
β-CD	37	9	36
γ-CD	42	11	47
NaX	11	7	46
NaY	9	3	39

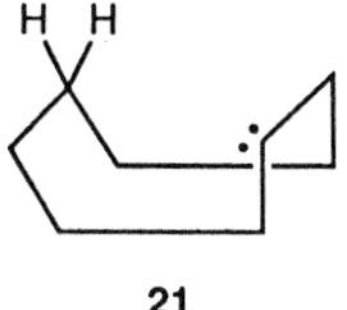

21

Figure 9. Required conformation for 1,5 CH insertion.

limit bicyclo[3.3.0]octane formation when azi-cyclooctane (**20**) is photolyzed within α-CD (Table 2).[60]

The occurrence of transannular C-H insertion becomes less probable as the cavity within the molecular reaction vessel gets smaller. This can be likened to tying ones shoelaces in a telephone booth; the smaller the booth, the harder the task.

III. CONCLUSIONS

The hosts studied in this chapter modify certain reaction pathways by providing the order or restricting the mobility needed by the guest molecules therein to react. By reducing the frequency of intermolecular reactions involving carbenes, interesting results for intramolecular reactions can be coaxed to occur. Supramolecular complexation serves as a facile way to modify the potential energy surface (PES) that a reaction follows. This is not unlike the effect that functional group substitution has on many other reaction mechanisms and outcomes. However, here, the reactant is not chemically *but* physically modified. It is as if the chemist has the ability to manually orchestrate the diverse happenings of a chemical reaction.

ACKNOWLEDGMENTS

We are indebted to the Petroleum Research Fund administered by the American Chemical Society for financial support, and also to the former American Maize-Products Company and Wacker Chemie for providing the cyclodextrins used in our studies.

REFERENCES

1. (a) *Comprehensive Supramolecular Chemistry*; Lehn, J.-M., Ed.; Pergamon: New York, 1995, Vols. 1–10; (b) Cram, D.J.; Cram, J.M. *Container Molecules and Their Guests*, Royal Society of Chemistry, Cambridge, 1994; (c) Cram, D.J. *Angew. Chem. Int. Ed. Engl.* **1988**, *27*, 1009; (d) Smith, D. R. *Chem. Ind.* **1994**, 14; (e) Dowden, J.; Kilburn, J.D.; Wright, P. *Contemp. Org. Synth.* **1995**, *2*, 289.

2. (a) *Advances in Carbene Chemistry*; Brinker, U.H., Ed., JAI, Greenwich, 1994; Vol. 1; (b) *Houben-Weyl, Methoden der Organischen Chemie*; Regitz, M., Ed., Thieme Verlag: Stuttgart, 1989; Vol. E19b; (c) Jones, W.M.; Brinker, U.H. in: *Pericyclic Reactions*, Vol. 1, Marchand, A. P.; Lehr, R. E. Eds.; Academic: New York, 1977; p. 109; (d) Kirmse, W. *Carbene Chemistry*, 2nd

ed., Academic, New York, 1971; (e) *Carbenes*; Jones, M., Jr.; Moss, R.A. Eds.; Wiley: New York, 1973, 1975; Vols. 1 and 2; (f) Jones, M., Jr. *Sci. Am.* **1976**, *234*(2), 101.

3. (a) Shin, S.H.; Keating, A.E.; Garcia-Garibay, M.A. *J. Am. Chem. Soc.* **1996**, *118*, 7626; (b) Keating, A.E.; Shin, S.H.; Houk, K.N.; Garcia-Garibay, M.A. *J. Am. Chem. Soc.* **1997**, *119*, 1474; (c) Shin, S.H.; Cizmeciyan, D.; Keating, A.E.; Khan, S.I.; Garcia-Garibay, M.A. *J. Am. Chem. Soc.* **1997**, *119*, 1859.

4. (a) Platz, M.S. in: *Kinetics and Spectroscopy of Carbenes and Biradicals*; Platz, M. S., Ed.; Plenum: New York, 1990, p. 143; (b) Chapman, O.L.; Johnson, J.W.; McMahon, R.J.; West, P.R. *J. Am. Chem. Soc.* **1988**, *110*, 501; (c) Tomioka, H.; Ozaki, Y.; Koyabu, Y.; Izawa, Y. *Tetrahedron Lett.* **1982**, *23*, 1917.

5. (a) Moss, R.A.; Ho, G.-J.; Liu, W.; Sierakowski, C. *Tetrahedron Lett.* **1992**, *33*, 4287; (b) Dix, E.J.; Herman, M.S.; Goodman, J.L. *J. Am. Chem. Soc.* **1993**, *115*, 10424; (c) Dix, E.J.; Goodman, J.L. *Res. Chem. Intermed.* **1994**, *20*, 149; (d) Tomioka, H. *JAERI-Conf.* **1995**, *95-020*, 33.

6. Harris, K.D.M. *Chem. Br.* **1993**, 132.

7. Jackson, J.E.; Platz, M.S. in: *Advances in Carbene Chemistry*, Brinker, U. H., Ed.; JAI: Stamford, CT, 1994; Vol. 1, p. 89.

8. (a) Skell, P.S.; Woodworth, R.C. *J. Am. Chem. Soc.* **1956**, *78*, 4496; (b) Woodworth, R.C.; Skell, P.S. *J. Am. Chem. Soc.* **1959**, *81*, 3383; (c) Su, M.-D. *J. Phys. Chem.* **1996**, *100*, 4339.

9. (a) Forrester, A.R.; Sadd, J.S. *J. Chem. Soc., Chem. Commun.* **1976**, 631; (b) Forrester, A.R.; Sadd, J.S. *J. Chem. Soc., Perkin Trans. 2*, **1982**, 1273.

10. Hoffmann, R.; Zeiss, G.D.; Van Dine, G.W. *J. Am. Chem. Soc.* **1968**, *90*, 1485.

11. (a) Villiers, M.A. *Compt. Rend. Acad. Sci. Paris* **1891**, *112*, 536; (b) Schardinger, F. *Wien. Klin. Wochenschr.* **1904**, *17*, 207.

12. Dodziuk, H. *Chemical Intelligencer* **1996**, 2(2), 56.

13. Tee, O.S.; Bozzi, M.; Clement, N.; Gadosy, T.A. *J. Org. Chem.* **1995**, *60*, 3509.

14. Rideout, D.C.; Breslow, R. *J. Am. Chem. Soc.* **1980**, *102*, 7816.

15. (a) Purdy, W.C. *Chem. Rev.* **1992**, *92*, 1457; (b) Easton, C.J. *Chem. Soc. Rev.* **1996**, 163.

16. Breslow, R. *Science* **1982**, *218*, 532.

17. Hamilton, A.D. *J. Chem. Educ.* **1990**, *67*, 821.

18. (a) Saenger, W. *Angew. Chem. Int. Ed. Engl.* **1980**, *19*, 344; (b) Wenz, G. *Angew. Chem. Int. Ed. Engl.* **1994**, *33*, 803.

19. Cramer, F.; Henglein, F.M. *Chem. Ber.* **1957**, *90*, 2561.

20. Tee, O.S.; Gadosy, T.A.; Giorgi, J.B. *Can. J. Chem.* **1996**, *74*, 736.

21. Cronstedt, A.F. *K. Sven. Vetenskapsakad. Handl., Stockholm* **1756**, *17*, 120.

22. Manly, R.; Holmes, J. *New Scientist* **1989**, *121* (1657), 39.

23. *Zeolite Molecular Sieves: Structure Chemistry and Use*; Breck, D.W., Ed.; Wiley: New York, 1974.

24. Smoot, A.L.; Lindquist, D.A. *J. Chem. Educ.* **1997**, *74*, 569.

25. Ramamurthy, V. in: *Photochemistry in Organized and Constrained Media*, Chapter 10, Ramamurthy, V., Ed.; VCH: New York, 1991.

26. Ramamurthy, V.; Corbin, D.R.; Turro, N.J.; Zhang, Z.; Garcia-Garibay, M.A. *J. Org. Chem.* **1991**, *56*, 255.

27. Lowenstein, W. *Amer. Mineral.* **1954**, *39*, 92.

28. Garcia-Garibay, M.A.; Zhang, Z.; Turro, N.J. *J. Am. Chem. Soc.* **1991**, *113*, 6212.

29. Primet, M.; Garbowski, E.; Mathieu, M.V.; Imelik, B. *J. Chem. Soc., Faraday Trans. 1*, **1980**, *76*, 1942.

30. (a) Demarco, P.V.; Thakkar, A.L. *Chem. Commun.* **1970**, 2; (b) MacNicol, D.D. *Tetrahedron Lett.* **1975**, 3325; (c) Chung, W.-S.; Turro, N.J.; Silver, J.; le Noble, W.J. *J. Am. Chem. Soc.* **1990**, *112*, 1202; (d) Ivanov, P.M.; Salvatierra, D.; Jaime, C. *J. Org. Chem.* **1996**, *61*, 7012.

31. (a) Qi, Z.H.; Mak, V.; Diaz, L.; Grant, D.M.; Chang, C.-J. *J. Org. Chem.* **1991**, *56*, 1537; (b) Chang, C.-J.; Choi, H.-S.; Wei, Y.-C.; Mak, V.; Knevel, A.M.; Madden, K.M.; Carlson, G.P.; Grant, D.M.;

Diaz, L.; Morin, F.G. in: *Biotechnology of Amylodextrin Oligosaccharides*; Chapter 19, Friedman, R.B., Ed., ACS Symposium Series 458, American Chemical Society: Washington, DC, 1991.

32. (a) Inoue, Y.; Okuda, T.; Kuan, F.-H.; Chujo, R. *Carbohydr. Res.* **1984**, *129*, 9; (b) Hall, L.D.; Lim, T.K. *J. Am. Chem. Soc.* **1986**, *108*, 2503; (c) Gidley, M.J.; Bociek, S.M. *J. Am. Chem. Soc.* **1988**, *110*, 3820; (d) Ripmeester, J.A. *J. Inclusion Phenom.* **1988**, *6*, 31.

33. (a) Fyfe, C.A.; Thomas, J.M.; Klinowski, J.; Gobbi, G.C. *Angew. Chem. Int. Ed. Engl.* **1983**, *22*, 259; (b) Thomas, J.M.; Klinowski, J.; Ramdas, S.; Hunter, B.K.; Tennakoon, D.T.B. *Chem. Phys. Lett.* **1983**, *102*, 158; (c) Haase, J.; Pfeifer, H.; Oehme, W.; Klinowski, J. *Chem. Phys. Lett.* **1988**, *150*, 189; (d) Fyfe, C.A.; Gies, H.; Kokotailo, G.T.; Pasztor, C.; Strobl, H.; Cox, D.E. *J. Am. Chem. Soc.* **1989**, *111*, 2470; (e) Fyfe, C.A.; Feng, Y.; Gies, H.; Grondey, H.; Kokotailo, G.T. *J. Am. Chem. Soc.* **1990**, *112*, 3264.

34. (a) Hamilton, J.A.; Sabesan, M.N. *Acta Crystallogr., Sect. B* **1982**, *38*, 3063; (b) Harata, K.; Uekama, K.; Otagiri, M.; Hirayama, F.; Ohtani, Y. *Bull. Chem. Soc. Jpn.* **1985**, *58*, 1234; (c) Takeo, K.; Kuge, T. *Agr. Biol. Chem.* **1970**, *34*, 568.

35. (a) Pokrovskii, V.A.; Mosin, V.V. *Teor. Eksp. Khim.* **1987**, *23*, 62; (b) Nibbering, N.M.M. *Adv. Mass Spectrom.* **1989**, *11A*, 101; (c) *Continuous-Flow Fast Atom Bombardment Mass Spectrometry*; Caprioli, R.M., Ed.; Wiley: New York, 1990; p. 1; (d) Pocsfalvi, G.; Liptak, M.; Huszthy, P.; Bradshaw, J.S.; Izatt, R.M.; Vekey, K. *Anal. Chem.* **1996**, *68*, 792.

36. Bakó, I.; Jicsinszky, L. *J. Inclusion Phenom. Mol. Recognit. Chem.* **1994**, *18*, 275.

37. Gelb, R.I.; Alper, J.S. *J. Phys. Org. Chem.* **1995**, *8*, 825.

38. Kano, K.; Ishimura, T.; Negi, S. *J. Inclusion Phenom. Mol. Recognit. Chem.* **1995**, *22*, 285.

39. Margheritis, C.; Sinistri, C. *Z. Naturforsch., A* **1996**, *51*, 950.

40. Moss, R.A.; Pilkiewicz, F.G. *J. Am. Chem. Soc.* **1974**, *96*, 5632.

41. (a) Komiyama, M.; Hirai, H. *Bull. Chem. Soc. Jpn.* **1981**, *54*, 2053; (b) Komiyama, M.; Hirai, H. *J. Am. Chem. Soc.* **1983**, *105*, 2018.

42. Chang, C.D.; Chu, C.T-W. *J. Catal.* **1982**, *74*, 203.

43. Van Hooff, J.H.C. *J. Catal.* **1983**, *79*, 242.

44. Anderson, M.W.; Klinowski, J. *Nature* **1989**, *339*, 200.

45. *Chemistry of Diazirines*; Liu, M.T.H., Ed.; CRC: Boca Raton, 1987; Vols. 1 and 2.

46. Regitz, M.; Maas, G. *Diazo Compounds*; Academic Press: New York, 1986.

47. (a) Abelt, C.J.; Pleier, J. *J. Org. Chem.* **1988**, *53*, 2159; (b) Abelt, C.J.; Lokey, J.S.; Smith, S.H. *Carbohydr. Res.* **1989**, *192*, 119; (c) Smith, S.H.; Forrest, S.M.; Williams, D.C., Jr.; Cabell, M.F.; Acquavella, M.F.; Abelt, C.J. *Carbohydr. Res.* **1992**, *230*, 289.

48. Brinker, U.H.; Buchkremer, R.; Kolodziejczyk, M.; Kupfer, R.; Rosenberg, M.; Poliks, M.D.; Orlando, M.; Gross, M.L. *Angew. Chem. Int. Ed. Engl.* **1993**, *32*, 1344.

49. Kupfer, R.; Poliks, M.D.; Brinker, U.H. *J. Am. Chem. Soc.* **1994**, *116*, 7393.

50. (a) Isaev, S.D.; Yurchenko, A.G.; Stepanov, F.N.; Kolyada, G.G.; Novikov, S.S.; Karpenko, N. *Zh. Org. Khim.* **1973**, *9*, 724; (b) Bayley, H.; Knowles, J.R. *Biochemistry* **1980**, *19*, 3883.

51. Bally, T.; Matzinger, S.; Truttmann, L.; Platz, M.S.; Morgan, S. *Angew. Chem. Int. Ed. Engl.* **1994**, *33*, 1964.

52. Fuchs, J.J. U.S. Patent 3 287 354, 1966; *Chem. Abstr.* **1967**, *66*, 65453.

53. Rosenberg, M.G.; Kam, S.M.; Brinker, U.H. *Tetrahedron Lett.* **1996**, *37*, 3235.

54. Wilt, J.W.; Wagner, W.J. *J. Org. Chem.* **1964**, *29*, 2788.

55. SYBYL 6.1 Molecular Modeling Package, TRIPOS Inc.

56. Nickon, A.; Stern, A.G.; Ilao, M.C. *Tetrahedron Lett.* **1993**, *34*, 1391.

57. Liu, M.T.H. *Acc. Chem. Res.* **1994**, *27*, 287.

58. (a) Bradley, G.F.; Evans, W.B.L.; Stevens, I.D.R. *J. Chem. Soc. Perkin Trans. 2*, **1977**, 1214; (b) For reviews see: Sydnes, L.K.; Brinker, U.H. in: *Houben-Weyl, Methoden der Organischen Chemie*, Vol. E 19b, Regitz, M., Ed.; Thieme Verlag: Stuttgart, 1989; p. 601; Baron, W.J.; DeCamp, M.R.; Hendrick, M.E.; Jones, Jr., M.; Levin, R.H.; Sohn, M.B. in: *Carbenes*; Jones, Jr., M.; Moss, R.A., Eds.; Wiley: New York, 1973, Vol. 1.

59. (a) Wojnarovits, L.; Szondy, T.; Szekeres-Bursics, E.; Foldiak, G. *J.Photochem.* **1982**, *18*, 273; (b) Wojnarovits, L. *J. Chem. Soc., Perkin Trans. 2*, **1984**, 1449; (c) Cope, A.C.; Brown, M.; Woo, G.L. *J. Am. Chem. Soc.* **1965**, *87*, 3107; (d) Friedman, L.; Shechter, H. *J. Am. Chem. Soc.* **1961**, *83*, 3159.
60. Kupfer, R.; Brinker, U.H. *Liebigs Ann.* **1995**, 1721.

THERMOCHEMISTRY OF CARBENES

Peter Chen

I. INTRODUCTION

If one were to ask a typical organic chemist to predict the bond dissociation energy for a C–H bond in an ordinary molecule, for example, the C–H in isobutyronitrile to form the 2-cyano-2-propyl radical (the product of pyrolysis of AIBN, a common

Advances in Carbene Chemistry
Volume 2, pages 45–75
Copyright © 1998 by JAI Press Inc.
All rights of reproduction in any form reserved.
ISBN: 1-55938-837-4

radical initiator), the ensuing thought process would be something like this (presuming of course that the chemist did not know the number offhand):

The C–H bond in acetonitrile is worth about 93 kcal/mol. The two methyl groups should weaken the C–H bond in isobutyronitrile relative to acetonitrile by about the same amount as the difference in C–H bond strengths between ethane and isobutane, which is 101–96 ~ 5 kcal/mol. That means the isobutyronitrile ought to be about 88 kcal/mol.

The answer is just about right; the actual bond dissociation energy is 87 kcal/mol. There are several assumptions implicit in the reasoning. Our chemist assumes that all C–H bonds β to a cyano group, barring special effects, are about the same as the C–H bond in acetonitrile. He then assumes that the two methyl groups weaken the bond by the same increment as in another pair of reference compounds—ethane and isobutane in this case. Furthermore, he assumes that the various structural changes affect the final bond energy in an additive fashion so that the important contributions can be simply summed up. The chain of reasoning illustrated here, with its attendant assumptions, is implicit in much of the intuitive understanding of chemical structure and reactivity that organic chemists use every day. One assumes that a phenyl group in one molecule is more or less like a phenyl group in another molecule, with separate corrections for other factors—e.g., steric effects, ring size, etc.—and that the properties of a large molecule, $\Delta H_{f,298}$ for example, are the sum of contributions from each component. The decomposition of a molecule into parts does not necessarily correspond to any real reaction, but the formal reaction can be calibrated or parameterized with a set of primary reference compounds and reactions. A representation of the assumptions gathered together under the label of additivity can be made with the help of isodesmic or homodesmotic reactions,[1] which form the basis for various decomposition schemes.

The qualitative notion of separate, additive contributions from fragments of molecules has been formalized in numerical schemes such as Benson's group equivalent,[2] which for molecules within the structural types with good reference data, give predictions of $\Delta H_{f,298}$ and $S_{f,298}$ with an absolute accuracy of ±2 kcal/mol and ±2 eu. Benson group equivalents for various radical groups have been derived,[3] meaning only that radical thermochemistry, or equivalently, the bond dissociation enthalpy for an ordinary molecule, is well-described by additivity, as was illustrated in the isobutyronitrile example above. The success of group equivalents, and even the organic chemist's intuition, attest to the overall validity of the assumptions implicit in any additivity picture. However, if one takes a radical, and breaks a second bond at the same site to form a carbene, the bond-strength, or again equivalently, $\Delta H_{f,298}$[carbene], often deviates strongly from expectations. A striking example is the comparison of the two sequential C–H bond dissociation enthalpies in formaldehyde. Going from formaldehyde to formyl radical and H· one needs 88 kcal/mol.[4] Formyl radical[5] has CO and CH bond lengths of 1.18 and 1.13 Å, respectively, and an HCO angle of 125°, which are not too different from the corresponding values in formaldehyde, i.e. the carbon in formyl is still sp^2-hybridized as in formaldehyde. Nevertheless, the bond dissociation of formyl to CO and H· costs only 16 kcal/mol.[6] Because it is the first bond energy in formaldehyde that is normal, and the second that is weak, one can attribute the very low bond strength in formyl to an unexpected thermodynamic stability of CO rather than instability in formyl. Clearly, there is an additional effect at work here in the thermochemistry of divalent carbon compounds, i.e. carbenes, that the usual additivity scheme has left out. The goal of this chapter is to suggest a solution to the problem of $\Delta H_{f,298}$[carbene] which not only identifies the missing factor, but also deals with it in a semiquantitative (or better) way.

The material that follow is divided into three parts. To start, a general valence–bond promotion energy model will be proposed as both a rationalization of known thermochemistry and a predictive tool linking internal properties of fragments to overall properties of molecules. Next, thermochemical data on several families of carbenes will be reviewed with an emphasis on the validation of the promotion energy model. Limitations of the model will also become evident. Lastly, recent work on an extreme example of bond weakening by a promotion energy will be reviewed.

II. QUALITATIVE PICTURE

The operative picture[7] used in this chapter derives most directly from a model first proposed[8] (and largely ignored) by J.P. Simons in 1965, and resurrected[9] in 1986 by E.A. Carter and W.A. Goddard III. The basic picture suggested that the C=C bond in an olefin that would formally cleave into a pair of singlet ground-state carbenes ought to be weakened by the sum of the singlet–triplet splittings in the two constituent carbenes. As originally proposed, the model applied to halogen-substituted olefins and carbenes such as tetrafluoroethylene dissociating to a pair of difluorocarbenes. A qualitative diagram correlating valence–bond structures across a planar dissociation coordinate connects the ground state of an olefin with a pair of triplet carbenes. The energy difference there is assumed to be that for an ordinary C=C bond, usually taken to be about 172 kcal/mol. However, halogen substitution on a carbenic center,[10] through π donation, lowers the closed-shell singlet state of the carbene below the open-shell triplet. The change in energetic ordering of carbene state does not affect the correlation, so for a dissociation to a pair of singlet carbenes, there is necessarily a curve crossing and a reduction in the actual C=C bond strength by a valence promotion energy. The model is shown pictorially in Figure 1.

It is important to acknowledge that the triplet is not really the "valence state". However, the valence state, by definition a nonstationary state formed as a linear combination of spectroscopic states, should contain a predominant contribution from the lowest energy state with open-shell character, which is, naturally, the triplet. Taking therefore the approximate identification of the triplet state of the

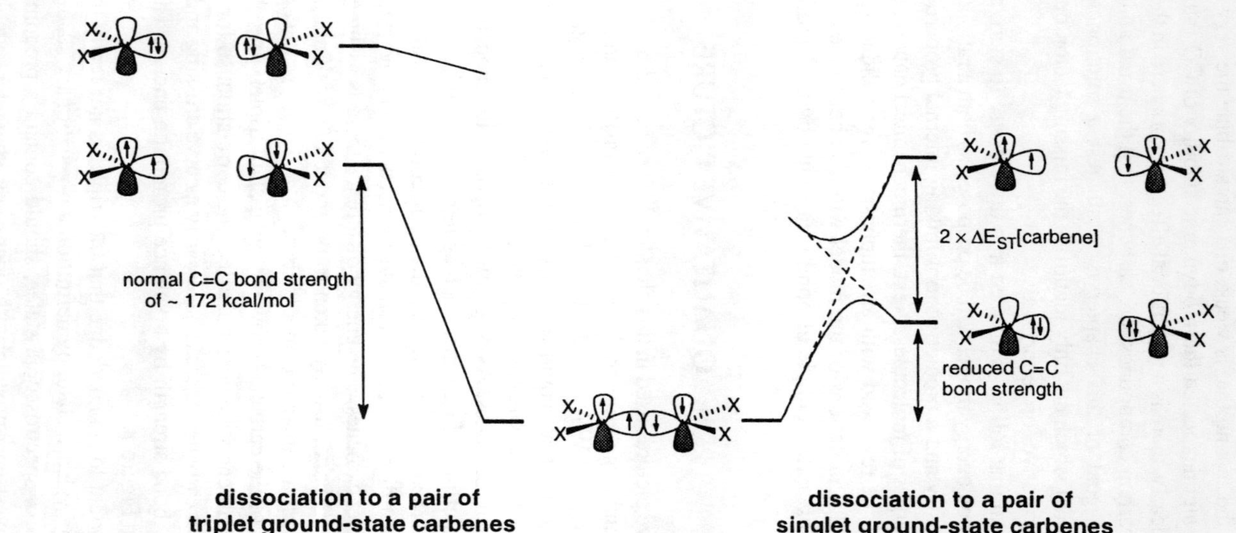

Figure 1. Valence bond promotion energy rationalization for a weakened C=C bond in olefins that are formally cleaved to a pair of singlet ground-state carbenes. The valence promotion energy is identified with the singlet–triplet splitting of the carbene.

carbene with its valence state, the valence promotion energy becomes simply the singlet–triplet splitting of the carbene (Figure 2). Accordingly, the actual C=C bond strength for an olefin going to a pair of singlet ground-state carbenes should be [$\Delta H_{diss} \sim [172 - (2 \times \Delta E_{ST})]$. In essence, one has related an overall property of the molecule, a bond strength, to properties intrinsic to fragments, singlet–triplet splittings. The reduced bond strength in the olefins, in this picture, is caused by a lower than expected $\Delta H_{f,298}$[carbene]. The deviation of $\Delta H_{f,298}$[carbene] from expectations, i.e. additivity, is just ΔE_{ST}, so conceptually the model is compatible with additivity in that one has simply identified one more correction that needs to be done.

The extension to C–H bond strengths is straightforward. A similar correlation diagram shows that the energy needed to break a C–H bond in a radical that forms H· and a singlet ground-state carbene ought to be $\Delta H_{diss} \sim (\Delta H_{diss}$[reference C–H] $- \Delta E_{ST}$). The reference C–H bond is simply a C–H bond of the same hydridization and substitution in an ordinary molecule. By this prediction, cleavage of a C–H bond to form a triplet ground-state carbene should show no unusual effects; looking at the ·CH$_3$ → :CH$_2$ + H· reaction, one finds $\Delta H_{diss} = 111$ kcal/mol,[11] which is perfectly normal for an sp^2-hybridized carbon. There is, accordingly, no reason that a bond α to a radical center is necessarily weakened; it all depends on whether or not the resultant carbene is a ground state singlet or triplet.

As in any qualitative model, there are pitfalls of which one should be aware. Principal among these is the question of geometry change. The valence promotion energy is defined for the carbene ground state, carbene valence state, and carbenic fragment in the larger molecule all having the same geometry. Given that the

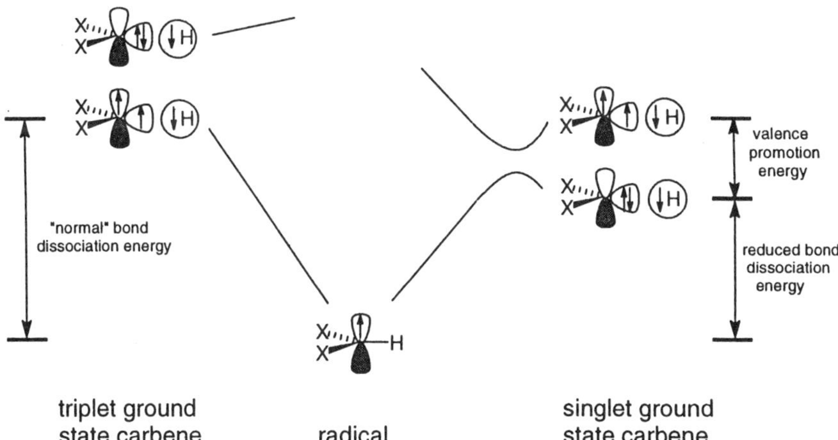

Figure 2. Valence bond promotion energy rationalization for the weakening of the C-H bond in a radical that formally cleaves to form a singlet ground-state carbene.

singlet–triplet splitting in carbenes has been shown to strongly depend upon geometry, in particular on the X–C–X angle at the carbenic center,[10] one concludes that the singlet–triplet splitting of the carbene is a good approximation to the promotion energy only if the geometry of the carbene (and carbenic fragment) is somehow constrained. Presumably, a further correction to the singlet–triplet splitting for the relative energies needed to distort the carbene into the correct geometry could be added in cases where constraint is not possible, but this begins to chip away at the simplicity of the model, which is, after all, its greatest attractiveness.

To show that the correlation of bond strength reductions to single–triplet splittings is really not a new idea at all, one needs only to examine a figure from Herzberg's monograph on diatomic molecules, published in 1950. Herzberg[12] applies Heitler's spin–valence theory[13] from 1934 to explain the anomalously weak bond in diatomic BeH by the correlation of the BeH ground electronic state to an open-shell Be (^{3}P) and H·. However, the ground state of a beryllium atom is Be (^{1}S), a closed-shell singlet. Accordingly, there is a curve crossing, and a reduction in the Be–H bond strength by the singlet–triplet splitting in atomic beryllium. The conceptual scheme is absolutely identical to that proposed above for carbenes. Other models closely related to the valence bond promotion energy scheme are the DSSE picture by Walsh,[14] the VB curve-crossing formalism by Pross and Shaik,[15] and the promotion energy arguments by Trinquier and Malrieu.[16] The several models vary in details, but the overall trends are always the same.

III. EXPERIMENTAL THERMOCHEMISTRY

A. Halocarbenes

Considering the simplicity of the halocarbenes, it is genuinely surprising that reliable thermochemical data has come out for CF_2 and CCl_2 only in the last several years. If one refers to a fairly recent reinvestigation[17] of $\Delta H_{f,298}[CCl_2]$ which reviews the literature up to 1985, one can see experimentally determined values by usually reliable methods ranging from 39 to 60 kcal/mol. Considering that CCl_2 is the "oldest" carbene, in the sense that it was the first carbene discussed as such in the literature,[18] one cannot help but be astonished that even its basic thermochemistry was not known reliably by the mid-1980s. The basic problems lay in experimental difficulties which, in some respects, are unavoidable for the carbenes because of the large geometry change upon ionization.

Looking again at CCl_2 as a representative case, most of the prior data could be grouped according to three experimental methods. A measured appearance potential for CCl_2^+ by either dissociative ionization or an ion–molecule reaction, combined with an ionization potential, $IP_{ad}[CCl_2]$, should give a reliable $\Delta H_{f,298}[CCl_2]$. Referring to Figure 3, one sees that the bond dissociation energy of tetrachloroethylene, and hence $\Delta H_{f,298}[CCl_2]$, is determined by a thermochemical cycle,

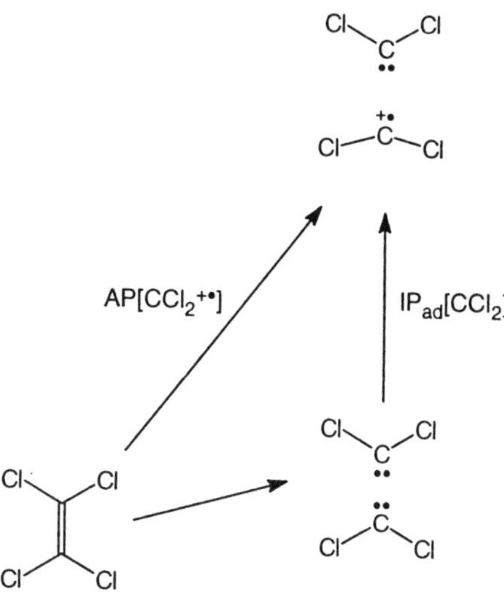

Figure 3. Thermochemical cycle connecting for the heat of formation of dichloro-carbene in terms of appearance and ionization potentials.

$$2 \times \Delta H_f[CCl_2] = \Delta H_f[Cl_2C{=}CCl_2] + AP[CCl_2^{+\bullet}] - IP_{ad}[CCl_2]$$

which ties the desired $\Delta H_{f,298}[CCl_2]$ to the heat of formation of a stable molecule, $Cl_2C{=}CCl_2$, and two spectroscopically measurable quantities. In practice, a number of stable reference molecules could be used in the experiment, and the values for $\Delta H_{f,298}[CCl_2^+]$ derived from the different measurements were reasonably consistent. However, measurement of $IP_{ad}[CCl_2]$ proved difficult, not the least because the experiment actually required the generation in situ of the reactive intermediate. Furthermore, the large change in both C–Cl bond length and Cl–C–Cl bond angle made the determination of an adiabatic ionization potential problematic. The accepted value[19] was $IP_{ad}[CCl_2] = 9.76$ eV, which, as we shall see, was seriously in error. A more recent determination[20] gave 9.10 ± 0.10 eV. Therefore, depending on the ionization potential that one chooses to believe, $\Delta H_{f,298}[CCl_2]$ can vary over a 15 kcal/mol range.

Bracketting with gas-phase proton transfer reactions, in principle, gives $\Delta H_{f,298}[CCl_2]$ to the precision set by ladder of compounds with known gas-phase basicities.[21] If one considers the reaction,

$$HCCl_2^+ + B \rightarrow CCl_2 + HB^+$$

and makes the usual (and experimentally justified) assumption that proton transfer will occur when it is thermodynamically favorable, then, $\Delta H_{f,298}[CCl_2]$ is bracketted when a pair of bases is found, only one of which takes a proton from $HCCl_2^+$.

In practice, the measurement is complicated in that the initial ion, $HCCl_2^+$, must be formed by some dissociative process, and could contain some internal energy. Unless the ions are carefully thermalized, the internal energy lowers the threshold for proton transfer, artifactually lowering the derived $\Delta H_{f,298}[CCl_2]$. The lowest reported value[17] for $\Delta H_{f,298}[CCl_2]$, 39 kcal/mol, was, in fact, done by pulsed ion cyclotron resonance spectroscopy using just this method, and probably suffered from incomplete thermalization. The same measurement, done at different times or by different groups, yielded numbers up to 16 kcal/mol higher.[22]

Other threshold techniques rely upon measurement of the onset for some reaction in which $\Delta H_{f,298}[CCl_2]$ is the only unknown besides the actual measured quantity. For example, using dissociative electron capture, the reaction,

$$CCl_4 + e^- \rightarrow CCl_2 + Cl_2^-$$

can be used to derive a value for $\Delta H_{f,298}[CCl_2]$, but again, independent measurements for the same reaction gave widely disparate results.[23]

A perusal of the values listed in ref. 17 clearly underlines the experimental difficulties inherent in the thermochemistry of carbenes. There are also some solution-phase techniques, but these are not discussed here because the substantial dipole moment of halocarbenes makes it likely that differential solvation phenomena would be non-negligible.

Two recent determinations by independent groups utilizing no common experimental techniques or auxiliary thermochemical data have arrived at a consistent value for $\Delta H_{f,298}[CCl_2]$. Paulino and Squires[24] determined the thermochemical threshold for collision-induced loss of Cl^- from CCl_3^-, which, in combination with the gas-phase acidity of chloroform, yields $\Delta H_{f,298}[CCl_2] = 52.1 \pm 3.4$ kcal/mol. A very similar result of $\Delta H_{f,298}[CCl_2] = 51.0 \pm 2.0$ kcal/mol comes from Kohn et al.[25] using the more conventional combination of appearance potential and ionization potential. The consistency of two completely independent measurements argues strongly for the derived number. We will take $\Delta H_{f,298}[CCl_2] = 51.0 \pm 2.0$ kcal/mol to be the correct heat of formation in the later discussion.

Because the latter measurement shares with earlier determinations the general method, it is instructive to examine it in more detail to highlight the inherent difficulties in appearance potential/ionization potential determinations of carbene thermochemistry. As mentioned above, the primary experimental problem comes in the determination of the adiabatic ionization potential of CCl_2. The problem in CCl_2 is completely general in that the determination of IP_{ad} for any singlet ground-state carbene will suffer from the same difficulties unless the carbene is geometrically constrained. Referring to Figure 4, one sees the large geometry change[25] in CCl_2 that occurs upon removal of an electron. Qualitatively, the increase in bond angle comes because removal of an electron from the in-plane sp^2 orbital lessens the energetic advantage of high s character in that orbital, while the decrease in bond length derives from increased π donation to the carbenic center upon

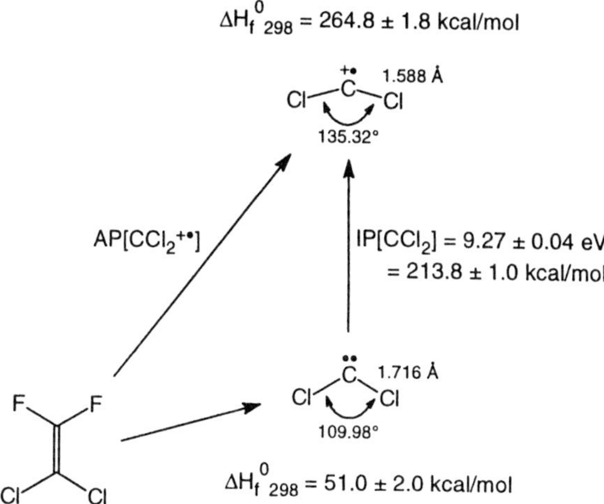

Figure 4. Thermochemical cycle and *ab initio* computed geometries for dichlorocarbene and its radical cation. The ionization potential, determined by photoelectron spectroscopy, gives $\Delta H_{f,298}[CCl_2]$.

ionization. Consequently, there is a large separation between the adiabatic and vertical ionization potentials for unconstrained singlet carbenes.

The poor Franck–Condon factor for ionization at threshold may make the adiabatic ionization potential unobservable. An examination of the threshold law,[26] i.e. the functional dependence of signal intensity on energy in excess of an ionization threshold, finds that, within a set of simplifying approximations,[27] integration of the photoelectron spectrum yields the mass spectrometric photoionization efficiency curve. Further integration of the photoionization efficiency curve yields the electron impact ionization efficiency curve. As is immediately evident from Figure 5, the adiabatic ionization potential in the simulated spectrum, which occurs just to the left of the first tick mark in the PES, becomes almost impossible to determine by the time one gets to the EIE. It is questionable that any deconvolution or extrapolation scheme[28] can pull IP_{ad} out from such a PIE or EIE curve. The solution, implemented in ref. 25, involves using photoelectron spectroscopy to take advantage of the best threshold law, and numerical computation of Franck–Condon factors based on geometries for CCl_2 and $CCl_2^{+\bullet}$ so that the entire photoelectron band profile can be included in a fit (Figure 6). By fitting the entire band, an adiabatic ionization potential can be extracted even when the intensity at threshold is so small as to be unobservable by direct methods. The resulting adiabatic ionization potential, $IP_{ad}[CCl_2] = 9.27 \pm 0.04$ eV, combined with $\Delta H_{f,298}[CCl_2^{+\bullet}] = 264.8 \pm 1.8$ kcal/mol from appearance potential measurements,[20] yields $\Delta H_{f,298}[CCl_2] = 51.0 \pm 2.0$ kcal/mol.

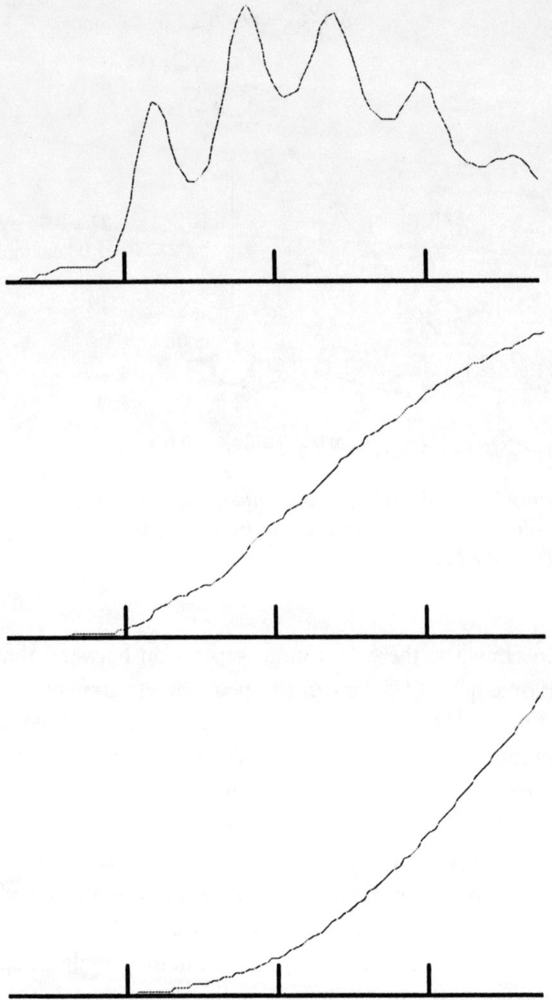

Figure 5.   Simulated photoelectron spectrum (PES, top trace) for a molecule with only modestly bad Franck–Condon factors at threshold over about 1 eV, mass spectrometric photoionization efficiency curve (PIE, middle trace), and mass spectrometric electron impact ionization efficiency curve (EIE, bottom trace). The adiabatic peak in the PES is the small peak to the left of the first tick mark.

A reliable heat of formation for CCl_2 can be used directly to check the qualitative bond strength picture proposed by Carter and Goddard. Correlation-consistent singlet–triplet splittings for CCl_2 of 25.9 kcal/mol and 20.9 kcal/mol have been reported by Carter et al[29] and Shin et al,[30] respectively, with the latter figure to be

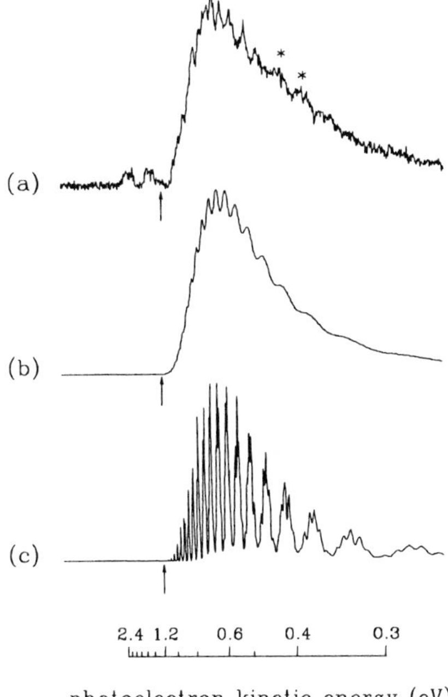

Figure 6. 10.49 eV vacuum-uv laser photoelectron spectrum of CCl_2 (trace a) and Franck–Condon simulations for 40 me V FWHM (trace b) and 6 me V (trace c) FWHM. The extrapolated position of the adiabatic ionization potential is marked with an arrow, and corresponds to $IP_{ad}[CCl_2] = 9.27 \pm 0.04$ eV. From Ref. 25.

considered somewhat better. While the one direct experimental study[31] of carbene singlet–triplet gaps by negative ion photoelectron spectroscopy did not report ΔE_{ST} for CCl_2 due to spectral congestion, the estimate of $\Delta E_{ST}[CCl_2] \sim 21$ kcal/mol is completely consistent with the values for the other halocarbenes in that study. The prediction, therefore, for $\Delta H_{f,298}[CCl_2]$, in kcal/mol, is,

$$\Delta H_f[CCl_2] \approx [\Delta H_f[Cl_2C{=}CCl_2] + 172 - (2 \times \Delta E_{ST})]/2 = [-3 + 172 - (2 \times 21)]/2$$

$$= 64 \text{ kcal/mol}$$

which is higher than the measured $\Delta H_{f,298}[CCl_2] = 51.0 \pm 2.0$ kcal/mol by a substantial 13 kcal/mol. The slightly higher $\Delta E_{ST} = 26$ kcal/mol from Carter and Goddard reduces the discrepancy to 8 kcal/mol, if one can justify taking that number in favor of the later one. Adjustment of the carbene geometry to that of the CCl_2 fragment in tetrachloroethylene further raises ΔE_{ST} by another 3 kcal/mol, which is still not enough to bridge the difference.

While the direction and approximate magnitude of the deviation of $\Delta H_{f,298}[CCl_2]$ from additivity is correctly predicted by $\Delta E_{ST}[CCl_2]$, the remaining discrepancy does need to be understood. The central approximation in the Carter/Goddard picture is the identification of the valence state of a carbenic fragment with the triplet state of the carbene. The advantage of doing so has been mentioned previously: we relate an intrinsic, measurable property of an isolated fragment to a measurable property of the molecule containing that same fragment. However, the triplet state of CCl_2 is not really the valence state. The valence state, by definition, is a nonstationary state formed as a linear combination of spectroscopic states. Because the lowest triplet state of CCl_2 is the lowest energy spectroscopic state with predominantly open-shell character, its contribution to the valence state should be large and dominant. Hence, the qualitative picture works, identifying ΔE_{ST} with the promotion energy. However, contributions from other electronic states with open-shell character (which are higher in energy) are nevertheless present, which necessarily raises the energy of the valence state above that of the lowest triplet state of CCl_2. Accordingly, ΔE_{ST} is actually only a lower bound on the promotion energy, making the estimated heat of formation an upper bound on $\Delta H_{f,298}[CCl_2]$.

For CF_2, the experimental determinations of $\Delta H_{f,298}[CF_2]$ also show a disturbing spread, probably for many of the same reasons as in CCl_2. A summary[17] of data up to 1985 lists values for $\Delta H_{f,298}[CF_2]$ ranging from −34 kcal/mol to −56 kcal/mol. The most reliable value comes from Paulino and Squires[24] by negative collision-induced dissociation thresholds, which give $\Delta H_{f,298}[CF_2] = -39.4 \pm 3.4$ kcal/mol. The singlet–triplet splitting, $\Delta E_{ST}[CF_2]$ is much larger than in CCl_2, with correlation-consistent *ab initio* calculations[30] giving $\Delta E_{ST}[CF_2] = 57.1$ kcal/mol. Experimental values of 57 kcal/mol,[32] and > 50 kcal/mol,[31] are in good agreement. Accordingly, an estimate of $\Delta H_{f,298}[CF_2]$ is

$$\Delta H_f[CF_2] \approx [\Delta H_f[F_2C{=}CF_2] + 172 - (2 \times \Delta E_{ST})]/2 = [-158 + 172 - (2 \times 57)]/2$$

$$= -50 \text{ kcal/mol}$$

which, in contrast to the case of CCl_2, is somewhat lower than the experimentally determined value. The discrepancy, 11 kcal/mol this time, arises presumably because, as in CCl_2, the singlet–triplet splitting of CF_2 is, in reality, only an approximation to the promotion energy.

Similar calculations for other halocarbenes for which ΔE_{ST} is known by either calculation[30] or experiment[31] can, in principle, be done. However, reliable experimental thermochemistry for these species is lacking for a comparison. As will be later clear, *ab initio* methods which yield absolute thermochemistry are practical for small molecular species such as halocarbenes and should give numbers good enough to make comparisons to qualitative predictions. Some benchmark cases will be discussed in a later section. However, it is clear that the application of the Carter/Goddard scheme to halocarbenes allows prediction to within about 10 kcal/mol, which, it should be noted, is not bad for a qualitative, pictorial model.

B. Vinylidene

The valence isomer of aceylene, known as vinylidene, is a singlet carbene with rather large singlet–triplet splitting of 47 kcal/mol. The large value for ΔE_{ST} was determined spectroscopically by photodetachment of the vinylidene radical anion.[33] Accordingly, its thermochemistry should be amenable to the analysis that was used above for halocarbenes. One can ascertain that vinylidene is thermally accessible from acetylene by the rapid isotopic scrambling[34] of doubly labeled acetylene, which occurs at 850 °C after a contact time of a few seconds (Figure 7). The onset of perturbations 44 kcal/mol above the ground state in the high-resolution spectrum[35] of acetylene has been attributed to the onset of the 1,2-H shift converting acetylene to vinylidene, and would give $\Delta H_{f,298}[H_2C{=}C{:}] \sim 98$ kcal/mol. A thermochemical cycle for which error analysis is better defined uses the proton transfer reaction between the vinyl radical and either water or methanol to determine the proton affinity of the vinyl radical. Negative ion photoelectron spectroscopy of the resulting vinylidene radical anion establishes the electron affinity of vinylidene.[33,36] One therefore arrives at:

$$\Delta H_{f,298}[H_2C{=}C{:}] = \Delta H_{f,298}[H_2C{=}CH] + PA[H_2C{=}C^{-\bullet}] - \Delta H_{f,298}[H^+] +$$
$$EA[H_2C{=}C{:}]$$

$$= (71.6 \pm 0.8) + (384.7 \pm 3.4) - (367.2) + (11.3 \pm 0.1) = 100.4 \pm 3.5 \text{ kcal/mol}$$

Furthermore, *ab initio* calculations[37] at the CBS-QCI/APNO level gives $\Delta H_{f,298}[H_2C{=}C{:}] = 103$ kcal/mol, supporting the experimental determination. For the corrected additivity estimate, one would think that one starts with the heat of formation of ethylene,[11] breaks two equivalent, ethylenic C–H bonds,[36] subtracts the hydrogen atoms, and finally, subtracts the singlet–triplet splitting,

$$\Delta H_{f,est}[H_2C{=}C{:}] \approx \Delta H_f[H_2C{=}CH_2] + 2 \times BDE[H_2C{=}CH_2] - 2 \times \Delta H_f[H\cdot] -$$
$$\Delta E_{ST}[H_2C{=}C{:}]$$

$$= 12.5 + 2 \times 111.5 - 2 \times 52.1 - 47 = 84.3 \text{ kcal/mol}$$

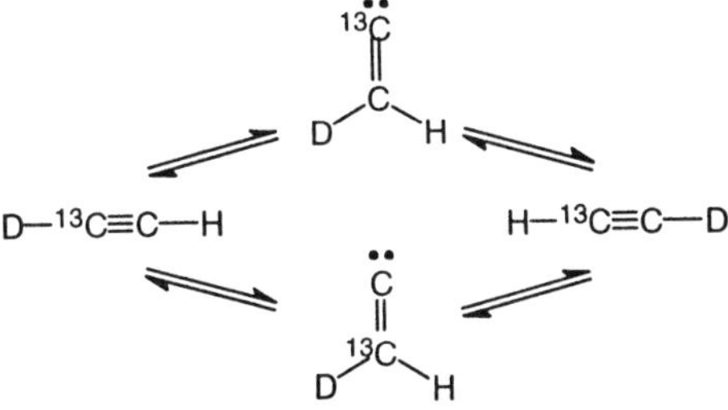

Figure 7. Isotopic scrambling in doubly-labeled acetylene as evidence for a vinylidene intermediate.

which lies 16 kcal/mol below the experimental determination of $\Delta H_{f,298}[H_2C{=}C{:}]$. This would give a discrepancy that is somewhat worse than that for the halocarbenes. The origin of the discrepancy is an incorrect calculation of the additivity estimate. The experimental C–H bond strength of ethylene is the enthalpy change going from ethylene, for which the C–H bonds are on sp^2-hybridized carbons, to the vinyl radical, for which the hybridization at the radical site is somewhere between sp^2 and sp, as judged by the geometry, where the CC–H angle is opened up significantly from 120°. The observed hyperfine structure in the ESR spectrum[38] of C_2H_3 is fit to a nonlinear $H_2C{=}C{-}H$ structure with rapid flipping of the lone hydrogen from one side to the other over a barrier[39] of approximately 2 kcal/mol. Because the hyperfine couplings are very sensitive to the extent of s character in the SOMO, the ESR data may be regarded as definitive. At the linear geometry, the carbon bearing the lone hydrogen is clearly sp-hybridized, so the uncorrected additivity estimate for the C–H bond strength in the vinyl radical should be the acetylenic,[36] rather than the ethylenic, C–H bond strength, less a few kcal/mol for the effect of nonlinearity. Making that change, one gets for the new, corrected additivity estimate.

$$\Delta H_{f,est}[H_2C{=}C{:}] = \Delta H_f[H_2C{=}CH_2] + BDE[H_2C{=}CH_2] + BDE[H{-}C{\equiv}C{-}H] -$$

$$2 \times \Delta H_f[H^{\bullet}] -$$

$$\Delta E_{ST}[H_2C{=}C{:}] - 2 \text{ kcal/mol}$$

$$12.5 + 111.2 + 132.8 - 2 \times 52.1 - 47 - 2 = 103.3 \text{ kcal/mol}$$

The new estimate, using the correct reference bond strengths, is in excellent agreement with the experimental thermochemistry, and underlines the importance in addifivity calculations of choosing the appropriate comparisons to make. It should be emphasized that the selection of reference bond strengths is not arbitrary, but, on the contrary, comes from some sort of physical evidence.

C. Cyclopropenylidene, Propadienylidene, Ethynylcarbene

The isomeric series of C_3H_2 carbenes, cyclopropenylidene, propadienylidene, and ethynylcarbene, provide a particularly good test bed for models of carbene thermochemistry.[40] Not only are the molecules small enough for practical computational studies, but they are small enough for gas-phase spectroscopy using relatively low-resolution techniques. They are furthermore better suited as tests for the qualitative models of $\Delta H_{f,298}[\text{carbene}]$ than the halocarbenes because the main geometric problem in the latter family of carbenes, variation of the X–C–X angle, is absent in the C_3H_2 series. Of the three bound isomers of C_3H_2 listed above, the first two are ground-state singlets, while the last is a ground-state triplet. What are the expectations? We can connect the three C_3H_2 isomers to two C_3H_4 compounds. Cyclopropenylidene is derived from cyclopropene, with $\Delta H_{f,298}[c\text{-}C_3H_4] = 66.0$

kcal/mol,[11] by the sequential cleavage of two C–H bonds. To make an additivity estimate, one needs $\Delta H_{f,298}[c\text{-}C_3H_3]$. Because the single report[41] of a bond dissociation energy in cyclopropene suffers from several systematic errors, we use the computed[42] (by the G-2 method, see below) value of $\Delta H_{f,298}[c\text{-}C_3H_3] = 118$ kcal/mol. For the geometry of the cyclopropenyl radical, one uses, in analogy to the example with vinyl radical above, the *ab initio* computed geometry that fits the ESR hyperfine structure.[43] The electron spin in that spectrum is strongly coupled to only one proton spin, and is fit by a nonplanar structure in which the unique C–H bond is bent about 50° out of the plane of the ring.[44] Accordingly, the carbon bearing the hydrogen is approximately sp^2-hybridized. Going from cyclopropenyl radical to cyclopropenylidene, one then needs to break that sp^2-hybridized bond, which nominally would be in the range of approximately 105–110 kcal/mol. One should note that the error in the corrected additivity estimate introduced by a poor choice of geometry for $c\text{-}C_3H_3$ would be in the individual C–H bond strengths from cyclopropene, but that the sum of the two sequential C–H bond strengths should be predicted well.

Similarly, propadienylidene and ethynylcarbene are connected to propyne, with $\Delta H_{f,298}[l\text{-}C_3H_4] = 44.3$ kcal/mol,[11] by cleavage of either one hydrogen from each end of the molecule, or removal of both hydrogens from one end. Using the approximate sp- and sp^2-hybridized C–H bond strengths[35] of 130 and 110 kcal/mol, and $\Delta H_{f,298}[H^\bullet] = 52.1$ kcal/mol, one gets the additivity estimates below.

The singlet–triplet splitting of cyclopropenylidene is very large,[45] computed to be ~60 kcal/mol. One rationalizes the large magnitude for the singlet–triplet gap by two effects. Of the two frontier orbitals in the carbene, the in-plane orbital has high s character (according to the Walsh model for strained rings), making it lower in energy relative to what it would have been in an acyclic system. Furthermore, the out-of-plane orbital, empty in the singlet, but singly occupied in the triplet, is not a nonbonding orbital, but rather a π^* antibonding orbital of the aromatic π system. The singlet is stabilized with respect to the triplet by both effects, which leads to a prediction of an internal separation of charge in the singlet ground state of cyclopropenylidene. The internal charge separation is represented by an ylid resonance structure which should contribute substantially. As proof, the dipole moment of cyclopropenylidene, as measured by microwave spectroscopy,[46] is $\mu = 3.4$ D, an extraordinarily large value for a neutral hydrocarbon. The singlet–triplet splitting in propadienylidene is computed[47] to be ~35 kcal/mol, and recently measured[48] to be just under 30 kcal/mol, as would be expected for a vinylidene-type carbene. Ethynylcarbene, as mentioned above, is a ground-state triplet, as determined by computation[49] and ESR spectroscopy.[50]

Using the variation of the Carter/Goddard picture in which the singlet–triplet splitting in singlet carbenes represents the sum of the deviations from additivity for the sequential cleavage of two C–H bonds in the hydrocarbon producing the carbene, the corrected additivity estimates for the heats of formation in the three isomeric C_3H_2 carbenes are respectively: $\Delta H_{f,298}[\text{cyclopropenylidene}] \approx 175 - 60$

= 115 kcal/mol, $\Delta H_{f,298}$[propadienylene] = 160 − 30 = 130 kcal/mol, and $\Delta H_{f,298}$[ethynylcarbene] ≈ 140 kcal/mol, the latter indicating that a bond α to a radical center is not necessarily weakened (Figure 8). The test of the model with these three carbenes is especially favorable because the singlet–triplet splittings on the two singlet ground-state species are so large and there is little opportunity for geometry change.

For cyclopropropenylidene, c-C$_3$H$_2$, two sets of measurements can be envisioned by which the heat of formation can be measured. The first involves dissociative photoionization to measure an appearance potential for c-C$_3$H$_2^{+\bullet}$, combined with an ionization potential measurement on c-C$_3$H$_2$. The second set involves determination of the gas-phase acidity of the cyclopropenium cation, or conversely, the proton

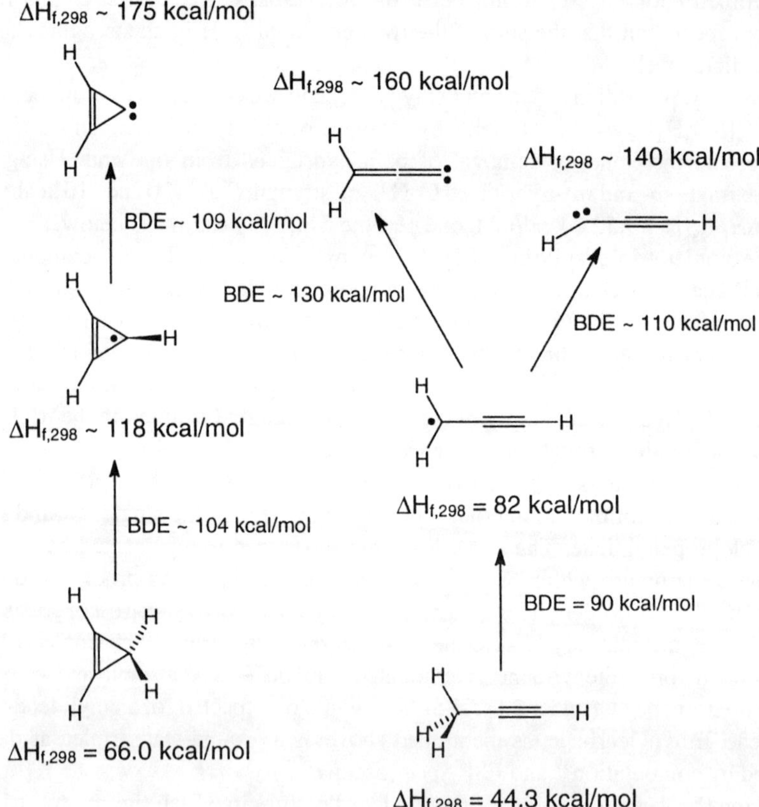

Figure 8. Uncorrected additivity estimates for the heat of formation for the three isomeric carbenes, starting from either cyclopropene or propyne. The reference bond energies were chosen by considering the hybridization of the carbon bearing the C–H bond to be broken.

affinity of cyclopropenylidene. Both cycles are shown on common diagram in Figure 9. With the heat for formation of c-$C_3H_3^+$ known from a variety of experiments, proton transfer equilibrium or bracketing would be a good way to determine $\Delta H_{f,298}[c\text{-}C_3H_2]$. While this latter experiment has not been performed, the proton affinity of c-C_3H_2 has been reported[51] to be at least 16 kcal/mol greater than that for NH_3. Using $PA[NH_3] = 204.0$ kcal/mol,[21] one obtains $PA[c\text{-}C_3H_2] \geq 220$ kcal/mol. From dissociative photoionization[52,53] comes $\Delta H_{f,298}[c\text{-}C_3H_3^+]. = 258.5 \pm 0.7$ kcal/mol, which sets a lower limit of $\Delta H_{f,298}[c\text{-}C_3H_2] \geq 113$ kcal/mol. This value, unfortunately, neither challenges nor confirms the prediction. An *ab initio* computed proton affinity[54] has been reported using the G-1 formalism which is reputed to have an accuracy of ± 2 kcal/mol. The computed proton affinity of $PA[c\text{-}C_3H_2]$ = 225 kcal/mol, gives $\Delta H_{f,298}[c\text{-}C_3H_2] = 118$ kcal/mol. With a series of additivity assumptions, the effective level of theory in a G-1 calculation is QCISD(T)/6-311 + G(2*df,p*)//MP2/6-31G,* which should predict proton affinities quite well.

The other thermochemical cycle involving appearance and ionization potentials has been successfully implemented.[40] The key technical innovation was a gas-phase source of reactive intermediates that produced the species cleanly and abundantly enough to utilize a low-resolution, low-selectivity spectroscopic technique such as photoelectron spectroscopy. The experimental photoelectron spectrum, taken with 10.49 eV vacuum–UV laser photoionization, is shown in Figure 10 with a Franck–

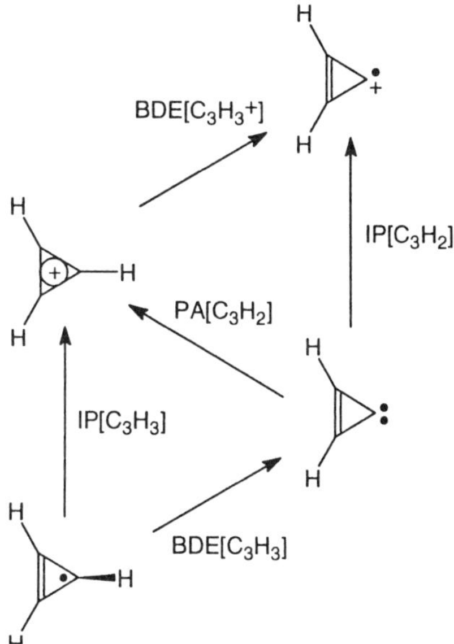

Figure 9. Two thermochemical cycles that could give the heat of formation for cyclopropenylidene. Measurement of either the ionization potential or the proton affinity is needed.

Condon simulation of the band profile based on *ab initio* computed geometries.[55] Not only does the simulation uniquely identify the structure as cyclopropenylidene, as opposed to one of its isomers, but it also determines with confidence the peak corresponding to the adiabatic ionization potential. The measured adiabatic ionization potential is $IP_{ad}[c\text{-}C_3H_2] = 9.15 \pm 0.03$ eV, or 211.0 ± 0.7 kcal/mol. For the cyclopropenylidene radical cation, $c\text{-}C_3H_2^{+\bullet}$, dissociative photoionization measurements[56] give its heat of formation as $\Delta H_{f,298}[c\text{-}C_3H_2^{+\bullet}] = 329 \pm 2$ kcal/mol. A SIFT atom transfer study[57] of the reaction, $c\text{-}C_3H_2^{+\bullet} + MH \rightarrow c\text{-}C_3H_3^{+} + M\cdot$ set an upper-bound of $\Delta H_{f,298}[c\text{-}C_3H_2^{+\bullet}] \leq 326 \pm 2$ kcal/mol, with the assumption of identical kinetic and thermochemical barriers to the reaction. The assumption is generally safe for gas-phase proton transfer reactions, but the generality for atom transfer reactions can be questioned. A small activation barrier for the reverse reaction would have the effect of increasing the upper-bound by a few kcal/mol, which would bring it into agreement with the value from dissociative photoionization. Accordingly, with $\Delta H_{f,298}[c\text{-}C_3H_2^{+\bullet}] = 329 \pm 2$ kcal/mol, one arrives at $\Delta H_{f,298}[c\text{-}C_3H_2] = 118 \pm 2$ kcal/mol, in excellent agreement with the value using the calculated proton affinity. The all-experimental $\Delta H_{f,298}[c\text{-}C_3H_2]$ also agrees

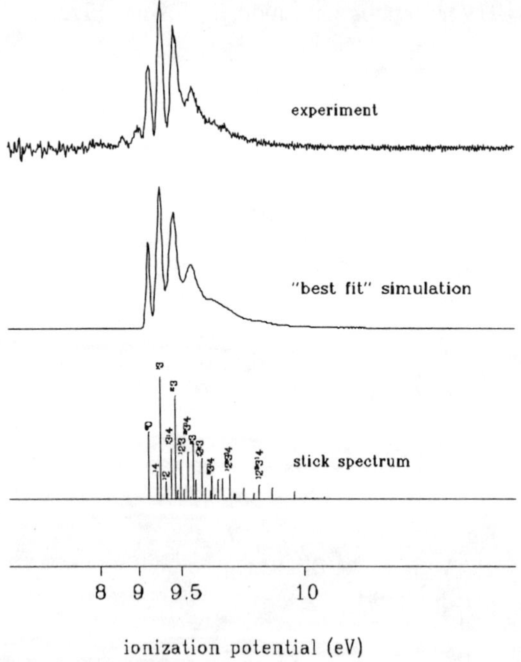

Figure 10. 10.49 eV vacuum-UV laser photoelectron spectrum of cyclo-propenylidene, and Franck–Condon simulations based on *ab initio* geometries as proof of chemical identity. From Ref. 55.

very well with the corrected additivity estimate of 115 kcal/mol, giving credence to the Carter/Goddard picture of carbene thermochemistry.

All *ab initio* estimates for absolute thermochemistry by the G-2 and CBS-QCI/APNO formalisms purport to predict heats of formation to within 2 kcal/mol or better. For cyclopropenylidene, further support for the experimentally determined heat of formation is provided by work using these two computational methodologies. The G-2 estimate is effectively a QCISD(T)/6-311 + G($3d2f$, $2p$)//MP2/6-31G* calculation, and predicts $\Delta H_{f,298}[c\text{-}C_3H_2]$ = 120 kcal/mol.[42] The CBS-QCI/APNO method[37] gives a comparable result.

For the other two isomers of C_3H_2, there exists some spectroscopic data,[40] e.g. $IP_{ad}[H_2C\!=\!C\!=\!C\text{:}]$ = 10.43 $\pm$ 0.02 eV, but there are insufficient auxiliary thermochemical measurements to complete a cycle to give heats of formation. However, for the purposes of the discussion here, in particular, for comparison to the corrected additivity estimates, a good quality *ab initio* computed relative energy should suffice. In calculations at the MP4/6-311++G(df, pd) level,[58] the two higher energy isomers, propadienylidene and ethynylcarbene, are predicted to lie 14.5 and 21.7 kcal/mol above cyclopropenylidene, respectively. Very similar results are found at the G-1 level,[54] which place propadienylidene and ethynylcarbene 12.5 and 23.2 kcal/mol above cyclopropenylidene, respectively. Using the latter estimates, and the experimental $\Delta H_{f,298}[c\text{-}C_3H_2]$ = 118 $\pm$ 2 kcal/mol, one arrives at $\Delta H_{f,298}[H_2C\!=\!C\!=\!C\text{:}]$ $\approx$ 131 kcal/mol and $\Delta H_{f,298}[HCCCH]$ $\approx$ 141 kcal/mol. These two heats of formation for the remaining two isomers of C_3H_2 very closely match the corrected additivity estimates made at the head of this section using canonical bond strengths and the singlet–triplet splitting of the carbene. Again, the agreement is very good and confirms the qualitative validity of the Carter/Goddard model.

D. Cyclopentadienylidene

Cyclopentadienylidene is a ground-state triplet, as confirmed by ESR[59] and quantum chemical calculations.[60] Matrix-isolation studies have also produced IR and UV/vis spectra[61] of the carbene. In $c\text{-}C_5H_4$, the two frontier orbitals, designated σ and π, resemble those in $c\text{-}C_3H_2$. However, in contrast to $c\text{-}C_3H_2$, in which the σ^2 configuration (relative to either ($\sigma^1\pi^1$ or pi sup 2) is strongly favored by both double occupancy of the lower energy frontier orbital, and by π aromaticity, the two effects work against each other in $c\text{-}C_5H_4$, effectively canceling out any particular stabilization of a singlet state for the carbene. Accordingly, the typical Hund's rule argument makes the carbene a ground-state triplet. One therefore predicts for $\Delta H_{f,298}[c\text{-}C_5H_4]$ the simple additivity value. Starting with cyclopentadiene, for which the heat of formation is known from conventional calorimetry, $\Delta H_{f,298}[c\text{-}C_5H_6]$ = 31 $\pm$ 1 kcal/mol, one breaks one C–H bond to go to cyclopentadienyl radical, and a second one at the same site to go to cyclopentadienylidene (Figure 11). Data for the cyclopentadienyl radical has been critically evaluated,[62] giving a recommended value of $\Delta H_{f,298}[c\text{-}C_5H_5]$ = 58 $\pm$ 2 kcal/mol, which does reflect some resonance stabilization of the radical. Breaking the second C–H bond, assuming a nominal sp^2 bond strength gives $\Delta H_{f,298}[c\text{-}C_5H_4] \approx 116$ kcal/mol as the additivity estimate. Given the triplet ground state of cyclopentadienylidene, one would expect the actual heat of formation to be close to the additivity value.

The experimental determination[63] combines gas-phase proton transfer bracketing and an electron photodetachment threshold measurement. The radical anion of the carbene was produced by dissociative electron attachment to diazocyclopentadiene. Its proton affinity was determined by proton transfer measurements against a series of alcohols, which when combined with the electron affinity from a photodetachment threshold, yielded $\Delta H_{f,298}[c\text{-}C_5H_4] = 112.3 \pm 4.7$ kcal/mol for the carbene. Again, the deviation from additivity is only on the order of a few kcal/mol, which can easily be attributed to the energy needed to accomodate small changes in bond lengths and angles. The successful prediction of ΔH_f in the case of cyclopentadienylidene indicates that the cyclic conjugated carbenes, such as cyclopropenylidene, are not all necessarily ground-state singlets with strongly lowered heats of formation. One would expect, however, that the next homologue, cycloheptatrienylidene, which is a ground-state singlet, should show a lower-than-expected heat of formation.

IV. THE MOST EXTREME TEST

So far, we have seen triplet carbenes with heats of formation that match additivity, and singlet carbenes whose heats of formation are lowered relative to additivity by their singlet–triplet splittings. In the latter case, either the C–H single bond in a radical which could give the singlet carbene, or the C=C double bond in an olefin that is the formal dimer of a pair of singlet carbenes, is weakened by a promotion energy which could be quite large. As an extreme test of the theoretical prediction, it should be possible to construct a singlet carbene with a singlet–triplet splitting

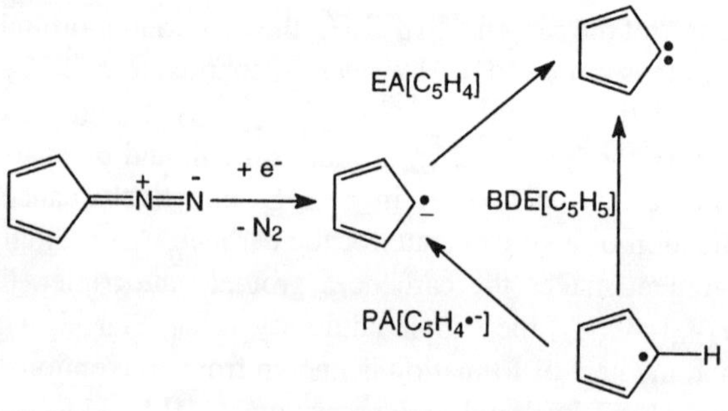

Figure 11. Preparation, by dissociative electron capture, of the radical anion of cyclopentadienylidene. Proton affinity and electron affinity measurents give the heat of formation of the carbene.

so large that the C=C bond in its formal dimer has a zero or negative bond dissociation enthalpy. For the halocarbenes and the purely hydrocarbon systems, the largest singlet–triplet splittings are $\Delta E_{\mathrm{ST}}[\mathrm{CF_2}]$ and $\Delta E_{\mathrm{ST}}[c\text{-}\mathrm{C_3H_2}]$, both approximately 60 kcal/mol. Certainly for tetrafluoroethylene, there is a C=C bond, and for the (unknown) formal dimer of cyclopropenylidene, there should be one as well.

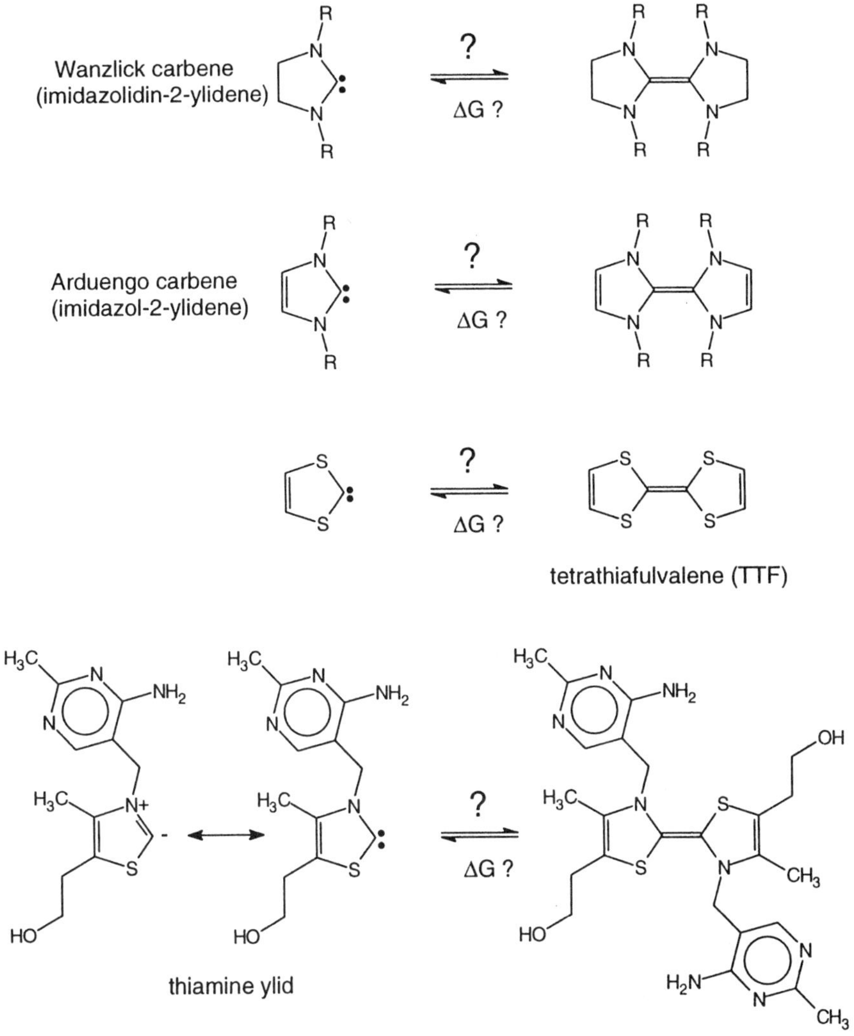

Figure 12. Singlet ground-state carbenes and their dimers. Heteroatom π-donors on the carbenic carbon, and small X–C–X bond angles, stabilize the singlet relative to the triplet. The interconversion between the carbenes and their olefinic dimers has not been observed for all cases.

Adding other better, heteroatom π donors to the carbenic center can increase ΔE_{ST}[carbene] even more. The choice of structures in Figure 12 also reflects the range of important applications for which such singlet carbenes or their dimers may be important. Thiamine ylid,[64] and the related thiazolium ylids are singlet carbenes with large singlet–triplet splittings, for example. Their catalytic activity in biochemical C–C forming processes[65] have inspired the use of similar catalysts in organic synthesis.[66,67] The rather low heat of formation of the sulfur-substituted carbenes is reflected in the thermal generation of dithiacarbenes from CS_2 and acetylenes,[68] for which the formal dimers are the tetrathiafulvalenes that find use as building blocks in organic metals. There have even been suggestions that interconversion between singlet carbenes and their dimers may be important to their chemistry, with the dimer playing a role beyond that of a reservoir for carbenes.[69] One can certainly appreciate the importance of knowing the C=C bond strength in the dimer.

Of all the structures, those with the largest singlet–triplet splittings should occur for the nitrogen-substituted systems. Decreasing the X–C–X angle should further favor the singlet over the triplet, meaning that we want a nitrogen-substituted carbene in a small ring. Two families of carbenes can be envisioned, depending on whether the five-membered ring is saturated or not. The former, the imidazolidin-2-ylidenes, were studied extensively by Wanzlick and others starting over 30 years ago.[70] The latter, the imidazol-2-ylidenes, were only recently reported by Arduengo.[71] Recognizing that there have been contributions by others, we nevertheless designate the two families of carbenes as Wanzlick and Arduengo carbenes as a matter of convenience in discussion.

A. Wanzlick Carbenes and Dimers

The imidazolidin-2-ylidenes were proposed by Wanzlick in 1962 as reactive intermediates in the pyrolysis of 2-trichloromethylimidazolidine.[70] Wanzlick expected a β elimination of HCl, producing the corresponding dichloro-olefin, but instead, isolated a biimidazolidinylidene, the formal dimer of the carbene (Figure 13). The carbene was subsequently trapped by reaction with olefins, forming cyclopropane derivatives. The various related structures were termed nucleophilic carbenes because of their preference for reaction with electron-poor substrates. Various efforts at the time to isolate the carbenes failed, but reactions of the dimers invariably led to carbene trapping products. The latter observation led to the suggestion that the thermal cleavage of the dimer's central C=C bond was facile, allowing an equilibrium between carbene and dimer. The question as to whether or not there existed such a thermal equilibrium between Wanzlick carbenes and their dimers was finally settled by Lemal,[72] who showed, in a clever crossover experiment, that there was no purely thermal interconversion (Figure 14). It was subsequently shown that the carbene chemistry of the dimers arose because of an electrophile-induced cleavage of the central C=C bond, releasing one mole of

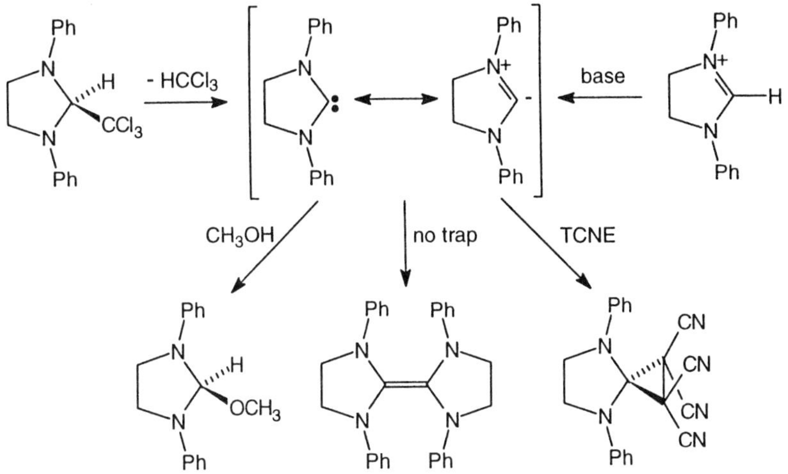

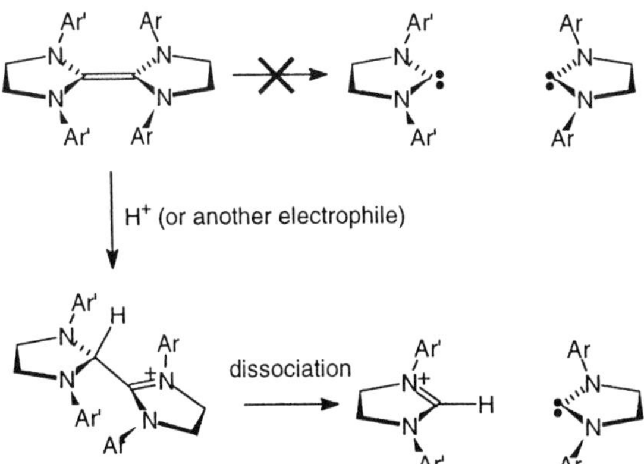

Figure 13. Generation and reactions of the *N,N*-diphenylimidazolidin-2-ylidene (Wanzlick carbene) by α-elimination or deprotonation. Trapping with methanol, tetracyanoethylene (TCNE) or dimerization are characteristic singlet carbene reactions.

carbene per mole of dimer, saying little about the actual C=C bond strength.[73] In particular, the interconversion, catalyzed by protonation of the dimer in protic solvents, explained most of Wanzlick's trapping results. Much other interesting work has been done on the nucleophilic carbenes. However, only that which is pertinent to the thermochemical question is reviewed here.

Figure 14. Electrophile-induced cleavage of the C=C bond in a Wanzlick dimer. The absence of crossover in the thermolysis of asymmetric Wanzlick dimers in the absence of electrophiles excludes the purely thermal dissociation.

The isolability of the formal dimers of the Wanzlick carbenes indicates that the dimers are at least kinetically, and possibly thermodynamically, stable relative to the separated fragments. That is to say that, if the C=C bond energy were to be negative, then there would have to be a significant activation barrier to cleavage to rationalize the isolability of the olefin. Alternatively, one could argue that the C=C bond energy is positive, in which case there is no mystery as to the isolability of the dimer. The experimental evidence actually does not permit a distinction to be made between the two possibilities. One would expect, on the basis of the *ab initio*-calculated singlet–triplet splitting for the imidazolidin-2-ylidene[74] of $\Delta E_{ST} \approx$ 70 kcal/mol, that the C=C bond dissociation enthalpy in the dimer would be small—perhaps only 30–40 kcal/mol—but nevertheless positive. The recently reported isolation and structure proof by X-ray crystallography of *N,N*-dimesitylimidazolidin-2-ylidene[75] has only highlighted the difficulty in drawing thermochemical conclusions from isolability or nonisolability of single compounds. The particular carbene was found to be stable against dimerization despite its very close resemblance to the *N,N*-diphenyl- and *N,N*-ditolylimidazolidin-2-ylidenes that would have been the ones found in Lemal's crossover experiments[72] if the carbenes in that case had, in fact, been produced by thermal dissociation. One can only say that the isolability of at least one Wanzlick carbene, and the formal dimers of Wanzlick carbenes, demonstrates the existence of a significant kinetic barrier along the dimerization/dissociation reaction coordinate. While this is consistent with the presence of an avoided crossing as predicted by theory, it says nothing about the thermochemistry.

B. Arduengo Carbenes and Dimers

The very surprising isolation of simply substituted imazol-2-ylidenes by Arduengo[71] in 1991 opened the way to a system for which there might possibly be a zero or negative C=C bond dissociation enthalpy (Figure 15). With its cyclic π system, the Arduengo carbenes have an even larger singlet–triplet splitting than their Wanzlick cousins. Heinemann and Thiel[74] computed $\Delta E_{ST} = 85$ kcal/mol, which means, by the Carter and Goddard picture, a C=C bond with approximately zero bond dissociation enthalpy. Although the formal dimer of benzannelated imidazolylidene has been reported,[76] no unsubstituted biimidazolylidene was known in the literature. The common assumption was that the bond energy was simply too negative. Recently, double deprotonation of alkyl-bridged bis-imidazolium salts, or sodium reduction of alkyl-bridged biimidazolium dications, has been reported (Figure 16)[77,78] to yield either the biimidazolylidene (the dimer), or the bis-imidazolylidene (the pair of carbenes), depending on the temperature and/or bridge length. The presence of a genuine double bond between the two carbenic fragments in the bis-trimethylene doubly-bridged dimer is confirmed by the X-ray crystal structure (Figure 17),[77] which shows a C=C bond length of 1.34 Å, and the ^{13}C NMR chemical shift[77,78] for what would be the carbenic carbon. In the ^{13}C

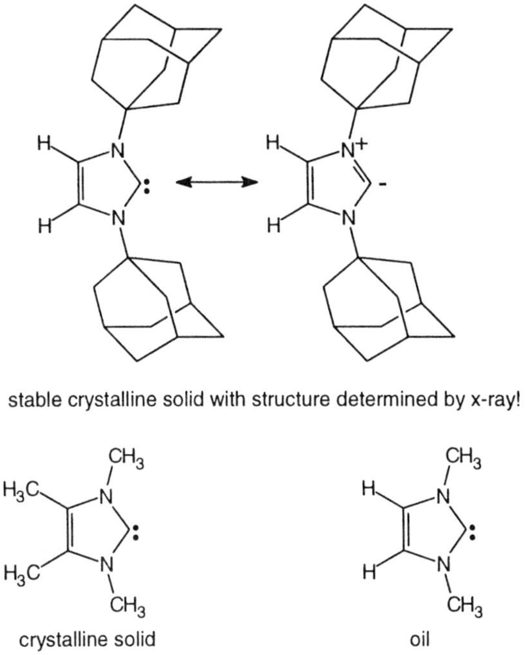

Figure 15. Arduengo carbenes prepared by deprotonation of the corresponding imidazolium salts. The carbenes are isolable.

NMR spectra of all Arduengo carbenes examined to date, that carbon appears at 210–220 ppm, while for the doubly bridged dimer it shows up at 127.5 ppm. Both the bond length and chemical shift would be highly unusual for a pair of carbenes, but quite unexceptional for an olefin.

Having established the structure, one has three key observations pertaining to the thermochemistry: (1) all unbridged or monobridged biimidazolylidenes, if formed, spontaneously dissociate to a pair of carbenes under all conditions employed, but (2) the bis-trimethylene doubly bridged biimidazolylidene is stable against dissociation even up to 100 °C and (3) the bis-tetramethylene doubly bridged biimidazolylidene exists predominantly as the olefin at –33 °C, and as a pair of carbenes at room temperature. From the three observations, one can claim that $\Delta G_{diss} < 0$ for the unbridged and singly bridged dimers, $\Delta G_{diss} > 0$ for the bis-trimethylene doubly bridged dimer, and $\Delta G_{diss} \approx 0$ for the bis-tetramethylene doubly bridged dimer. Furthermore, the interconversion of carbenes and dimers in one case indicates that there is no insurmountable kinetic barrier other than what is imposed by the thermochemistry. If one assumes that ΔH_{diss} for the olefinic carbene dimer is approximately constant, then the differences in ΔG_{diss} arise from differences in ΔS_{diss}. The entropy change upon cleavage of the C=C bond will always be positive,

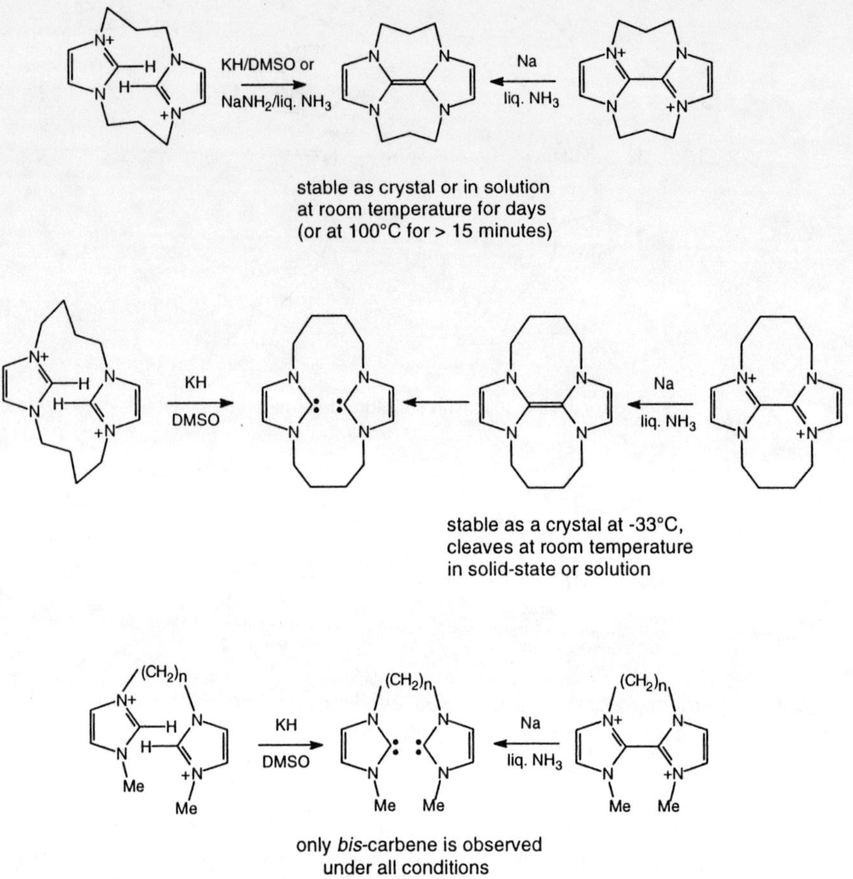

Figure 16. Preparation and chemical behavior of bridged biimidazolylidenes (Arduengo dimers) and bis-imidazolylidenes (a pair of Arduengo carbenes). Only the bis-trimethylene, doubly-bridged dimer is bound at room temperature.

with an upper-bound of ~35 eu (for cleavage of the unbridged dimer). The observation of at least one case with $\Delta G_{diss} > 0$ means that, by assumption, $\Delta H_{diss} > 0$ for all of the biimidazolylidenes. Because $T\Delta S$ at room temperature for dissociation of the unbridged dimer is about 10 kcal/mol, one can place another bound of $\Delta H_{diss} < 10$ kcal/mol. In fact, it should be much less, with a reasonable estimate being $\Delta H_{diss} = 4 \pm 3$ kcal/mol.

Recalling that the prediction for the bond enthalpy of an olefin that formally dissociates into a pair of singlet carbenes is $[172 - (2 \times \Delta E_{ST})]$, which for the dimer of an Arduengo carbene is approximately zero, one sees here a striking confirmation of the quantitative predictive power of a qualitative, pictorial model for bond strengths. The Arduengo carbenes and their dimers constitute the most extreme case

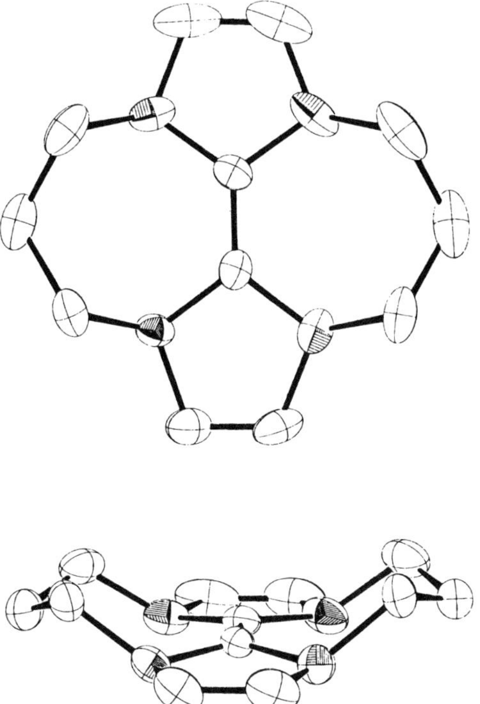

Figure 17. Two views of the X-ray crystal structure of a doubly-bridged biimida-zolylidene. The four nitrogen atoms are shown with cutaway ellipsoids. The central C=C bond length of 1.34 Å, and the only minor pyramidalization of the olefinic carbons is indicative of a genuine double bond.

of a bond weakened by promotion energy effects that can be imagined. Further-more, the relatively normal bond length and ^{13}C chemical shift in the dimer supports the notion that there is a small, but finite, barrier caused by the avoided crossing along the dissociation reaction coordinate, as predicted by the valence promotion energy picture.

A single report[79] of a covalent, ketene-like adduct of the *N,N*-diadamantylimidazol-2-ylidene and carbon monoxide is a challenge to theory. Because the singlet–triplet splitting in CO is even larger than that in the Arduengo carbene, such an adduct should not be bound at all (Figure 18). Subsequent attempts to reproduce the work have been unsuccessful,[80] but it is always more difficult to prove the "nonex-istence," as opposed to the existence, of a molecule or phenomenon. Further studies should clarify the issues.

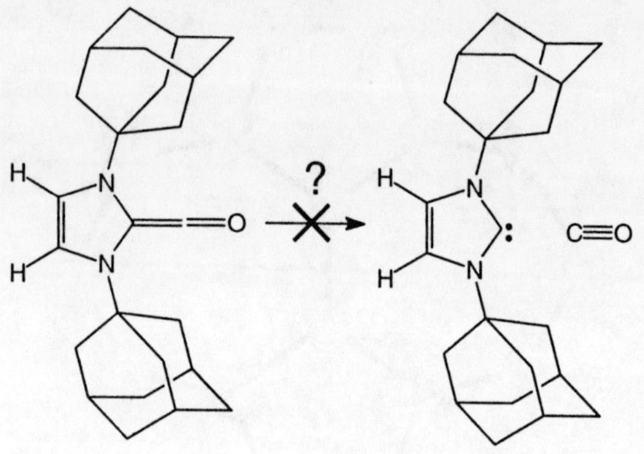

Figure 18. An adduct of an Arduengo carbene and CO is reported to be bound and isolable, despite theoretical arguments against the existence of the bond. A judgment on the report awaits further confirmation.

V. CONCLUSION

The availability of reliable experimental thermochemistry of some model carbenes has allowed a quantitative test of the predictions of a simple model for carbene heats of formation. It is quite evident that the singlet–triplet splitting in singlet ground-state carbenes serves as a good guide for when and by how much ΔH_f[carbene] deviates from additivity estimates. In favorable cases, the prediction is accurate to within a few kcal/mol, although in the more general case, more generous bounds of about 10 kcal/mol pertain. Nevertheless, the correlation demonstrates that the organic chemist's intuition that overall molecular properties are somehow determined by the properties of the molecule's constituent parts is, in a broad sense, correct for predicting trends and patterns.

REFERENCES

1. A cogent discussion of the extraction of energetic contributions due to strain or aromaticity may be found in: J.F. Liebman, A. Greenberg, *Strained Organic Molecules*; Academic Press: New York, 1978, Chapter 1.
2. Benson, S.W.; Cruickshank, F.R.; Golden, D.M.; Haugen, G.R; O'Neal, H.E.; Rodgers, A.S.; Shaw, R.; Walsh, R. *Chem. Rev.* **1996**, *69*, 279.
3. O'Neal, H.E.; Benson, S.W. in *Free Radicals*, vol. II, Kochi, J.K., Ed.; Wiley: New York, 1973; pp. 338–340.
4. Berkowitz, J.; Ellison, G.B.; Gutman, D. *J. Phys. Chem.* **1994**, *98*, 2744.
5. Harmony, M.D.; Laurie, V.W.; Kuczkowski, R.L.; Schwendeman, R.H.; Ramsay, D.A. *J. Phys. Chem. Ref. Data* **1979**, *8*, 619.
6. Data from ref. [4] and: D.R. Stull, E.F. Westrum Jr., G.C. Sinke, *The Chemical Thermodynamics of Organic Compounds*, Wiley: New York, 1969.

7. Blush, J.A.; Clauberg, H.; Kohn, D.W.; Minsek, D.W.; Zhang, X.; Chen, P. *Acc. Chem. Res.* **1992**, *25*, 385.

8. Simons, J.P. *Nature* **1965**, *205*, 1308.

9. Carter, E.A.; Goddard III, A. *J. Phys. Chem.* **1986**, *90*, 998; E.A. Carter, W.A. Goddard III, *J. Phys. Chem.* **1987**, *91*, 4651; E.A. Carter, W.A. Goddard III, *J. Chem. Phys.* **1988**, *88*, 3132.

10. Gleiter, R.; Hoffmann, R. *J. Am. Chem. Soc.* **1968**, *90*, 5457; Baird, N.C.; Taylor, K.F. *J. Am. Chem. Soc.* **1978**, *100*, 1333; Minkin, V.I.; Simkin, B.; Gluhovtsev, M.N. *Russ. Chem. Rev.* **1989**, *58*, 622; Liebman, J.F.; Simons, J. in *Molecular Structure and Energetics*, vol. 1, J.F. Liebman, A. Greenberg, Ed.; VCH Publishers Inc.: Deerfield Beach, FL, 1986, Chap. 3.

11. Using data from: Seetula, J.A.; Russell, J.A.; Gutman, D. *J. Am. Chem. Soc.* **1990**, *112*, 1347; Bunker, P.R.; Sears, T.J. *J. Chem. Phys.* **1985**, *83*, 4866.

12. Herzberg, G. *Molecular Spectra and Molecular Structure. 1. Spectra of Diatomic Molecules*, 2nd ed., Van Nostrand Reinhold: New York, 1950, pp. 357–359.

13. Murrell, J.N.; Kettle, S.F.A.; Tedder, J.M. *The Chemical Bond*, 2nd ed.: Wiley: New York, 1985; pp. 247–249.

14. Walsh, R. *Acc. Chem. Res.* **1981**, *14*, 246; Walsh, R. *Pure Appl. Chem.* **1987**, *59*, 69; Grev, R.S.; Schaefer III, H.F.; Baines, K. *J. Am. Chem. Soc.* **1990**, *112*, 9458.

15. Pross, A.; Shaik, S.S. *Acc. Chem. Res.* **1983**, *16*, 363.

16. Trinquier, G.; Malrieu, J.-P. *J. Phys. Chem.* **1990**, *94*, 6184.

17. Lias, S.G.; Karpas, Z.; Liebman, J.F. *J. Am. Chem. Soc.* **1985**, *107*, 6089.

18. Hine, J.J. *J. Am. Chem. Soc.* **1950**, *72*, 2438.

19. Shapiro, J.S.; Lossing, F.P. *J. Phys. Chem.* **1968**, *72*, 1552.

20. Rademann, K.; Jochims, H.-W.; Baumgärtel, H. *J. Phys. Chem.* **1985**, *89*, 3459.

21. Lias, S.G.; Liebman, J.F.; Levin, R.D. *J. Phys. Chem. Ref. Data* **1984**, *13*, 695.

22. Levi, B.A.; Taft, R.W.; Hehre, W. *J. Am. Chem. Soc.* **1977**, *99*, 8454; Lias, S.G.; Ausloos, P. *Int. J. Mass Spec. Ion Phys.* **1976**, *22*, 135; Ausloos, P.; Lias, S.G. *J. Am. Chem. Soc.* **1978**, *100*, 4594.

23. Scheunemann, H.-U.; Illenberger, E.; Baumgärtel, H. *Ber. Gunsenges. Phys. Chem.* **1980**, *84*, 580; Verhaart, G.J.; Van Sprang, H.A.; Brongersma, H.H. *Chem. Phys.* **1980**, *83*, 389.

24. Paulino, J.A.; Squires, R.R. *J. Am. Chem. Soc.* **1991**, *113*, 5573.

25. Kohn, D.W.; Robles, E.S.J.; Logan, C.F.; Chen, P. *J. Phys. Chem.* **1993**, *97*, 4936.

26. A discussion of threshold laws may be found in: McDowell, C.A. *Mass Spectrometry*, McGraw-Hill: New York, 1963, pp. 506–588.

27. Geltman, S. *Phys. Rev.* **1956**, *102*, 171.

28. Detailed procedures for deconvolution of thermal energy in mass spectrometric threshold measurements are given by: Chupka, W.A. *J. Chem. Phys.* **1971**, *54*, 1936.

29. Carter, E.A.; Goddard III, W.A. *J. Chem. Phys.* **1988**, *88*, 1752.

30. Shin, S.K.; Goddard III, W.A.; Beauchamp, J.L. *J. Phys. Chem.* **1990**, *94*, 6963; *J. Chem. Phys.* **1990**, *93*, 4986.

31. Murray, K.K.; Leopold, D.G.; Miller, T.M.; Lineberger, W.C. *J. Chem. Phys.* **1988**, *89*, 5442.

32. Koda, S. *Chem. Phys. Lett.* **1978**, *55*, 353; *Chem. Phys.* **1986**, *66*, 383.

33. Ervin, K.M.; Ho, J.; Lineberger, W.C. *J. Chem. Phys.* **1989**, *91*, 5974.

34. Brown, R.F.C.; Eastwood, F.W.; Jackman, G.P. *Aust. J. Chem.* **1978**, *31*, 579.

35. Chen, Y.; Jonas, D.M.; Kinsey, J.L.; Field, R.W. *J. Chem. Phys.* **1989**, *91*, 3976.

36. Ervin, K.M.; Gronert, S.; Barlow, S.E.; Gilles, M.K.; Harrison, A.G.; Bierbaum, V.M.; DePuy, C.H.; Lineberger, W.C.; Ellison, G.B. *J. Am. Chem. Soc.* **1990**, *112*, 5750.

37. Montgomery, J.A.; Ochterski, J.W.; Petersson, G.A. *J. Chem. Phys.* **1994**, *101*, 5900.

38. Fessenden, R.W.; Schuler, R.H. *J. Chem. Phys.* **1963**, *39*, 2147.

39. Drago, R.S.; Petersen Jr., H. *J. Am. Chem. Soc.* **1967**, *89*, 5774.

40. Clauberg, H.; Chen, P. *J. Am. Chem. Soc.* **1991**, *113*, 1445; Clauberg, H.; Minsek, D.W.; Chen, P. *J. Am. Chem. Soc.* **1992**, *114*, 99. The heat of formation for $c\text{-}C_3H_2^{\bullet+}$ used in this review is 3 kcal/mol

higher than that in the original publications because it has come to our attention that the atom-transfer experiment from which the number came may have underestimated $\Delta H_f[c\text{-}C_3H_2^{\bullet+}]$.

41. DeFrees, D.J.; McIver, R.T.; Hehre, W.J. *J. Am. Chem. Soc.* **1980**, *102*, 3335. The authors reported that the cyclopropene radical cation reacted with cyclopropene in a fast ion-molecule reaction. They therefore worked under such low density conditions that thermalization of the cyclopropene radical cations was likely to have been incomplete. Furthermore, the cyclopropene radical cation has been shown to rapidly isomerize to a linear structure, which would invalidate any thermochemical conclusions based on the icr experiment. See ref. [53].

42. Shea, D.A.; Chen, P., unpublished results.

43. Closs, G.L.; Redwine, O.D. *J. Am. Chem. Soc.* **1986**, *108*, 506.

44. Chipman, D.M.; Miller, K.E. *J. Am. Chem. Soc.* **1984**, *106*, 6236; Byun, Y.G.; Saebo, S.; Pittman, C.U. *J. Am. Chem. Soc.* **1991**, *113*, 3689.

45. Lee, T.J.; Bunge, A.; Schaefer, H.F. *J. Am. Chem. Soc.* **1985**, *107*, 137; Bofill, J.M.; Farras, J.; Olivella, S.; Sole, A.; Vilarrasa, J. *J. Am. Chem. Soc.* **1988**, *110*, 1694.

46. Kanata, H.; Yamamoto, S.; Saito, S. *Chem Phys. Lett.* **1987**, *140*, 221.

47. Maier, G.; Reisenauer, H.P.; Schwab, W.; Carsky, P.; Hess, B.A.; Schaad, L.J. *J. Am. Chem. Soc.* **1987**, *109*, 5183.

48. Robinson, M.S.; Polak, M.L.; Bierbaum, V.M.; DePuy, C.H.; Lineberger, W.C. *J. Am. Chem. Soc.* **1995**, *117*, 6766.

49. DeFrees, D.J.; McLean, A.D. *Astrophys. J.* **1986**, *308*, L31.

50. Bernheim, R.A.; Kempf, R.J.; Gramas, J.V.; Skell, P.S. *J. Chem. Phys.* **1965**, *43*, 196.

51. McAllister, T.; Nicholson, A.J. *J. Chem. Soc. Faraday Trans.* **1981**, *77*, 821.

52. Matthews, C.S.; Warneck, P. *J. Chem. Phys.* **1969**, *51*, 854.

53. Parr, A.C.; Elder, F.A. *J. Chem. Phys.* **1968**, *49*, 2659; Parr, A.C.; Jason, A.J.; Stockbauer, R. *Int. J. Mass Spec. Ion Phys.* **1978**, *26*, 23; Parr, A.C.; Jason, A.J.; Stockbauer, R. *Int. J. Mass Spec. Ion Phys.* **1980**, *33*, 243.

54. MacLagan, R.G.A.R., *THEOCHEM* **1992**, *258*, 175.

55. Clauberg, H.; Chen, P. *J. Phys. Chem.* **1992**, *96*, 5676.

56. Smith, D.; Adams, N.G. *Int. J. Mass Spec. Ion Phys.* **1987**, *76*, 307; Dannacher, J.; Heilbronner, E.; Stadelmann, J.-P.; Vogt, J. *Helv. Chem. Acta* **1979**, *62*, 2186.

57. Prodnuk, S.D.; DePuy, C.H.; Bierbaum, V.M. *Int. J. Mass Spec. Ion Proc.* **1990**, *100*, 693. Note that we use $\Delta H_f[c\text{-}C_3H_3^+] = 258.5 \pm 0.7$ kcal/mol value from photoionization mass spectroscopy to derive an upper limit on $\Delta H_{f,298}[c\text{-}C_3H_2^{+\bullet}]$.

58. DeFrees, D.J.; McLean, A.D., *Astrophys. J.* **1986**, *308*, L31.

59. Wasserman, E.; Barash, L.; Trozzolo, J.; Murray, R.W.; Yager, W.A. *J. Am. Chem. Soc.* **1964**, *86*, 2304.

60. Kassee, M.Z.; Nimlos, M.R.; Downie, K.E.; Waali, E.E. *Tetrahedron* **1985**, *41*, 1579; Bofill, J.M.; Bru, N.; Farras, J.; Olivella, S.; Sole, A.; Villarrasa, J. *J. Am. Chem. Soc.* **1988**, *110*, 3740; Collins, C.L.; Davy, R.D.; Schaefer, H.F. *Chem. Phys. Lett.* **1990**, *171*, 259.

61. Baird, M.S.; Dunkin, I.R.; Hacker, N.; Poliakoff, M.; Turner, J.J. *J. Am. Chem. Soc.* **1981**, *103*, 5190.

62. Puttemans, J.P.; Smith, G.P.; Golden, D.M. *J. Phys. Chem.* **1990**, *94*, 3226.

63. McDonald, R.N.; Bianchina Jr., E.J.; Tung, C.C. *J. Am. Chem. Soc.* **1991**, *113*, 7115.

64. Sugimoto, H.; Hirai, K. *Tetrahedron Lett.* **1985**, *26*, 883.

65. Breslow, R. *J. Am. Chem. Soc.* **1958**, *80*, 3719; Kluger, R. *Chem. Rev.* **1987**, *87*, 863.

66. Stetter, H. *Angew. Chem. Int. Ed. Engl.* **1976**, *15*, 639.

67. Lappert, M.F.; Maskell, R.K. *J. Chem. Soc. Chem. Comm.* **1982**, 580.

68. Hartzler, H.D. *J. Am. Chem. Soc.* **1973**, *95*, 4379.

69. Castells, J.; López-Calahorra, F.; Geijo, F.; Pérez-Dolz, R.; Bassedas, M. *J. Heterocycl. Chem.* **1986**, *23*, 715; M.B. Doughty, G.E. Risinger, *Bioorg. Chem.* **1987**, *15*, 1.

70. Wanzlick, H.W. *Angew. Chem. Int. Ed. Engl.* **1962**, *1*, 75.

71. Arduengo III, A.J.; Harlow, R.L.; Kline, M. *J. Am. Chem. Soc.* **1991**, *113*, 361; Arduengo III, A.J.; Dias, H.V.R.; Harlow, R.L.; Kline, M. *ibid.* **1992**, *114*, 5530; Arduengo III, A.J.; Dias, H.V.R.; Dixon, D.A.; Harlow, R.L.; Klooster, W.T.; Koetzle, T.F. *ibid*, **1994**, *116*, 6812; Regitz, M. *Angew. Chem. Int. Ed. Engl.* **1996**, *35*, 725.

72. Lemal, D.M.; Lovald, R.A.; Kawano, K.I. *J. Am. Chem. Soc.* **1964**, *86*, 2518; Winberg, H.E.; Carnahan, J.E.; Coffman, D.D.; Brown, M. *J. Am. Chem. Soc.* **1965**, *87*, 2055.

73. Hoffmann, R.W. *Angew. Chem. Int. Ed. Engl.* **1968**, *7*, 754; Lemal, D.M. in *The Chemistry of the Amino Group*, S. Patai, Ed.; Wiley: New York, 1968, Chap. 12.

74. Heinemann, C.; Thiel, W. *Chem. Phys. Lett.* **1994**, *217*, 11.

75. Arduengo III, A.J.; Goerlich, J.R.; Marshall, W.J. *J. Am. Chem. Soc.* **1995**, *117*, 11027.

76. Shi, Z.; Thummel, R.P. *Tetrahedron Lett.* **1995**, *36*, 2741.

77. Taton, T.A.; Chen, P. *Angew. Chem. Int. Ed. Engl.* **1996**, *35*, 1011.

78. Z. Shi, V. Goulle, R.P. Thummel, *Tetrahedron Lett.* in press.

79. Lyashchuk, S.N.; Skrypnik, Y.G. *Tetrahedron Lett.* **1994**, *35*, 5271.

80. Dixon, D.A.; Arduengo III, A.J.; Dobbs, K.D.; Khasnis, D.V. *Tetrahedron Lett.* **1995**, *36*, 645.

THE CARBENE ROUTE TO BRIDGEHEAD DOUBLE BONDS

Maitland Jones, Jr.

Twenty-five years ago we embarked on a series of examinations of reactions of carbenes in the gas phase. One of our schemes involved attaching a divalent carbon to the bridgehead position of a preformed bicyclic molecule. It seemed to us that such a carbene would have no way to avoid carbon–carbon insertion to give a bridgehead double bond, provided that we could avoid carbon–hydrogen insertion. We knew that simple carbenes preferred the carbon–hydrogen insertion reaction to the carbon–carbon insertion when both reactions were possible. For example, *tert*-butylcarbene gives roughly 90% 1,1-dimethylcyclopropane and only 10% 2-methyl-2-butene (Scheme 1).[1]

We reasoned that we could make carbon–hydrogen insertion unpalatable by keeping the bridges of the bicyclic molecule relatively short, and thus holding the carbon–hydrogen bonds at some distance from the carbene. Generation of the reactive species in the gas phase would make intermolecular reactions unlikely.

Advances in Carbene Chemistry
Volume 2, pages 77–96
Copyright © 1998 by JAI Press Inc.
All rights of reproduction in any form reserved.
ISBN: 1-55938-837-4

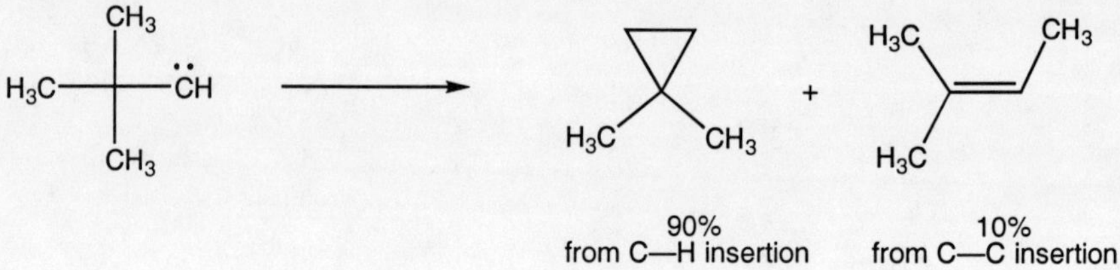

90%
from C—H insertion

10%
from C—C insertion

Scheme 1.

With the hydrogens removed from the action, as it were, the carbene would be left with no possible intramolecular reaction other than carbon–carbon insertion. We further reasoned that thermodynamics would generally favor the alkene product, although it was clear that as the double bond became more and more twisted, the difference in energy between carbene and alkene would surely diminish. The rearrangement of *tert*-butylcarbene to 2-methyl-2-butene is exothermic by several tens of kilocalories, but the thermochemical situation is far from certain if the alkene is required to be severely twisted. Eventually, the energies of the carbenes and the alkenes should become comparable. So, already we have an interesting question: will the equilibrium shown in Scheme 2 be a source of carbenes? Could we turn around our original idea of making bridgehead alkenes from carbenes and use the twisted alkenes as sources of carbenes?

In this chapter I will not attempt to augment the excellent recent reviews of the chemistry of bridgehead alkenes;[2] rather, I will concentrate on the route we developed to twisted alkenes. I will summarize our early experiments, but will emphasize our recent work on adamantene. I will also highlight some exciting new work appearing from Japan. I hope as well to raise some questions and suggest some areas in which new work is needed.

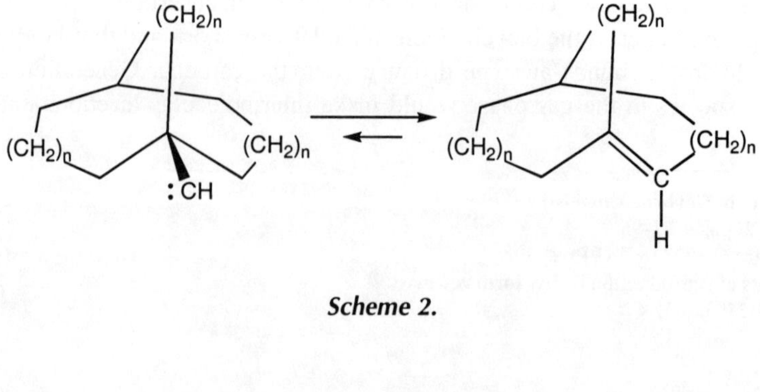

Scheme 2.

The first person to attempt to generate a bridgehead carbene was probably Jim Wilt who decomposed tosylhydrazone **1** in solution, but was unable to isolate products easily attributable to carbene **2**.[3] Wolfgang Kirmse made the related carbene **3**, and did isolate dimeric products that could have resulted from reactions of a carbene or bridgehead alkene.[4] We generated **2** in the gas phase, hoping to avoid the solvent reactions that apparently plagued Wilt and, perhaps, Kirmse. Of course, the choice of **2** carries problems with it, as ring expansion can take place in two ways (bond a or bond b) to give either bicyclo[2.2.2]oct-1-ene (**4**) or bicyclo[3.2.1]oct-1-ene (**5**) (Scheme 3). In the event, we isolated neither hoped-for alkene, but instead found **6** as the sole product (Scheme 4). We had not anticipated that the cyclohexenes contained within both **4** and **5** would fail to survive the reaction conditions, or the exothermicity of their formation, and would undergo reverse Diels–Alder reactions to give the observed triene. As shown in Scheme 4, a labeling experiment resolves the issue of which bond participates in ring expansion in the favor of bond a.[5]

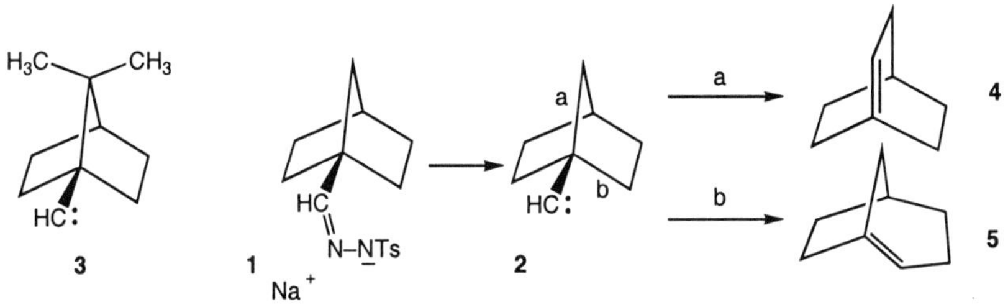

Scheme 3.

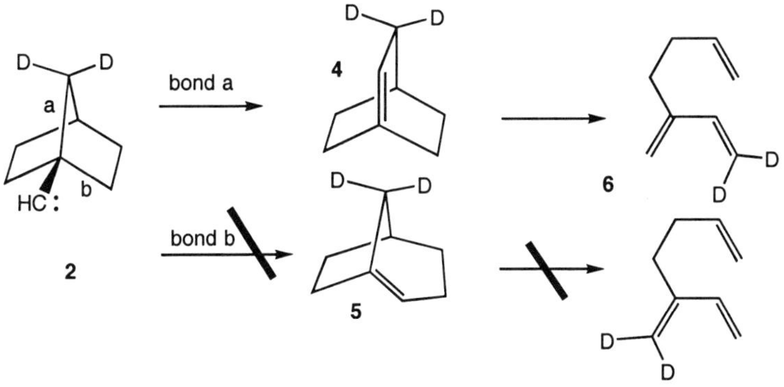

Scheme 4.

Rearrangement takes place under kinetic control to give the less stable of the two possible bridgehead alkenes. Carbocations behave in exactly the same way, as it is the short bridge that migrates when there is a choice.[6] This similarity in mode of ring expansion is not surprising, as in each intermediate the reaction must begin with the interaction of an empty $2p$ orbital with an adjacent bond.

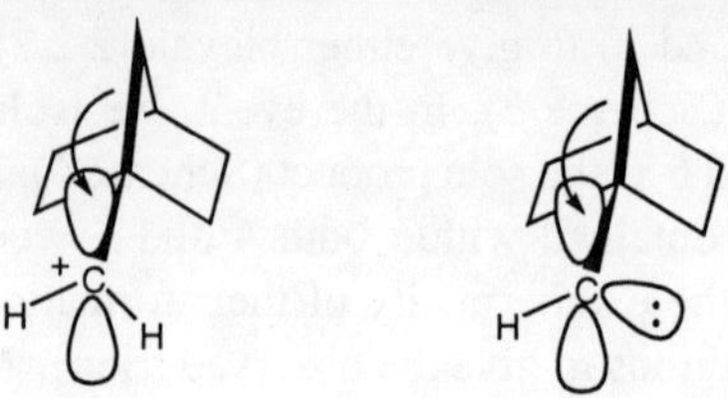

That **4** was not isolable is scarcely surprising. Schleyer and Maier have categorized bridgehead alkenes in terms of their olefinic strain (OS), defined as the difference in strain energy between the bridgehead alkene in its minimum energy conformation and the related polycyclic alkane in its minimum energy conformation.[7] Three broad categories emerged from Schleyer and Maier's study: isolable alkenes with OS ≤ 17 kcal/mol, detectable alkenes with OS between 17 and 21 kcal/mol, and unstable alkenes with OS ≥ 21 kcal/mol. Alkene **4**, with a calculated OS of 40.4 kcal/mol fits solidly in the third category and could not have been expected to survive without special efforts to isolate it at low temperature. Despite the failure of this method to give an isolable bridgehead alkene, these results do confirm the basic worth of the idea. Or do they? It was clear from the start that we envisioned this reaction as proceeding through the carbene. But the carbene is not an obligatory intermediate in this reaction; it is easy to write a mechanism that uses

Scheme 5.

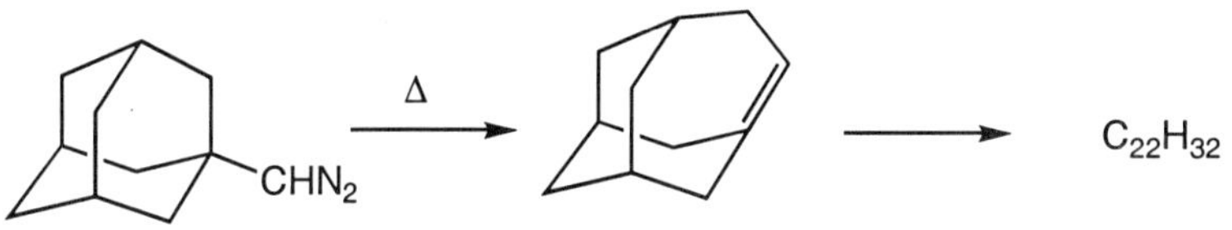

Scheme 6.

the diazo compound **7** rather than the carbene **2** (Scheme 5). Indeed, there is much recent[8] (and some old[9]) work that strongly indicates that diazo compounds are very active in intramolecular chemistry. We shall return to this point several times.

We next decomposed 1-adamantyldiazomethane (**8**) in the hope of obtaining homoadamant-3-ene, **9**, a bridged *trans*-cycloheptene with a calculated OS value of 20.2 kcal/mol. We thermally decomposed **8** and obtained dimers of the formula $C_{22}H_{32}$ (Scheme 6). When the apparatus was modified so that the products could be examined before warm up, the dimers were replaced with **9** which was stable until about –20 °C, at which point dimerization set in.[10] Although this work makes the case for synthesis of bridgehead alkenes unassailable, of course it makes no progress in the resolution of one central question: is the carbene or the diazo compound the active ingredient in the reaction?

Another bridged *trans*-cycloheptene, **10**, has been synthesized by photolysis of diazo compound **11** (Scheme 7). It, like **9**, was characterized by spectroscopy, and is stable at low temperature.[11]

This method is also capable of generating extremely unstable alkenes. The cubylcarbene–homocubene system provides a nice example. Eaton and Hoffmann first reported that 9-phenyl-1(9)-homocubene (**12**) could be made by the thermal and photochemical decomposition of cubylphenyldiazomethane.[12] Eaton and Appell later showed that the conversion of carbene **13** into **12** was reversible (Scheme 8).[13]

We and the Platz group concentrated on the parent system, and used the carbon–carbon insertion of homocub-9-ylidene (**14**) and of cubylcarbene (**15**) to generate

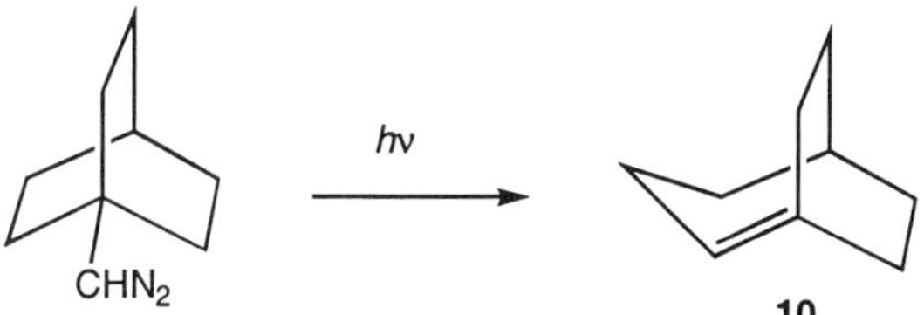

Scheme 7.

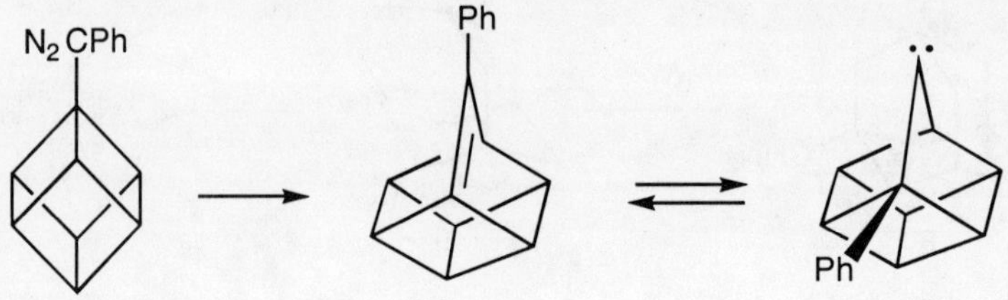

Scheme 8.

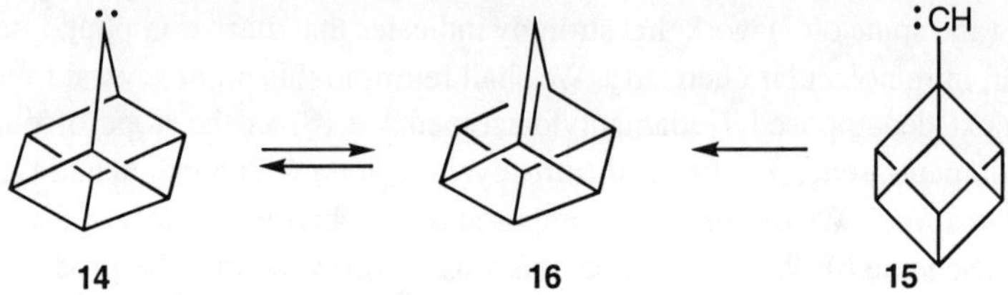

Scheme 9.

homocub-1(9)-ene, **16** (Scheme 9).[14] Generation of **15** actually led to the trapping products of **14**, which was the clue that **16** and **14** were close in energy. A variety of kinetic and chemical trapping experiments allowed us to evaluate the equilibrium constant as close to 1 at 20 °C.[14] Borden and Hrovat later used GVB-PP(1) calculations with both 3-21G and 6-31G* basis sets to estimate that the energy difference between **14** and **16** was on the order of 1 kcal/mol, in excellent agreement with the experimental findings.[15]

In this case it can be shown that the carbene *is* capable of ring expansion. Ning Chen devised a synthesis of the polycycle **17** and irradiated it in benzene/CH_3OD to give **18** and **19** (Scheme 10), the products of trapping of *both* **14** and **16**. The dependence of the ratio of **18** and **19** on methanol concentration showed that the carbene was the first formed intermediate in this case, and thus must be able to undergo rearrangement to the alkene. This information neither shows that simpler bridgehead carbenes such as **2** *are* capable of ring expansion, nor that diazo compounds are *not* capable of the reaction.

Adamantene, **21**, is another bridgehead alkene with a high OS of 39.5 kcal/mol.[7] One obvious potential precursor is diazo compound **20**, or some related nitrogenous compound such as the tosylhydrazone, diazirine, or aziridyl imine. A potential problem is that **20** presents two different bonds for ring expansion. Which will it be? Migration of bond a would lead to adamantene, **21**, but movement of bond b would lead to the related bridgehead alkene **22** (Scheme 11). Clearly, the chemistry of these species is complicated enough without having to deal with the sorting out of reactions of two bridgehead alkenes reacting at the same time. However, given

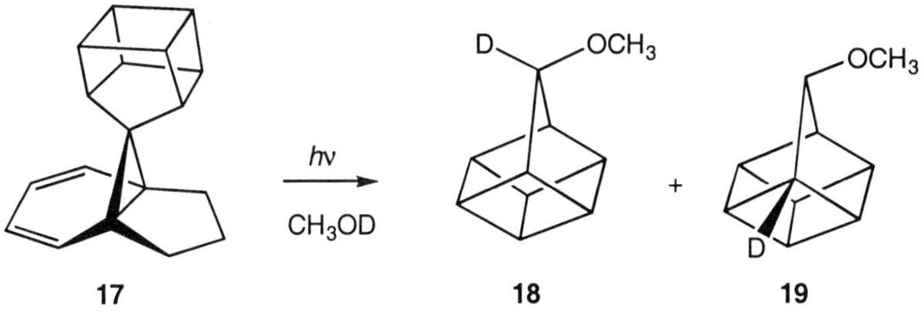

Scheme 10.

that **2** rearranged through exclusive migration of the shorter bridge, it seemed likely that **20** would behave similarly, and lead to the sought-for **21**. Some years ago,[16] we showed that the anticipated ring expansion of **20** did take place, as we isolated the Diels–Alder adduct **23** (Scheme 12), a compound already known from attempts to make adamantene in solution.[17]

The motivation for pursuing adamantene further was many-fold. First, the early work had shown tantalizing hints of diradical behavior in **21**, and we hoped to confirm and, perhaps, expand that work. Second, and at least as important, it must be admitted that our gas-phase route to adamantene was far from optimal: far better to have a method that worked in solution, and at low temperature. We thought that

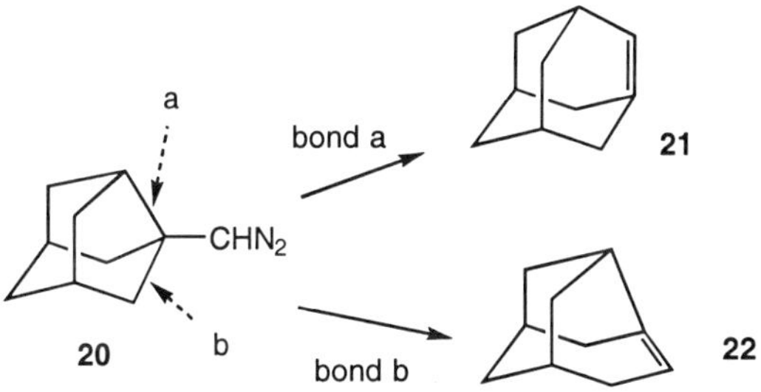

Scheme 11.

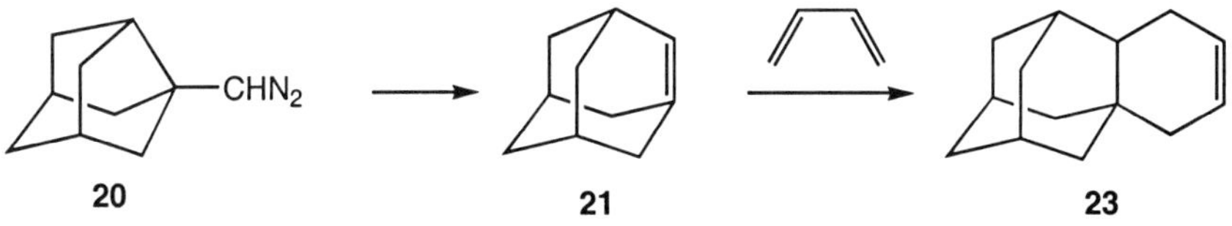

Scheme 12.

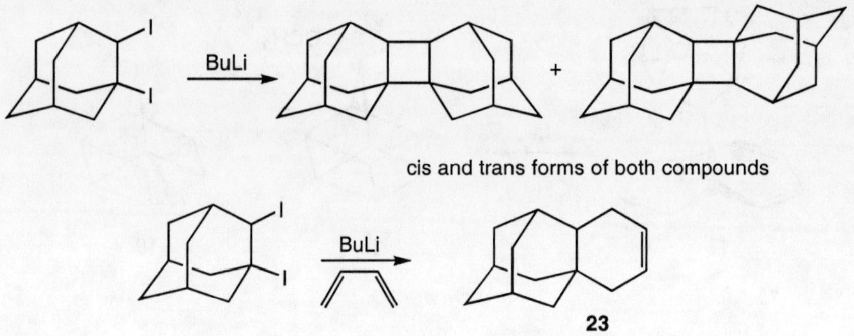

cis and trans forms of both compounds

23

Scheme 13.

such a method might already exist in the alkyllithium-induced decompositions of 1,2-dihaloadamantenes discovered by Tony McKervery[18] and Dieter Lenoir.[19] Both the dimers of adamantene and the product of butadiene trapping (**23**) were isolated from such decompositions (Scheme 13). The problem with those results was that no one could be quite sure that the reactive intermediate was really adamantene. Haloadamantanes can be efficiently coupled by alkyllithium reagents, for example,[20] and dimer formation could be no more than a double coupling of this kind. Trapping with butadiene is more persuasive, but even it is suspect. A mechanism involving an organometallic-mediated cycloaddition could be devised, and the related Diels–Alder addition to furan has had an on-again, off-again history.[21] So, we felt that one could not really credit any of these results as definitively proving the intermediacy of adamantene, despite our largely intuitive feeling that adamantene was probably involved in the organometallic reactions.

Our strategy was to find new reactions diagnostic of adamantene and show that the intermediate in the organometallic reactions also underwent them. In this way we hoped to forge a stronger link between the gas-phase chemistry and the solution reactions. In particular, adamantene is formally a cyclohexene and one might wonder why the products of reverse Diels–Alder reaction were not observed, as, for example, they were when **4** was generated from **7** or carbene **2**. Accordingly,

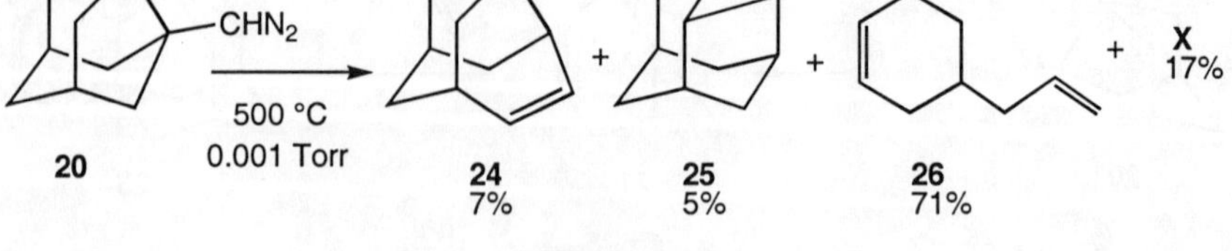

20

500 °C
0.001 Torr

24
7%

25
5%

26
71%

X
17%

Scheme 14.

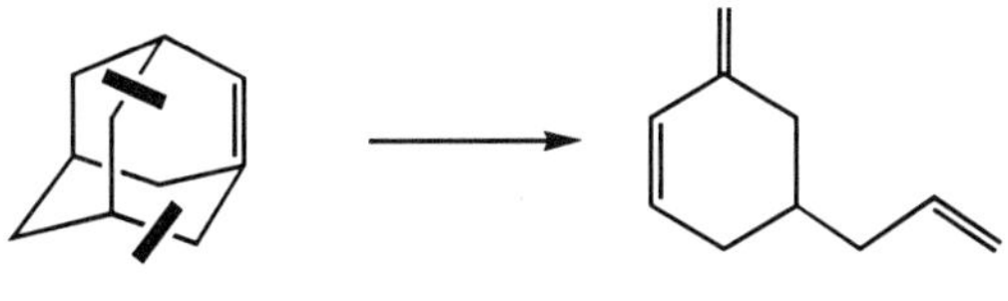

Scheme 15.

we decomposed the tosylhydrazone leading to **20** at higher temperature and lower pressure than before and were rewarded by not just one monomeric product, but with four (Scheme 14). Two of them, **24** and **25**, were known, the third, **26**, we made by an authentic synthesis, and the fourth, **X**, remains of unknown structure, alas.

Compound **26** is the result of a reverse Diels–Alder reaction of adamantene (Scheme 15), but the sources of **24** and **25** are more problematical.

The presence of **24** and **25** implicates carbenes **27** and **28**. As we have seen, strained alkenes such as homocubene can revert to carbenes.[13,14] Adamantene has three possible reversals: bond *a* gives **29**, bond *b* gives **27**, and participation of the carbon–hydrogen bond *c* leads to **28** (Scheme 16).

Of course, the presence of the unknown **X** is disturbing. Might it signal ring expansion in the original starting material **20** or carbene **29** in the "wrong" sense? We have come to expect ring expansion through the short bridge, but the presence of **X** makes this assumption tenuous. Although we were unable to determine the structure of **X**, we were able to show that it is not **30**, the putative product of reverse Diels–Alder reaction in **22** which would be formed by ring expansion of the longer bridge (Scheme 17). Accordingly, there is no need to revise our earlier findings that ring expansion took place primarily, if not exclusively, through the short bridge.

Now, equipped with a set of products that must come from adamantene, we set out to test whether the McKervey route, the treatment of 1,2-diiodoadamantane **31**

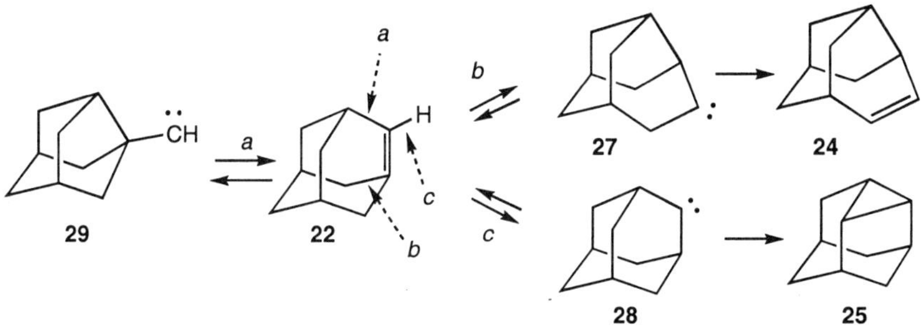

Scheme 16.

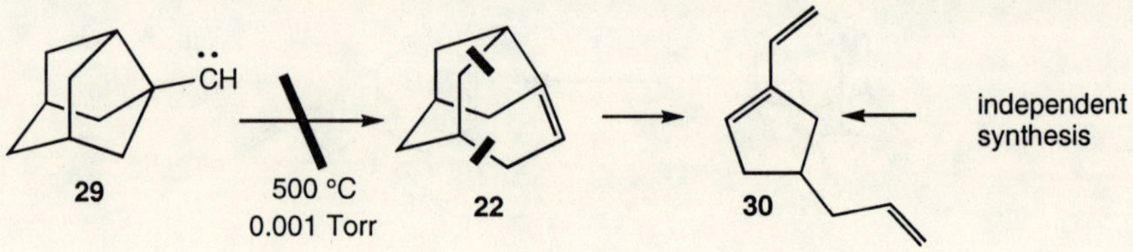

Scheme 17.

with butyllithium, could produce the same products. As there was little hope of
seeing them under low temperature "McKervey conditions," we tried to mimic
those conditions at higher temperature in the gas phase. To do this, we applied Udo
Brinker's technique of passing a precursor over methyllithium-coated quartz chips.[22]
We cannot take methyllithium to 500 °C, but even at our limit of ~100 °C we see traces
of the products of the unimolecular decomposition of adamantene, **24, 25, 26**, and
X (Scheme 18). By far the largest product, however, is adamantane. At 60%
conversion of starting material, 70% of the product is adamantane. Control experiments
established that adamantane is a product of adamantene, not the result of a pair of
sequential reductions of the two iodides. Thus, neither 1-iodoadamantane or
2-iodoadamantane is reduced under "Brinker conditions," whereas *trans*-2,3-diio-
dobicyclo[2.2.2]octane is smoothly converted into bicyclo[2.2.2]octene. It seems
that 1,2-diiodo compounds are converted into strained and unstrained alkenes under
the "Brinker tube" conditions, and that especially highly strained alkenes such as
adamantene are then further reduced, although normal alkenes are untouched.

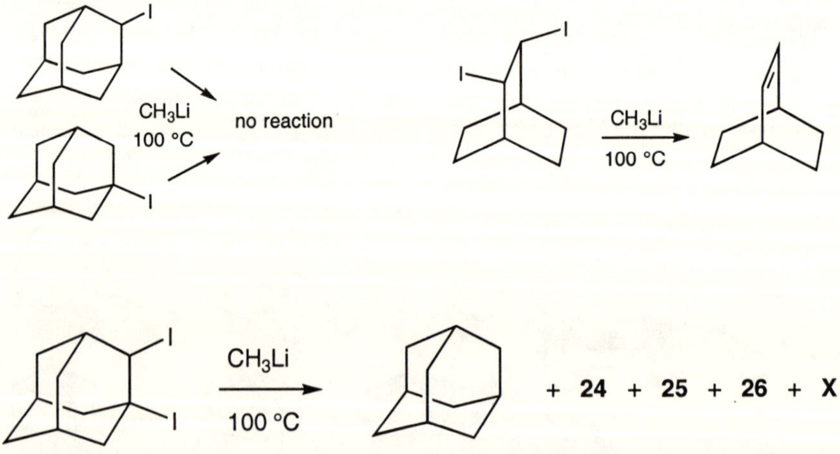

Scheme 18.

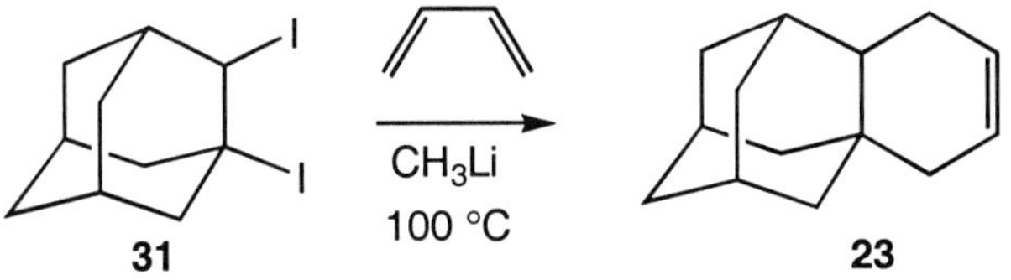

Scheme 19.

When a stream of 1,3-butadiene is used to sweep 1,2-diiodoadamantane into the methyllithium tube, adamantane is no longer produced and small amounts of the Diels–Alder adduct **23** are isolated instead (Scheme 19). It is surely true that we cannot explain every nuance of these reactions; the mechanism of the reduction of adamantene by methyllithium remains obscure, for example. Nonetheless, it seems certain to us that adamantene is formed under these rather odd conditions. The case for adamantene formation under the more conventional solution conditions is now greatly strengthened.[21]

There is an interesting but brief report on the bicyclo[1.1.1]carbinyl system.[23] Wiberg and Waddell have thermally decomposed tosylhydrazone **32** and observed two products, **33** and **34** along with another compound of unknown structure (Scheme 20).[23] The carbene **35** (or the related diazo compound) could be an intermediate leading to the bridgehead alkene **36**. However, under these conditions one must surely suspect (as Wiberg and Waddell clearly did) that protonation to give a carbocation **37** that could give **38** through ring expansion would be important (Scheme 21). Indeed, we might favor such an ionic, non-carbene route had Wiberg and Waddell not pointed out that carbocations such as **38** generally open to monocyclic products, none of which were observed.[24]

A related carbene, **39**, as well as substituted versions, has been investigated jointly by the Platz and Kirmse groups under similar, alcoholic conditions (Scheme 22).[25] The question is whether or not **39** undergoes carbon–carbon insertion to give **40**, the parent compound of **36**, which was possibly formed in the reactions of

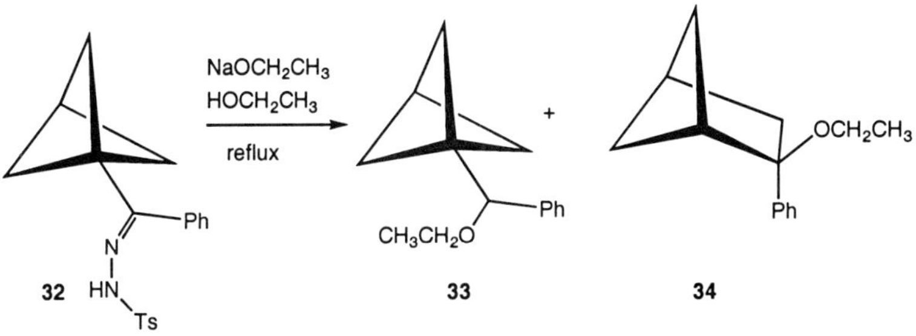

Scheme 20.

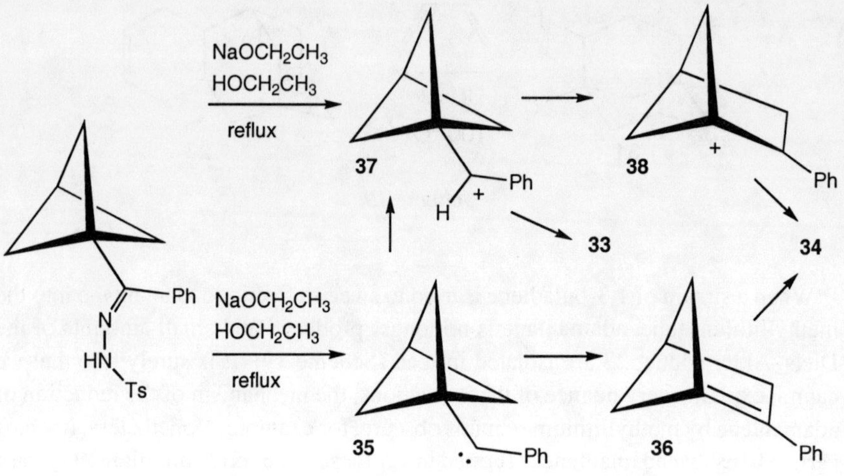

Scheme 21.

Wiberg and Waddell. Platz and Kirmse's conclusion was that carbocations such as **41**, formed by protonation of the diazo compound or the carbene, were the likely intermediates. There is no necessary conflict between the two investigations, as Wiberg and Waddell's cation would have been the bridgehead cation **38**, whereas Platz, Kirmse, and coworkers were dealing with **41**.

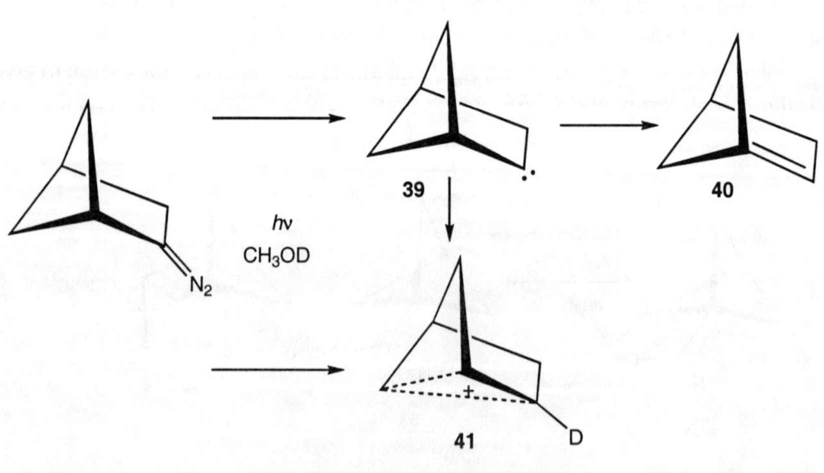

Scheme 22.

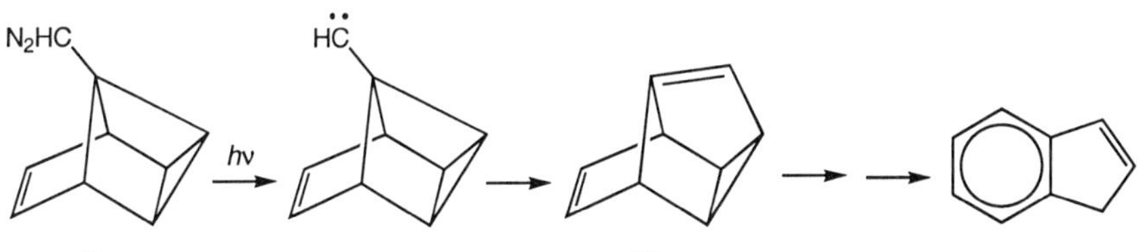

Scheme 23.

Another highly strained bridgehead alkene (**42**) has been made by the diazo compound–carbene route by Stapersma et al.[26] They decomposed diazo compound **43** at low temperature and were doubtlessly disappointed to find indene as the major product (Scheme 23). There was no evidence for addition to the double bond, which would have led to a [4.4.4.4]fenestrane derivative. Once again, the carbene is only a presumed, not an obligatory intermediate.

A major remaining problem involves the mechanism of ring expansion. As noted several times earlier, the carbenes which are so often written are not obligatory intermediates. The precursor diazo compounds, or the isomeric diazirines could be the real sources of the bridgehead alkenes. Even the observation that carbene **14** is capable of ring expansion (to **16**) does not place **14** on the reaction pathway from diazo compound to product alkene. One's intuitive feeling must surely be that carbenes can undergo ring expansion. After all, a variety of non-diazo sources leads from *tert*-butylcarbene to 2-methyl-2-butene, albeit as the minor product.[1] We have used the methyllithium tube in an attempt to generate carbene **29** from 1,1-diiodide **44** (Scheme 24). The results are qualitatively similar to those of the decomposition of 1,2-diiodoadamantane: small amounts of the products of the unimolecular decomposition of adamantene (**24**, **25**, **26**, and **X**) accompany larger amounts of adamantane. It seems that carbene **29** can undergo ring expansion. Once again, this observation does not place **29** on the reaction pathway from diazo compound to bridgehead alkene; it merely shows that if this carbene were formed, it would undergo ring expansion.

In at least one case it appears that carbenes are not the major sources of ring-expanded materials. When aziridylimine **45** was photolyzed in methanol/benzene, both the carbene and the bridgehead alkene were trapped to give **18** and **19** (Scheme 25). The isolation of both **18** and **19** means little: either **14** or **16** could be

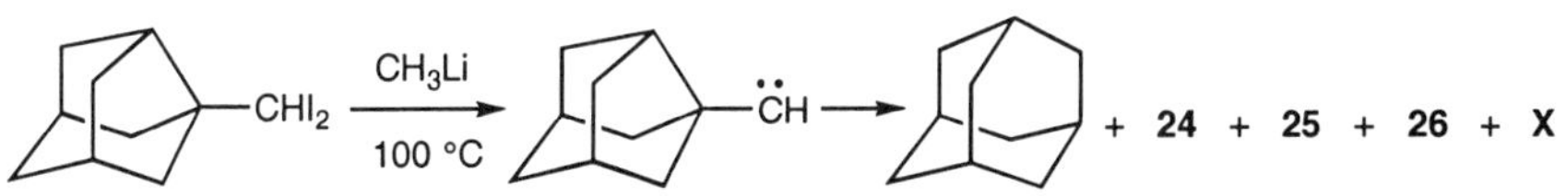

Scheme 24.

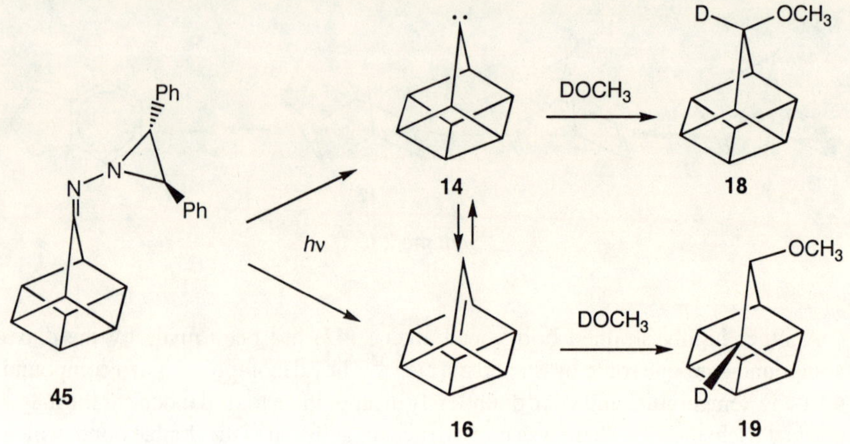

Scheme 25.

formed first only to rearrange to the other. However, when the concentration of CH_3OD is varied, it can be seen that as $[CH_3OD]$ increases, so does the yield of **19**. Compound **19** results from trapping of the bridgehead alkene, which must be the first formed intermediate from **45**, or the related diazo compound. Direct ring expansion, without carbene formation, must be the dominant reaction when **45** is photolyzed. Whether this observation is general, and can be extended to other bridgehead diazo compounds remains to be seen.

A resolution of the question could involve the application of the Platz–Jackson technique in which pyridine is used to scavenge carbenes.[27] The UV spectrum of the ylide formed from reaction of pyridine with the carbene is monitored; when the signal for the ylide reaches a steady maximum, all carbenes formed are presumably effectively scavenged by pyridine. If the ring-expanded product persists in such a situation, it must issue from some source other than the carbene. If it vanishes as the carbene is scavenged by pyridine, it almost certainly owes its existence to carbene chemistry.

A second problem concerns the reactions of disubstituted diazo compounds (or their surrogates) to generate bridgehead alkenes of increased stability. Presumably, as the size of R increases, the kinetic stability of the bridgehead alkene would also increase and one might be able to "dial a reactivity" by simply varying the size of R in a sensible way.

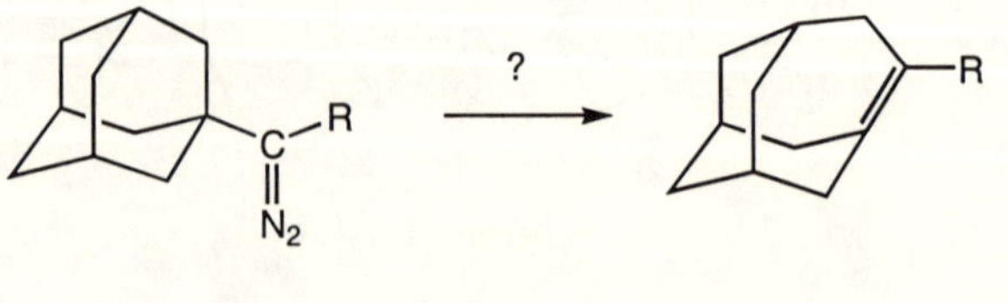

One difficulty is that not just any R will do; if intramolecular reaction of the related carbene with R is easy (and if the carbene is the intermediate in ring expansion, which is a real question), one could expect only the uninteresting products of reaction with R. We had used diadamantyldiazomethane to avoid this problem, but apparently overshot the mark as the resulting 4-(1-adamantyl)homoadamant-3-ene (**46**) is amazingly stable (Scheme 26).[28] This compound resists rearrangement up to +185 °C. Its stability is even more remarkable when one realizes that compared to the parent compound homoadamant-3-ene (**9**) itself, **46** is thermodynamically much less stable.

Bridgehead alkenes in general rehybridize, pyramidalizing one end of the carbon–carbon double bond, thus gaining overlap as the interorbital angle deviates increasingly from 90°. This distortion is much easier to do with R = H rather than the bulky adamantyl group. For example, it is calculated that the dihedral angle between the orbitals making up the "double" bond is 42.3° in **46** but only 26° in **9**.[29] Therefore, **46**, with by far the more twisted double bond, is substantially less stable thermodynamically than **9**. However, **46** is surrounded by especially high barriers thanks to the appended adamantane cage, and is kinetically much more stable than **9**.

Clearly, there should be room to study substituted bridgehead double bonds by selecting R groups intermediate in size between the reactivity-stifling adamantyl group and the too-small hydrogen. Moreover, this stabilizing technique should be applied to other bridgehead alkenes. It is a pleasure to report that these long-overdue experiments are finally underway in two groups in Japan, and that promising results are beginning to appear.

Shoji Eguchi and his coworkers at Nagoya have studied the decompositions of a phenyl- and carbalkoxy-substituted adamantyldiazomethane.[30] In each case a bridgehead homoadamantene of stability intermediate between **46** and **9** was obtained. For example, pyrolysis of diazo compound **47** at 400 °C led to **48** and **49** (Scheme 27). Compound **48** is logically attributed to carbon–hydrogen insertion by carbene **50**, formed by phenylcarbene rearrangement,[31] but diketone **49** would appear to come from the reaction of bridgehead alkene **51** with oxygen.

Photolysis was largely ineffective in producing carbenes from **47**, but catalytic decomposition with $Rh_2(OAc)_4$ led to rapid decomposition and formation of **51** at

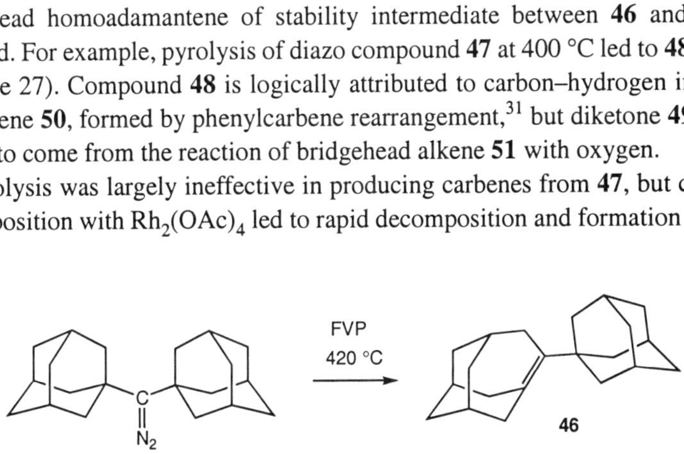

Scheme 26.

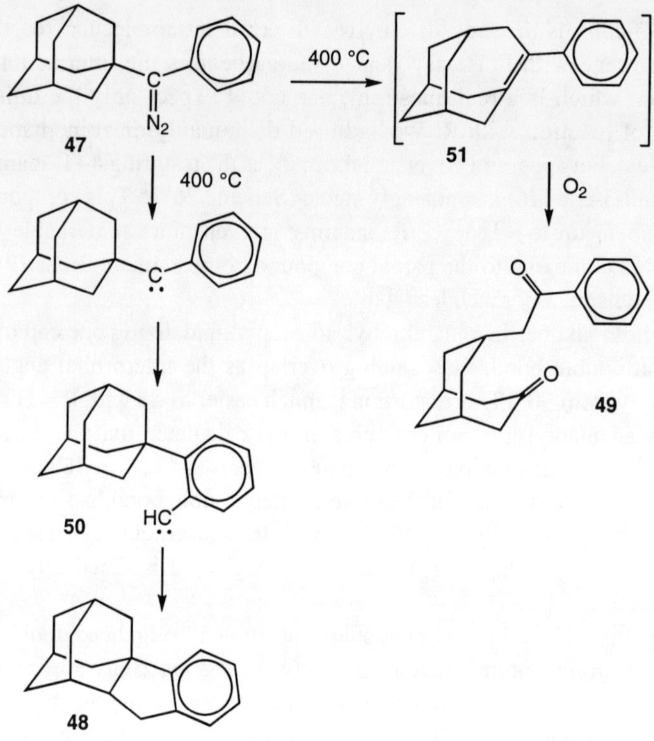

Scheme 27.

low temperatures at which its decomposition could be monitored. More than half of bridgehead alkene **51** survived in solution at room temperature after 12 hours.

Diazo compound **52** was decomposed in similar fashion (Scheme 28). Rhodium-catalyzed decomposition of **52** was rapid, and led to **53** which could be trapped by added ethyl alcohol (and other reagents). Bridgehead alkene **53** was less stable than **51**, but was still far more stable than **9**. For example, half of it survived after one hour at room temperature.

None of this work, beautiful though it is, speaks to the question of the mechanism of bridgehead alkene formation. Is a carbene involved? We don't know, although it surely seems most unlikely in the rhodium-catalyzed decompositions. The thermal reactions are more likely to involve carbenes. In the pyrolysis of the phenyl-substituted diazo compound **47** the carbene is surely formed, as the phenyl-carbene rearrangement to **50** demonstrates. A resolution of the question of whether the carbene is involved in formation of the bridgehead alkene awaits further work.

Takeuchi and his group in Kyoto have used an adamantyl group to stabilize adamantene.[32] Although they were not able to isolate 2-(1-adamantyl)adamantene (**54**), they were able to trap it both intramolecularly and intermolecularly. Photolysis

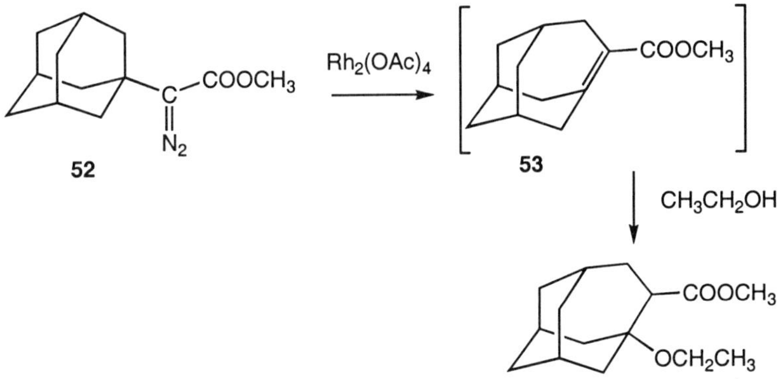

Scheme 28.

of diazo compound **55** in hexane at 14 °C led to **56** in 35% yield in addition to unreacted diazo compound (Scheme 29). Bridgehead alkene **54** seems a necessary intermediate, and there is a number of potential routes from **54** to **56**. That favored by the authors is shown in Scheme 30, and features an alkene-to-carbene rearrangement much like that shown by adamantene itself. The absence of **57** seems curious, as the related product is formed from reverse carbon–carbon insertion in adamantene to give carbene **27**. When methanol was added to the photolysis medium, **56** was still isolated along with **58**, and several products formed by protonation of diazo compound **55**. Ether **58** could be shown not to result from protonation, however, as its yield actually increases as the methanol concentration decreases. Its likely source is **54**. Further work on adamantenes substituted with smaller alkyl groups is in progress in Kyoto.[32]

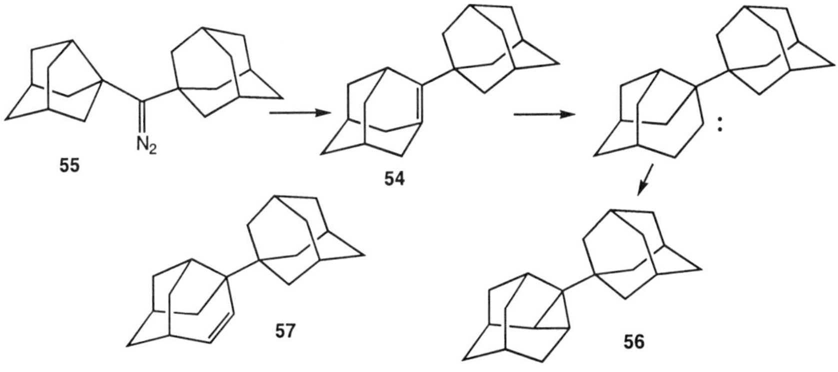

Scheme 29.

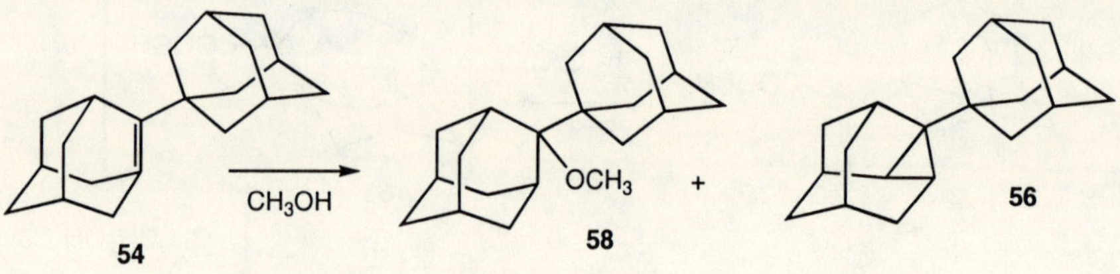

Scheme 30.

What problems remain? Plenty, but let me just mention two quite obvious large areas. First of all, the question of the intermediacy of carbenes in the ring expansion is not even settled. There are a few individual facts; this carbene can ring expand; that one can't, and so on. What is needed is a general, broad investigation of the ring expansion process under several reaction conditions. One nice way to do this is to use non-nitrogenous sources of carbenes, so that the question of diazo compound intervention is avoided. One possibility is compound **59**. But even when

we know in general if simple bridgehead substituted carbenes can ring-expand to bridgehead alkenes, we will not know if they in fact undergo the reaction when diazo compounds are irradiated. Here, we need experiments of the Jackson–Platz type.

Second, there need to be more investigations of the kind underway in Nagoya and Kyoto. It is perhaps presumptuous of me to say that, since we at Princeton have been more guilty than anyone in not carrying out such an investigation, but nonetheless, it is true. It should be possible to make a wide variety of bridgehead alkenes of varying twist angle and stability through the ring expansion reaction.

Finally, the connection between solution and the gas phase should be strengthened. It appears that our original fears that intermolecular solution reactions would out-compete the intramolecular chemistry were too pessimistic. There should be a stronger effort at trying these reactions in solution, especially from non-diazo precursors.

ACKNOWLEDGMENTS

The National Science Foundation and the Petroleum Research Fund have been steadfast supporters of our work in this area over two decades. My heartfelt thanks go to them. My coworkers over the years have contributed both hard work and clever thinking to the experiments described in these pages. So, HC Glick, Igor Likhotvorik, Tony Wolf, Dave Martella, Motoo Fukushima, Si Sellers, Theo Klebach, and Fred Hollowood, many, many thanks for both science and friendship. My collaborators, Paul Schleyer, Matt Platz, and Udo Brinker have given more than their share to our efforts, and I thank them for it. I must single out two people, Ning Chen and Nanying Bian, for their help on the carbene-bridgehead double bond problem. What has been most impressive to me is not so much their impressive intellectual prowess, but rather, their willingness to do the nasty, hard work necessary to burrow through some barrier after clever thought has failed to find some route around it. Thank you very much.

REFERENCES

1. Glick, H.; Likhotvorik, I.R.; Jones, M., Jr. *Tetrahedron Lett.* **1995**, *32*, 5715.

2. Warner, P.M. *Chem. Rev.* **1989**, *89*, 1067; Luef, W.; Keese, R. *Topics in Stereochemistry* Vol. 20, Eliei, E.L.; Wilen, S.H., Eds.; Wiley: New York, 1991, p 231; See also: Szeimies, G. *Advances in Strain in Organic Chemistry*, Halton, B., Ed.; JAI Press: Stamford, CT, 1992.

3. Wilt, J.W.; Schneider, C.A.; Dabek, H.F., Jr.; Kraemer, J.F. *J. Org. Chem.* **1966**, *31*, 1543.

4. Kirmse, W., private communication, 1982.

5. Wolf, A.D.; Jones, M., Jr. *J. Am. Chem. Soc.* **1973**, *95*, 8209.

6. Fort, R.C., Jr. in *Carbonium Ions*, Olah, G.A.; Schleyer, P.v.R., Eds.; Wiley: New York, 1973, Vol. IV, p 1783.

7. Maier, W.F.; Schleyer, P.v.R. *J. Am. Chem. Soc.* **1981**, *103*, 1891.

8. Modarelli, D.A.; Morgan, S.C.; Platz, M.S. *J. Am. Chem. Soc.* **1992**, *114*, 7034; White, W.R.; Platz, M.S. *J. Org. Chem.* **1992**, *57*, 2841; Celebi, S.; Leyva, S.; Modarelli, D.A.; Platz, M.S. *J. Am. Chem. Soc.* **1993**, *115*, 8613; Moss, R.A.; Ho, G.-J. *J. Phys. Org. Chem.* **1993**, *6*, 126; Tomioka, H.; Kitagawa, H.; Izawa, Y. *J. Org. Chem.* **1979**, *44*, 3072; Seburg, R.A.; McMahon, R.J. *J. Am. Chem. Soc.* **1992**, *114*, 7183; Yamamoto, N.; Bernardi, F.; Bottoni, A.; Olivucci, M.; Robb, M.A.; Wilsey, S. *J. Am. Chem. Soc.* **1994**, *116*, 2064; Fox, J.M.; Scacheri, J.E.G.; Jones, K.G.L.; Jones, M., Jr.; Shevlin, P.B.; Armstrong, B.; Sztyrbicka, R. *Tetrahedron Lett.* **1992**, *33*, 5021.

9. Frey, H.M. *Pure and Appl. Chem.* **1964**, *9*, 527, and references therein; Mansoor, A.M.; Stevens, I.D.R. *Tetrahedron Lett.* **1966**, 1733; Chang, K.-T.; Shechter, H. *J. Am. Chem. Soc.* **1979**, *101*, 5082.

10. Martella, D.J.; Jones, M., Jr.; Schleyer, P.v.R.; Maier, W.F. *J. Am. Chem. Soc.* **1979**, *101*, 7634.

11. Gudipati, M.S.; Radziszewski, J.G.; Kaszynski, P.; Michl, J. *J. Org. Chem.* **1993**, *58*, 3668.

12. Eaton, P.E.; Hoffmann, K.-L. *J. Am. Chem. Soc.* **1987**, *109*, 5285; Eaton, P.E.; White, A. *J. Org. Chem.* **1990**, *55*, 1321.

13. Eaton, P.E.; Appell, R.B. *J. Am. Chem. Soc.* **1990**, *112*, 4055.

14. Chen, N.; Jones, M., Jr.; White, W.R.; Platz, M.S. *J. Am. Chem. Soc.* **1991**, *113*, 4981.

15. Hrovat, D.A.; Borden, W.T. *J. Am. Chem. Soc.* **1992**, *114*, 2719.

16. Martella, D.J.; Jones, M., Jr.; Schleyer, P.v.R. *J. Am. Chem. Soc.* **1978**, *100*, 2896.

17. Burns, W.; McKervey, M.A. *J. Chem. Soc., Chem. Commun.* **1974**, 858; Burns, W.; Grant, M.A.; McKervey, M.A.; Step, G. *J. Chem. Soc., Perkin Trans.* **1976**, *1*, 234; Gillespie, D.G.; Walker, B.J. *Tetrahedron Lett.* **1977**, 1673; Lenoir, D.; Kornrumpf, W.; Fritz, H.P. *Chem. Ber.* **1983**, *116*, 2390.

18. Grant, D.; McKervey, M.A.; Rooney, J.J.; Samman, N.G.; Step, G. *J. Chem. Soc., Chem. Commun.* **1972**, 1186.
19. Lenoir, D. *Tetrahedron Lett.* **1972**, 4049; Lenoir, D.; Firl, J. *Liebigs Ann. Chem.* **1974**, 1467.
20. van Zorge, J.A.; Strating, J.; Wynberg, H. *Recl. Trav. Chim. Pays-Bas* **1970**, *89*, 781.
21. Bian, N.; Jones, M., Jr. *J. Am. Chem. Soc.* **1995**, *117*, 8957. The introduction gives a history of the reaction with furan.
22. Brinker, U.H.; Ritzer, J. *J. Am. Chem. Soc.* **1981**, *103*, 2116; Fukushima, M.; Jones, M., Jr.; Brinker, U.H. *Tetrahedron Lett.* **1982**, 3211.
23. Wiberg, K.B.; Waddell, S.T. *J. Am. Chem. Soc.* **1990**, *112*, 2194.
24. Wiberg, K.B.; Lowry, B.R. *J. Am. Chem. Soc.* **1963**, *85*, 3188.
25. Kirmse, W.; Meinert, T.; Modarelli, D.A.; Platz, M.S. *J. Am. Chem. Soc.* **1993**, *115*, 8918.
26. Stapersma, J.; Rood, I.D.C.; Klumpp, G.W. *Tetrahedron* **1982**, *38*, 3051.
27. Platz, M.S.; Modarelli, D.A.; Morgan, S.; White, W.R.; Mullins, M.; Celebi, S.; Toscano, J.P. *Prog. React. Kinetics* **1994**, *19*, 3.
28. Sellers, S.F.; Klebach, T.C.; Hollowood, F.; Jones, M., Jr. *J. Am. Chem. Soc.* **1982**, *104*, 5492.
29. Ermer, O., Bochum, private communication quoted in ref. [28].
30. Ohno, M.; Itoh, M.; Umeda, M.; Furuta, R.; Kondo, K.; Eguchi, S. *J. Am. Chem. Soc.* **1996**, *118*, 7075.
31. McMahon, R.J.; Abelt, C.J.; Chapman, O.L.; Johnson, J.W.; Kreil, C.L.; LeRoux, J.-P.; Mooring, A.M.; West, P.R. *J. Am. Chem. Soc.* **1987**, *109*, 2456.
32. Okazaki, T.; Isobe, H.; Kitagawa, T.; Takeuchi, K. *Bull. Chem. Soc. Jpn.* **1996**, *69*, 2053.

HIGH RESOLUTION FLUORESCENCE SPECTROSCOPY:
DETERMINATION OF THE ZERO-FIELD SPLITTING PARAMETERS OF EXCITED TRIPLET STATES OF AROMATIC CARBENES

Eva Migirdicyan, Boleslaw Kozankiewicz, and Matthew S. Platz

Advances in Carbene Chemistry
Volume 2, pages 97–132
Copyright © 1998 by JAI Press Inc.
All rights of reproduction in any form reserved.
ISBN: 1-55938-837-4

ABSTRACT

This chapter describes the fluorescence spectroscopy of Shpolskii matrix isolated aromatic carbenes and their conformational isomerism. The triplet–triplet fluorescence and excitation spectra of dibenzocycloheptadienylidene (DBC), diphenylcarbene (DPC), and 2-naphthylphenylcarbene (2-NPC) in n-hexane and n-heptane at 4.2 to 20 K have been investigated by site-selective laser experiments. Different spectra attributed to the carbenes in different geometrical conformations have been obtained for DPC and for 2-NPC. The attributions are confirmed by quantum chemical calculations using the CS-INDO-CIPSI method. In the case of 2-NPC, the photochemical transformation of the pseudo Z/*cis* into the pseudo E/*trans* conformer has been studied in the n-heptane matrix by both EPR and fluorescence spectroscopy.

The fluorescence decays of DBC, DPC, and 2-NPC at 4.2 to 10 K are nonexponential and significantly modified in the presence of a magnetic field. This is attributed to the fluorescence from the different sublevels of the first excited triplet state T_1 at a rate faster than the rate of spin-lattice relaxation.

Previous EPR experiments indicate that the zero-field splitting (ZFS) parameter $D(T_0)$ of DBC, DPC, and 2-NPC in the ground state T_0 are all around 0.4 cm^{-1}. However, the magnetic resonance techniques cannot be used to determine the ZFS parameter $D(T_1)$ in the first excited triplet T_1 whose component lifetimes are generally lower than 150 ns because the concentration of carbenes in the T_1 state is too low. This is why the $D(T_1)$ values of DBC, DPC, and 2-NPC have been estimated either from the magnetic field effect on the fluorescence decays or by hole-burning experiments. In all the cases, $D(T_1)$ is significantly smaller than $D(T_0)$. This is explained by the substantial delocalization of the π electron onto the aromatic rings in the T_1 state.

The hole-burning technique, which has been applied for the first time to a triplet–triplet transition, has led to the precise determination of the D and E parameters of the pseudo E/*trans* 2-NPC in the T_0 and T_1 states:

$$D(T_0) = 0.50 \text{ cm}^{-1} \qquad\qquad E(T_0) = 0.017 \text{ cm}^{-1}$$
$$D(T_1) = 0.038 \text{ cm}^{-1} \qquad\qquad E(T_1) = 0.005 \text{ cm}^{-1}$$

I. INTRODUCTION

The modern era of carbene chemistry can be dated to Jack Hine's pioneering studies[1] of the alkaline hydrolysis of chloroform

$$CHCl_3 + HO^- \rightleftharpoons {}^-CCl_3 + H_2O$$

$$^-CCl_3 \rightleftharpoons :CCl_2 + Cl^-$$

In his classic book *Divalent Carbon*,[2] Hine was quick to point out that Dumas[3] and Geuther[4] had postulated the existence of carbenes as reactive intermediates as early as in the 19th century.

Nevertheless, it was Hine's mechanistic studies that directly inspired Doering and Hoffmann[5] to intercept the postulated intermediate: CCl_2, with cyclohexene.

$$CHCl_3 \; + \; \text{(cyclohexene)} \xrightarrow{KOC(CH_3)_3} \text{(7,7-dichlorobicyclo[4.1.0]heptane)}$$

This dramatic demonstration of the synthetic potential of carbenes greatly stimulated mechanistically inclined scientists. Physical chemists, physical organic chemists, and spectroscopists began the quest for direct observation of the short-lived intermediates. In 1959 Herzberg issued the first report of his studies of the optical spectroscopy of methylene.[6] By working in the gas phase, Herzberg lengthened the lifetime of the parent carbene by minimizing the rate of bimolecular reactions. Thus, it was possible to obtain the absorption spectra of singlet and triplet methylene in this manner.

Attention soon turned to matrix spectroscopy. In this technique, a suitable precursor is dissolved or cocrystallized in an inert solvent host and then frozen to cryogenic temperatures, or it is deposited into an inert matrix at low temperature. Photolysis of the precursor can release a carbene intermediate. Relatively high concentrations of carbenes can be generated in this fashion because the low temperature and high viscosity of the medium prevents diffusion and bimolecular reactions.[7] In 1962 scientists at Bell Laboratories first used this method to detect a carbene (diphenylcarbene) by electron paramagnetic resonance (EPR) spectroscopy. This work was performed with randomly oriented samples.[8] Shortly thereafter, Brandon et al. detected diphenylcarbene in a single crystal.[9] These reports ushered in an era of rapid discovery:[10,11] several carbenes (including methylene) were subsequently studied by matrix EPR spectroscopy, including electron nuclear double resonance (ENDOR) techniques.[12] The fluorescence spectroscopy of arylcarbenes in low temperature matrices was also discovered at this time[13] and more recently, matrix infrared (IR) studies of carbenes have been reported.[7]

These studies profoundly enhanced our knowledge of carbenes. They demonstrated that many carbenes have triplet ground states and determined the bond angles at many divalent carbons.

Of course, many other techniques have been developed in subsequent years to study carbenes in solution. These include chemically induced dynamic nuclear polarization (CIDNP)[14] and flash photolysis methods with ever increasing time resolution.[15] Nevertheless, as we attempt to show, matrix spectroscopy remains an active and powerful tool for the study of carbenes.

In glassy matrices, the distribution of molecules isolated in a great variety of sites results in solute spectra consisting of broad bands. Such matrices have been widely used by scientists at Bell Laboratories[8,11,13] to study the EPR and fluorescence spectroscopy of carbenes. In contrast, crystalline matrices provide only a small number of distinct lattice sites that can be occupied by guest molecules or trapped species generated from them by photolysis. Substitutional solutions of aromatic carbenes oriented in single crystals have been investigated by optical, EPR, and ENDOR spectroscopy at the University of Chicago.[9,10,12,14] Such experiments are, however, limited to carbenes that have a similar volume and geometry as the host molecules.

Shpolskii and coworkers[16] have shown that sharp electronic spectra can be obtained for planar aromatic molecules dispersed in n-alkane polycrystalline matrices when the carbon skeleton chain length of the linear paraffin equals the longest dimension of the carbon skeleton of the aromatic guest. A typical Shpolskii spectrum consists of several sharp "quasi line" spectra, each one being associated with guest molecules located in a well-defined lattice site of the matrix. Each quasi line spectrum can be described in terms of a zero-phonon line that is predominant and a phonon band, since the coupling between the aromatic impurity and the n-alkane is generally very weak.[17]

In this chapter, we present the fluorescence spectra and decays of dibenzocyclo-heptadienylidene (DBC), diphenylcarbene (DPC), and 2-naphthyl-phenylcarbene (2-NPC) dispersed in n-hexane and n-heptane Shpolskii matrices. EPR experiments

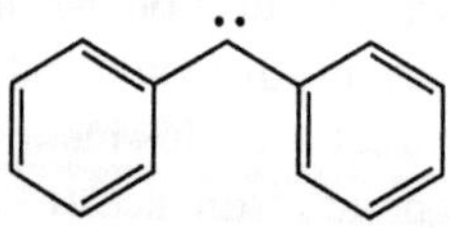

DPC

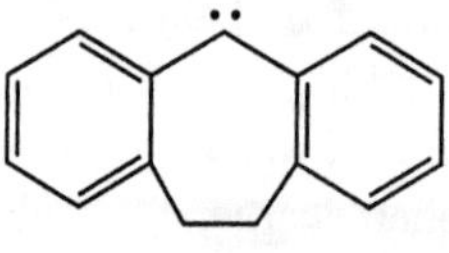

DBC

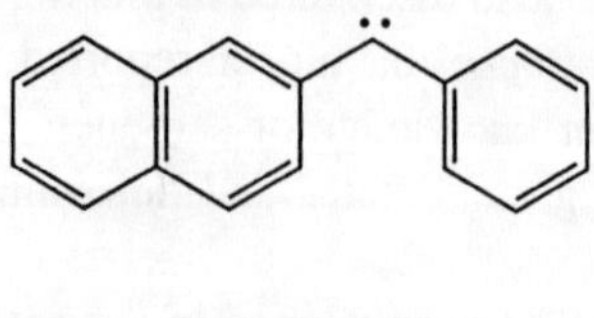

E / trans 2-NPC

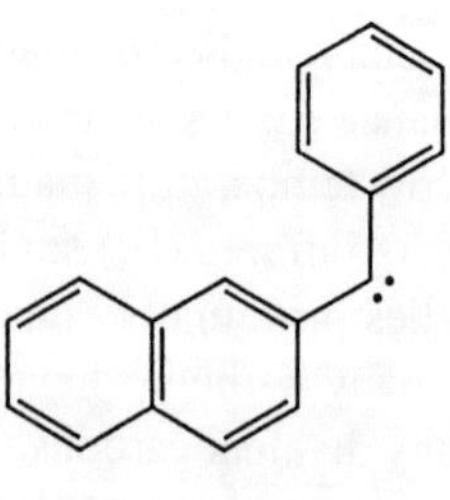

Z / cis 2-NPC

indicate that these species have a triplet ground state.[8,11] Among these carbenes, only DBC, which contains an ethano-bridge linking the two aromatic rings, has a rigid structure. The two other carbenes, DPC and 2-NPC, can exist in several planar and nonplanar conformations resulting from rotation of the aromatic ring around the single bond to the divalent carbon. Special emphasis will be placed on the study of conformational isomerism by fluorescence spectroscopy. From the magnetic field effect on the fluorescence decays, the ZFS parameters $D(T_1)$ in the first excited triplet T_1 of carbenes are determined. Finally, hole-burning experiments performed on 2-NPC in *n*-hexane and in *n*-heptane at 1.8 K are discussed. This is the first time that such investigations have been carried out on a triplet–triplet transition.

II. FLUORESCENCE SPECTRA OF AROMATIC CARBENES

A. Previous Studies in Organic Glasses and in Single Crystals

The fluorescence emitted from a triplet carbene was first observed by Trozzolo and Gibbons.[13] The UV photolysis of diphenyldiazomethane dispersed in a 2-methyltetrahydrofuran glass (2-MTHF) at 77 K gives rise to a broad emission with a maximum at 480 nm (see Figure 1), which disappears upon warming the frozen solution. On the basis of the linear correlation between the intensities of the EPR and the emission spectra, Trozzolo and Gibbons have attributed the broad band at 480 nm to the fluorescence of diphenylcarbene (DPC). This fluorescence corresponds to the spin-allowed transition from the first excited triplet state T_1 to the ground triplet state T_0. This assignment was strengthened by the subsequent time-resolved measurements of Ware and coworkers,[18] who found a single exponential fluorescence decay corresponding to a lifetime of 70 to 100 ns at 77 K, depending on the glass used. Similar results were obtained with substituted diphenylcarbenes.[13,18]

Electronic spectra exhibiting sharp features were observed by Anderson et al.[19] for DPC trapped in substitutional sites of benzophenone single crystals at 2 to 20 K. The absorption and fluorescence spectra obtained near the origin of the T_0–T_1 transition are shown in Figure 2. The spectra consist of a sharp zero-phonon line at 470.7 nm and broad phonon sidebands extending to high energy in absorption ($\lambda_{max} \approx 465$ nm) and to low energy in emission ($\lambda_{max} \approx 476$ nm). The intensity of the sharp zero-phonon line relative to the broad phonon band increases strongly with decreasing temperature down to 2 K, which indicates strong excitation-phonon coupling. This large coupling can be attributed to relatively large geometrical changes upon excitation, which could greatly distort the lattice. Particularly noteworthy is the similarity between the broad fluorescence band with $\lambda_{max} = 480$ nm observed[13] in the 2-MTHF glass at 77 K and the broad phonon fluorescence band with $\lambda_{max} = 476$ nm obtained[19] in the benzophenone matrix at 2 to 30 K. The 4-nm shift between the two bands is due, at least partly, to a matrix effect.

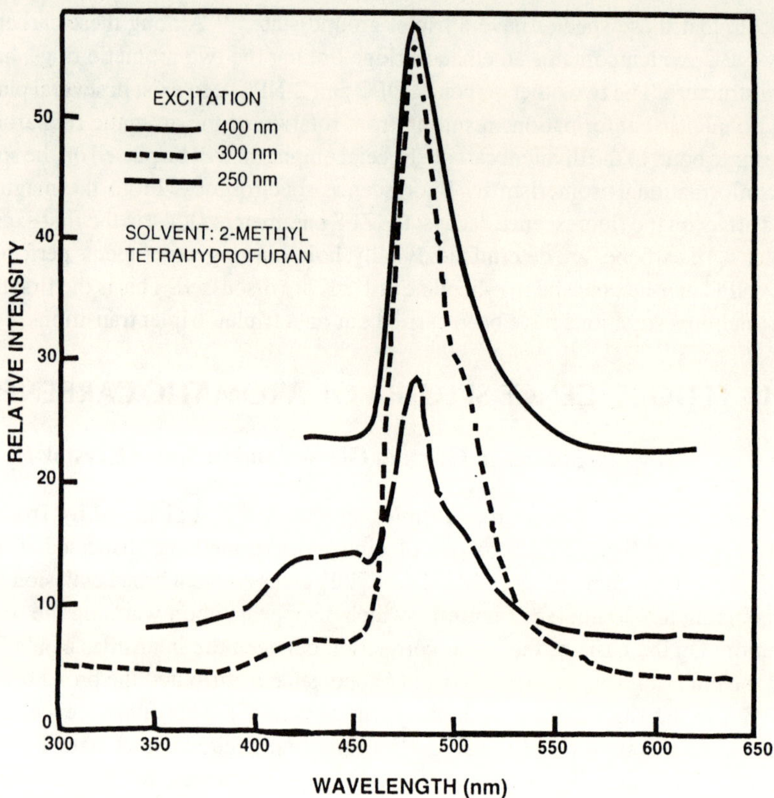

Figure 1. Luminescence of diphenylcarbene. (Reprinted with permission from *J. Am. Chem. Soc.* **1967**, *89*, 239).

B. Site-Selected Spectra in Shpolskii Matrices

Under conventional broad-band excitation, the fluorescence spectra of carbenes in Shpolskii matrices at 4.2 to 20 K generally present a multisite structure due to the distribution of guest molecules in several lattice sites. The spectra, with origins shifted by 1 to 2 nm from each other, can be attributed to geometrically equivalent carbenes incorporated in different crystallographic sites of the matrix. In contrast, the spectra with origins shifted by 10 to 12 nm from each other are generally due to different conformers of the carbene. The multisite structure can be considerably simplified by using narrow-band laser excitation, where only the fragments that have an absorption energy coincident with the laser frequency are selectively excited. The sharp electronic spectra presented here correspond to carbenes located in the most prominent sites of Shpolskii matrices either before or after annealing of the samples.

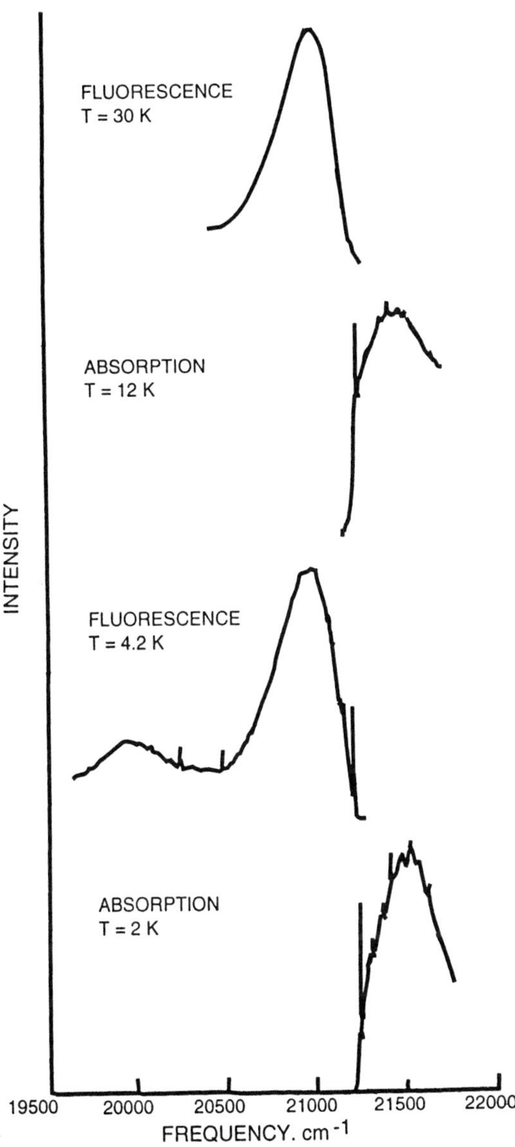

Figure 2. Absorption and emission for the origin region of the 470 nm ground triplet → excited triplet transition of diphenylcarbene in benzophenone at various temperatures. (Reprinted with permission from *J. Chem. Phys.* **1979**, *71*, 1559).

1. *Dibenzocycloheptadienylidene (DBC)*

The site-selected fluorescence (I, II) and excitation (III, IV) spectra of DBC[20] in *n*-hexane at 20 K are presented in Figure 3. The spectra are composed of sharp bands, indicating that the trapped carbene is either planar or nearly planar. However, DBC, which contains a seven-membered ring, cannot be strictly planar. Like 1,4-cycloheptadiene,[21] DBC will likely exist in two different nonplanar conformations, C_2 and C_s, whose fluorescence spectra are expected to lie in different spectral

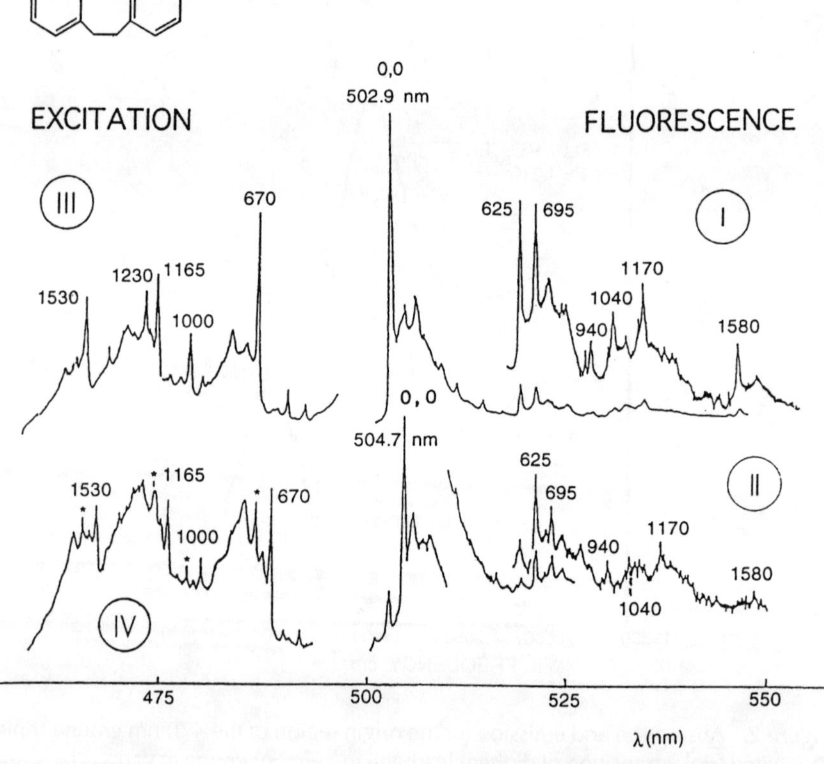

Figure 3. Site-selected fluorescence and excitation spectra of DBC in *n*-hexane at 20 K. The emissions I and II are respectively excited with 486.5 and 488 nm laser lines. The excitation spectra III and IV are obtained by monitoring the fluorescence origins at 502.9 and 504.7 nm, respectively, while scanning the dye laser. (Reprinted with permission from *J. Phys. Chem.* **1992,** *96,* 2486).

regions and to have different structures. However, the C_2 conformer, which is close to planarity, is predicted to be more stable than conformer C_s for steric reasons[22] and by π-orbital conjugation. Therefore the two fluorescence spectra I and II starting at 502.9 and 504.7 nm, which are sharp and have similar vibrational structures, can both be attributed to the nearly planar C_2 conformer trapped in two different sites of the *n*-hexane matrix. The analysis of these spectra is given in Ref. 20.

2. Diphenylcarbene (DPC)

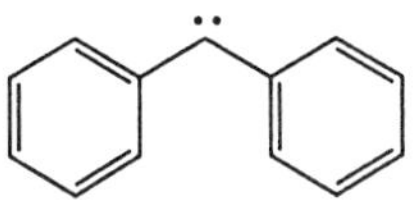

The fluorescence spectra of perprotonated DPC (DPC-h_{10}) in *n*-hexane at 77 K obtained at several concentrations c of diphenyldiazomethane, the DPC precursor, are displayed in Figure 4. The spectra consist of a broad band with a maximum at 482 nm, which is very similar to the fluorescence detected by Trozzolo and Gibbons,[13] and a sharper spectrum starting at 502.8 nm, which was not detected by these authors in their glassy matrix. The relative intensity of the sharp and broad spectra is concentration dependent. For $c = 10^{-2}$ M, only the sharp fluorescence is detectable. Its vibronic structure can be analyzed by using vibrational modes and frequencies of benzenoid compounds.[23] As c decreases, the broad band increases in intensity with respect to the sharp spectrum. This is true for experiments performed between 4.2 and 77 K on DPC-h_{10}, as well as on perdeuterated DPC (DPC-d_{10}).[24] These observations might be interpreted by triplet–triplet energy transfer between the two species responsible for the sharp and for the broad fluorescence spectra, the efficiency of which increases with c. The two fluorescence spectra have been attributed to two different geometrical conformations of DPC.

The low-energy sharp fluorescence is attributed to a DPC molecule with a geometry similar to that of DBC, the carbene with the ethano bridge, since DPC and DBC both have their fluorescence origins at 502.8 nm in the *n*-hexane matrix.[20] In this geometry, the two aromatic rings of DPC are not too far from being coplanar.

The broad blue-shifted fluorescence is attributed to a nonplanar DPC conformer in which the carbenic bond angle is 140° and the dihedral angles φ_1 and φ_2 of each benzene ring and the average plane are close to 30°. This geometry corresponds to that ($\varphi_1 = 26.5°$ and $\varphi_2 = 29.5°$) determined by Anderson and Kohler,[25] who used the ENDOR technique for DPC in benzophenone. Such an interpretation implies that DPC has nearly the same geometry in frozen *n*-hexane and in benzophenone single crystals, which is consistent with the fact that the broad fluorescence with $\lambda_{max} = 482$ nm observed in the *n*-hexane matrix (see Figure 4) is very similar to the broad structureless phonon band with $\lambda_{max} = 476$ nm previously observed[19] for DPC substitutionally incorporated in benzophenone single crystals (see Figure 2).

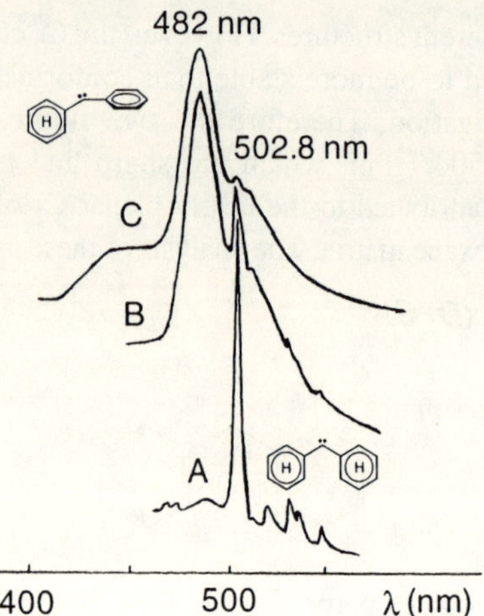

Figure 4. Concentration dependence of the fluorescence spectra of DPC-h_{10} in *n*-hexane at 77 K. The spectra are excited with 315 nm radiation. The concentration *c* of the DPC precursor, the diphenyldiazomethane, is $c = 10^{-2}$ M (curve A), $c = 10^{-4}$ M (curve B), and $c = 10^{-5}$ M (curve C).

These attributions have been confirmed by quantum chemical calculations using the CS-INDO (conformations spectra intermediate neglect of differential overlap) method,[26] which was specially designed for dealing with problems of conformation of conjugated molecules in their ground as well as excited states. The CI treatment itself was performed using the CIPSI (configuration interaction by perturbative selected iterations) procedure.

The energies and oscillator strengths of the singlet and triplet levels have been calculated[24] for DPC in three different geometries: the planar species of C_{2v} symmetry, the quasi-planar species of C_2 symmetry (having the same geometry as DBC[20]) and the nonplanar species of C_1 symmetry described by Anderson and Kohler.[25] The corresponding energy level diagram is displayed in Figure 5 and the data obtained for the triplet levels are collected in Table 1.

The vertical triplet–triplet transitions listed in Table 1 compare favorably with the experimental results. In particular, the first T_0–T_1 transition energy of 2.62 eV computed for the C_{2v} structure, i.e., $1^3B_1 - 1^3A_2$, is in good agreement with the energy of 2.49 eV (497 nm) corresponding to the average energy between the fluorescence and excitation origin bands of the sharp spectrum (see Figure 4 of Ref. 24). When the molecular symmetry is lowered, an hypsochromic shift is observed experimentally (2.60 eV) as well as theoretically (2.80 eV), as is usual for nonplanar structures. This is illustrated in Figure 5.

In addition, the calculations predict an important change in the nature of the main configuration included in the CI expansion of the first excited triplet state going

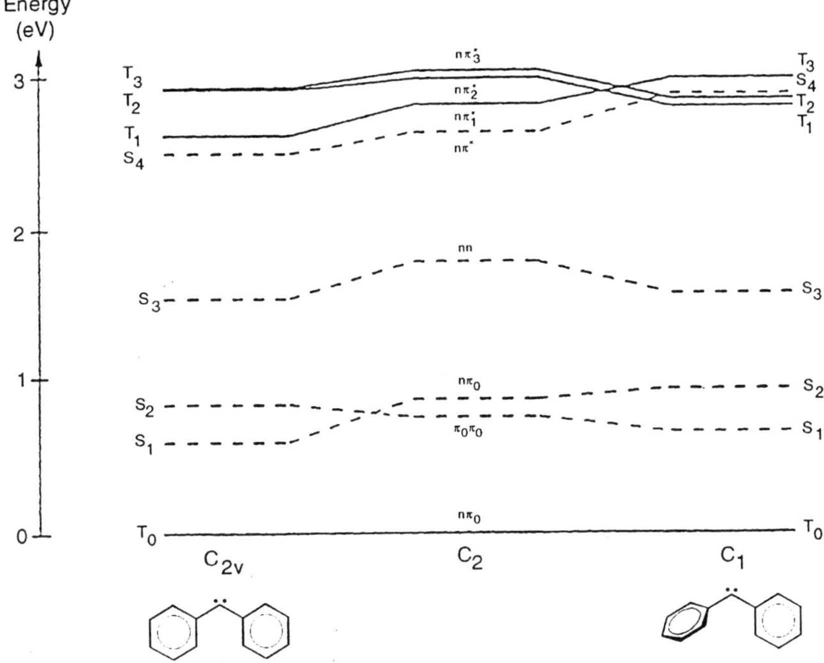

Figure 5. Energy level diagram of DPC computed for the C_{2v}, C_2, and C_1 symmetry species. (Reprinted with permission from *J. Phys. Chem.* **1993**, *97*, 13358).

from C_{2v} or C_2 to C_1 structures, namely $(n\pi_2^*)$ occupancy instead of $(n\pi_1^*)$. As a result, the oscillator strength of the first $T_0 - T_1$ transition ($f = 0.022$) in the nonplanar structure C_1 appears to be greatly reduced by comparison with the planar one ($f = 0.122$). This is again in good agreement with the data obtained from the fluorescence decays (*vide infra*).

Table 1. Calculated Transition Energy and Oscillator Strength of the Lower Triplet States of DPC Conformers

	C_{2v}			C_2			C_1		
State	Energy (eV)	f	State	Energy (eV)	f	State	Energy (eV)	f
1^3B_1	0.00		1^3B	0.00		1^3A	0.00	
1^3A_2	2.62	0.122	1^3A	2.82	0.095	4^3A	2.99	0.164
2^3B_1	2.92	0.024	2^3B	2.99	0.005	2^3A	2.80	0.022
2^3A_2	2.94	0.002	2^3A	3.03	0.016	3^3A	2.85	0.012

3. *2-Naphthylphenylcarbene (2-NPC)*

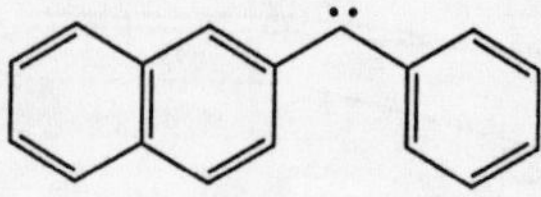

2-NPC is another species that exhibits conformational isomerism. This type of isomerism in triplet carbenes has been the subject of several earlier studies using EPR spectroscopy. For example, it was found that 1- and 2-naphthylcarbenes exist in two different conformations, which have different zero-field splitting (ZFS) parameters.[27] The measured D values were assigned to specific conformers on the basis of calculations that considered the relationship between the ZFS parameters D and π spin densities in the frame of the point spin model.[28,29] Similarly, EPR experiments[30] performed with 2-NPC in a 2-MTHF glass at 77 K indicated that two D values, $D = 0.4044$ cm^{-1} and $D = 0.3898$ cm^{-1}, can be obtained for the two different conformers of 2-NPC in the triplet ground state. The extension of this study combining EPR and fluorescence spectroscopy will now be described.[31]

The site-selected quasi-line fluorescence and excitation spectra of 2-NPC in *n*-hexane at 4.2 K are displayed in Figure 6. The good coincidence between the fluorescence and excitation origin bands (not shown on the figure) indicates that the sharp bands are indeed zero-phonon lines and, consequently, 2-NPC is a good candidate for hole-burning experiments (*vide infra*). The fluorescence origins at 600.1 and 588.8 nm are separated by more than 11 nm, which suggests that the two spectra correspond to different geometrical isomers.[31] Both spectra, which have slightly different vibrational structures, have been analyzed in Ref. 31.

When 2-NPC is introduced into the *n*-heptane matrix, the two sharp fluorescence spectra are again observed at 4.2 K with origin bands slightly shifted to 589.5 and 600.5 nm, but their relative intensity depends now on the excitation time. This is illustrated in Figure 7. The spectrum with origin at 600.5 nm, which is predominant in the early stages of excitation with the N$_2$ laser line at 337 nm (Figure 7A), decreases to the benefit of the spectrum with origin at 589.5 nm, as the photolysis of the sample progresses (Figure 7B). This process is reversed on annealing the sample in the dark at 85 K (Figure 7C). These observations are attributed to the photochemical (hν) and thermal (Δ) interconversion between the two conformers responsible for the spectra with the origins at 600.5 and 589.5 nm.

$$\text{conformer emitting at 600.5 nm} \underset{\Delta}{\overset{h\nu}{\rightleftharpoons}} \text{conformer emitting at 589.5 nm}$$

The light-induced isomerization that has been exclusively observed in the *n*-heptane matrix can also be monitored by EPR spectroscopy.[31] Figure 8 presents the low-field z transitions of 2-NPC in a 2-MTHF glass at 77 K (curve A) and in a

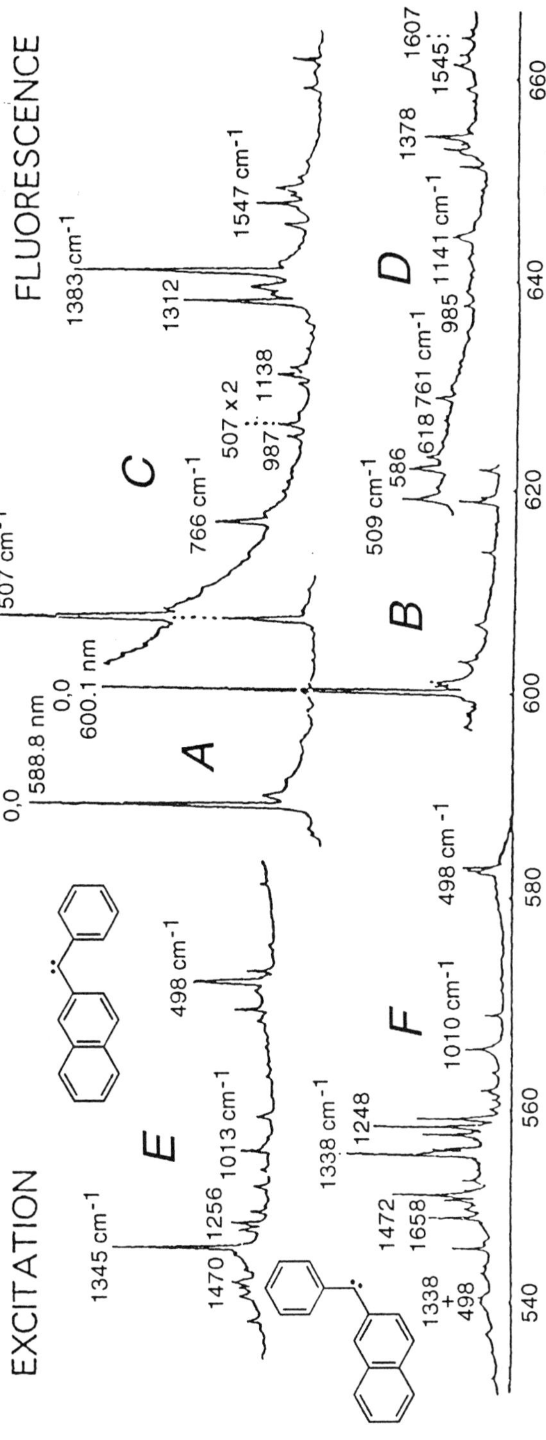

Figure 6. Selectively excited fluorescence (A, B, C, and D) and excitation (E and F) spectra of 2-NPC in annealed *n*-hexane at 4.2 K. The emissions A and B are excited with laser lines at 555.6 and 545.8 nm, respectively. The emissions C and D are excited with laser lines at 588.8 and 600.1 nm, respectively. The excitation spectra *E* and F are obtained by monitoring the fluorescence origins at 588.8 and 600.1 nm, respectively, while scanning the dye laser. (Reprinted with permission from *J. Phys. Chem.* **1994**, *98*, 10419).

109

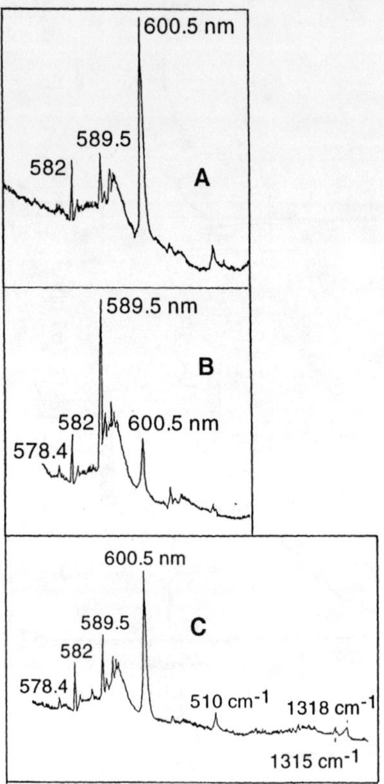

Figure 7. Fluorescence spectra of 2-NPC in *n*-heptane at 4.2 K excited with the N$_2$ laser line at 337 nm. The emissions A and B correspond to respectively 15 min and 5 hours excitation. The emission C is obtained after annealing the sample. (Reprinted with permission from *J. Phys. Chem.* **1994**, *98*, 10419).

polycrystalline *n*-heptane matrix at 77 K (curves B–D). The two broad bands at 975 and 1150 G in the glassy matrix become sharper and shifted to 990 and 1230 G, respectively, in the polycrystalline matrix. The effect of line-narrowing in crystalline matrices was already noted at Bell Laboratories for DPC.[32] The two bands of 2-NPC are attributed to the two different conformers. The band at 1230 G, which is predominant at short photolysis time (curve B), decreases in intensity during photolysis with radiation with wavelengths greater than 375 nm to the benefit of the band at 990 G (curves C and D). The correlation between the light-induced transformations in the EPR and fluorescence spectra establishes that the conformer responsible for the fluorescence origin at 600.5 nm has a low-field z transition at 1230 G and consequently a $D = 0.42 \pm 0.02 \text{ cm}^{-1}$ value in the ground state. The D value of the other conformer is $0.41 \pm 0.01 \text{ cm}^{-1}$.

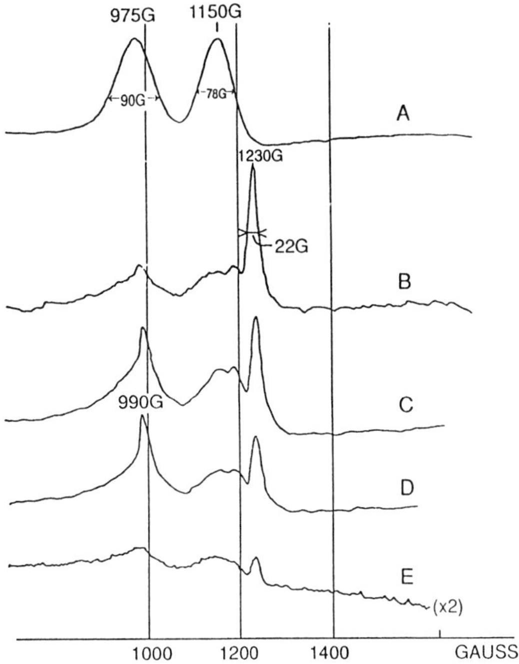

Figure 8. EPR spectra of the two conformers of 2-NPC in 2-methyltetrahydrofuran at 77 K (curve A) and in *n*-heptane at 77 K (curves B, C, D, and E). Spectrum B corresponds to sample B containing freshly produced 2-NPC obtained from photolysis of its precursor. Spectra C and *D* correspond to sample B exposed to radiations $\lambda > 375$ nm. Spectrum *E* is obtained by exposing directly the carbene precursor to radiations $\lambda > 375$ nm. (Reprinted with permission from *J. Phys. Chem.* **1994**, *98*, 10419).

In order to assign the two quasi line fluorescence spectra obtained in Shpolskii matrices to specific conformers, the energies and the ZFS parameters D of the triplet states of 2-NPC in several planar and nonplanar conformations have been calculated using the *ab initio* and CS-INDO-CIPSI methods.[33] The variable geometrical parameters are the apex angle α and both the dihedral angles Φ_1 (out of plane torsion centered on the carbenic atom for the phenyl group) and Φ_2 (rotation angle of the phenyl around C_1C_3 bond).

$$\alpha = (2\ 1\ 3)$$

$$\Phi_1 = (3\ 1\ 2\ 5)$$

$$\Phi_2 = (4\ 3\ 1\ 2)$$

Table 2. ROHF/3-21G Optimized Structures of 2-NPC in the Ground State and Corresponding Energies

	$\Phi_1{}^a$	$\Phi_2{}^a$	α^a	Energy (au)
E/*trans*	0	0	149.05	−647.025221
	30.77	30.63	135.24	−647.033220
	45	20.12	134.68	−647.032976
	90	1.19	133.57	−647.031422
	135	−19.33	135.18	−647.033090
	153.17	−33.45	135.56	−647.033507
Z/*cis*	180	0	148.91	−647.025513

Note: [a]All values in degrees, see text for definitions.

The results collected in Table 2 indicate that the conformers corresponding to the minimum energy in the ground state are not the planar species, but the nonplanar conformers such that $\Phi_1 = 30.77°$, called the pseudo E/*trans*, and $\Phi_1 = 153.17°$, called the pseudo Z/*cis*. Among the two conformers, the pseudo Z/*cis* has the lowest ground state energy. It is important to note that these geometries are very close to the one determined for DPC by the ENDOR technique.[25]

The T_0–T_1 transition energies together with the $D(T_0)$ values in the ground T_0 state, calculated for the pseudo E/*trans* and the pseudo Z/*cis* conformers are listed in Table 3. This table also contains the energies corresponding to the appropriate fluorescence origins of the two conformers (which are very close in the *n*-hexane or *n*-heptane matrices[31]) and the $D(T_0)$ values measured in the *n*-heptane matrix.[31] Correlation between the calculated and observed data establishes that the conformer with fluorescence origin at 588.8 nm in *n*-hexane (or 589.5 nm in *n*-heptane) is the pseudo E/*trans*, which has the lowest calculated and observed $D(T_0)$ value. Similarly, the conformer with the fluorescence origin at 600.1 nm in *n*-hexane (or 600.5 nm in *n*-heptane) is the pseudo Z/*cis*, which has the largest calculated and observed

Table 3. Calculated and Observed $T_0 - T_1$ Transition Energies and $D(T_0)$ Values of 2-NPC

	$T_0 - T_1$ Transition Energies (values in eV)		$D(T_0)$ (cm^{-1})	
	Calculated	Observed	Calculated	Observed
Pseudo E/*trans* $\Phi_1 = 30.77°$	2.05	2.10	0.3405	0.41 ± 0.01
Pseudo Z/*cis* $\Phi_1 = 153.17°$	2.01	2.06	0.3461	0.42 ± 0.02

$D(T_0)$ value. This attribution is consistent with the fact reported in Ref. 31 that the conformer with the largest $D(T_0)$ value fluoresces at the highest wavelength and is also consistent with the recent reinterpretation[34] of the EPR data previously obtained[30] for 2-NPC in the glassy 2-MTHF at 77 K.

III. FLUORESCENCE DECAYS AND ZERO-FIELD SPLITTINGS OF THE EXCITED TRIPLET STATES OF AROMATIC CARBENES

The fluorescence decays of DBC,[20] of DPC-h_{10},[24] and -d_{10}[24] and of 2-NPC[31] in *n*-hexane at 4.2 to 30 K excited either with 337 nm radiation from a N_2 laser (DBC and 2-NPC) or with radiation selected from a dye laser (DPC) have been measured on the fluorescence origin bands of the carbenes in the absence and in the presence of a magnetic field. They are given in Figures 9–13 on a semilogarithmic scale.

A. Decays in the Absence of a Magnetic Field

In the absence of a magnetic field, all the decays are nonexponential. They have been analyzed as a sum of three exponential decays with the same preexponential factor A,[35] since the three sublevels of the emitting triplet are equally populated by the excitation pulse.

$$I_{\text{calc}} = A \sum_{i=1}^{3} \exp(-t/\tau_i)$$

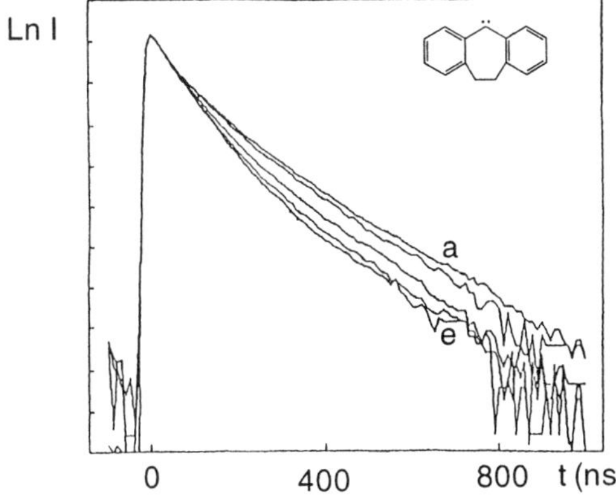

Figure 9. Fluorescence decays of DBC in *n*-hexane at 20 K in the absence (a) and in the presence of 45, 65, 105, and 145 G (b–e) magnetic fields. The decays are excited with the 337 nm nitrogen laser line and measured on the intense fluorescence origin band at 502.9 nm. (Reprinted with permission from *J. Phys. Chem.* **1992**, *96*, 2486).

Table 4. Component Lifetimes (ns) of the Fluorescence Decays of
Carbenes in the Absence of a Magnetic Field

$\tau_i(ns)$	DBC	DPC-h_{10} λ_{obs}		DPC-d_{10} λ_{obs}		2-NPC λ_{obs}	
		482 nm	502.8 nm	480 nm	500.8 nm	588.8 nm	600.1 nm
τ_1	35	26	4.2	36	3.9	2	3
τ_2	67	92	8.0	117	4.6	9	3
τ_3	140	163	26	186	29	18	19

This treatment leads to the component lifetimes τ_i of the fluorescence decays listed in Table 4. The results presented here for DPC-h_{10} at 20 to 30 K are somewhat different from those obtained for this carbene in glasses at 77 K by Ware and coworkers[18] who found a single exponential decay with a lifetime of 70 to 100 ns.

It is remarkable that the component lifetimes of DPC-h_{10} and -d_{10} measured on the broad fluorescence (attributed to the nonplanar C_1 species) are 6 to 12 times longer than those measured on the sharp fluorescence (attributed to the quasi-planar C_{2v} species). Besides the differences in nonradiative decay rates, the variations in τ_i have been explained by the oscillator strengths calculated for T_0–T_1 transitions,[24] which increase from 0.022 (C_1 species) to 0.122 (C_{2v} species).

The component lifetimes of DPC-d_{10} are slightly longer than those of DPC-h_{10} for the broad fluorescence as well as for the slow component of the sharp fluorescence. Such a small deuterium effect has already been observed for m-xylylene biradicals and interpreted in terms of the energy gap law on the rate of intersystem crossing (ISC) between T_1 and S_3, the closest singlet state below T_1.[35] The same interpretation holds for DPC since this carbene, like m-xylylene biradicals, has several singlet states S_1, S_2 ... S_n between the T_1 and T_0 triplet states.[24]

B. Decays in the Presence of a Magnetic Field

Upon application of a magnetic field, the fluorescence decay curves are generally modified, as shown in Figures 9–12. For DPC, however, the modification is detected (see Figures 10 and 11) at fields which are higher for the decays measured on the broad fluorescence ($H = 380$ to 970 G) than for the decays measured on the sharp fluorescence ($H = 28$ to 215 G). On the other hand, the magnetic field effect observed for 2-NPC is different in n-hexane and in n-heptane matrices. The decay of 2-NPC in n-hexane measured on the fluorescence origin at 588.8 nm remains nearly unchanged for fields up to 900 G, while the decay of the other 2-NPC conformer measured at 600.1 nm is significantly altered for fields of $H = 20$ to 76 G (see Figures 12 and 13). In n-heptane, however, the fluorescence decay is field-de-

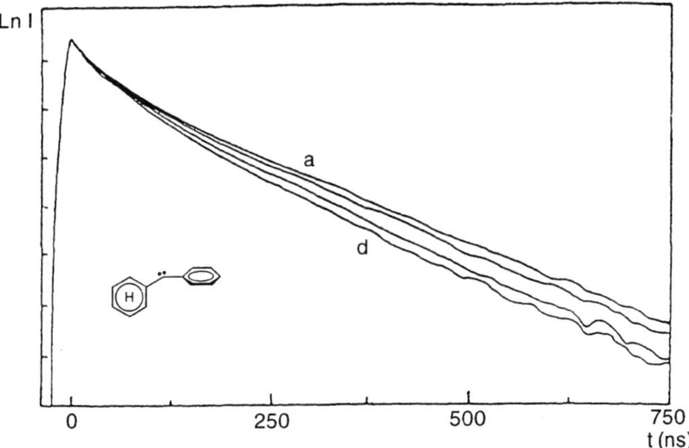

Figure 10. Fluorescence decays of DPC-h_{10} in *n*-hexane at 20 to 30 K in the absence (a) and in the presence of 380, 708, and 970 G (b–d) magnetic fields. The decays are excited with the 471 nm radiation emitted from the dye laser and measured on the broad fluorescence at 482 nm. (Reprinted with permission from *J. Phys. Chem.* **1993**, *97*, 13358).

pendent for the 589.5 nm line and field independent for the 600.5 nm line. This observation is surprising and has not yet been explained.

The nonexponential fluorescence decays of carbenes are attributed to the emission from the T_1 spin sublevels, at rates that are faster than the rate of spin-lattice relaxation between the different sublevels. The modification of the decay curves

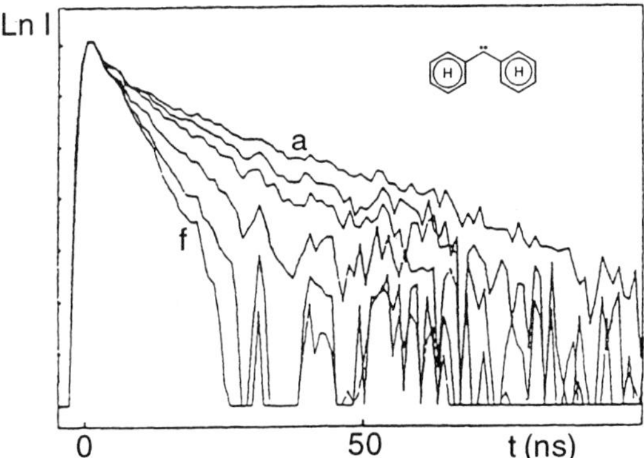

Figure 11. Fluorescence decays of DPC-h_{10} in *n*-hexane at 20 to 30 K in the absence (a) and in the presence of 28, 35, 65, 116, and 215 G (b–f) magnetic fields. The decays are excited with the 491 nm radiation emitted from the dye laser and measured on the sharp fluorescence at 502.8 nm. (Reprinted with permission from *J. Phys. Chem.* **1993**, *97*, 13358).

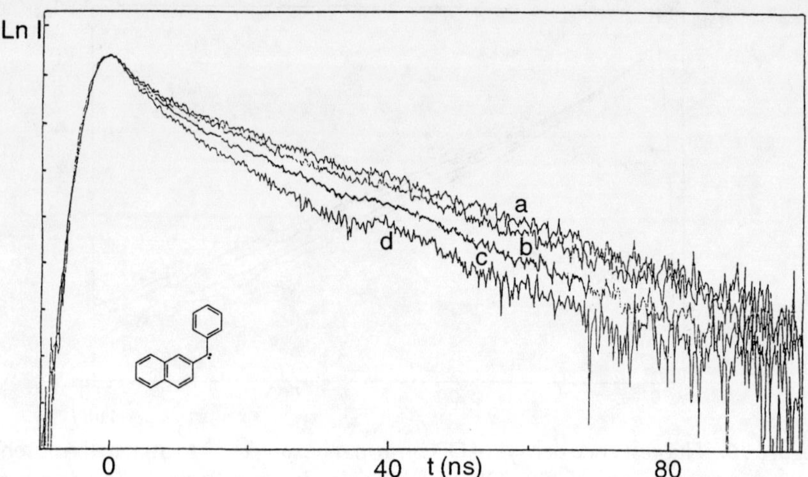

Figure 12. Fluorescence decays of the pseudo Z/*cis* 2-NPC in *n*-hexane at 4.2 K measured on the sharp origin band at 600.1 nm in the absence (a) and in the presence (b–d) of 20, 41, and 76 G magnetic fields. The decays are excited with the 337 nm radiation from the N_2 laser. (Reprinted with permission from *J. Phys. Chem.* **1994**, *98*, 10419).

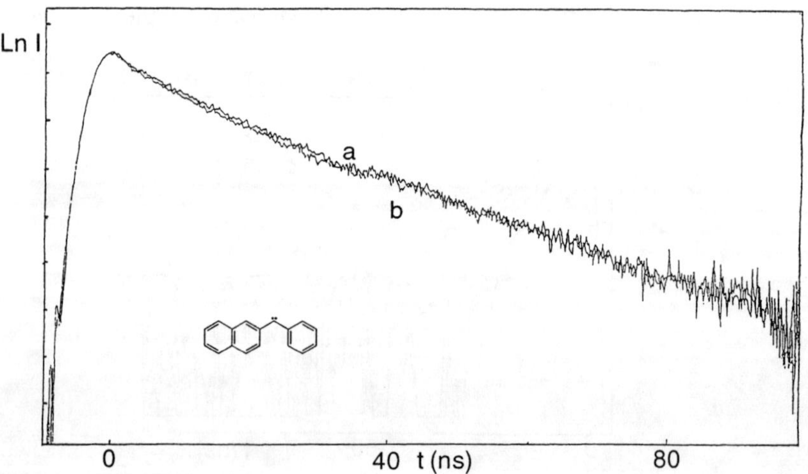

Figure 13. Fluorescence decays of the pseudo E/*trans* 2-NPC in *n*-hexane at 4.2 K measured on the sharp origin band at 588.8 nm in the absence (a) and in the presence (b) of a 900 G magnetic field. The decays are excited with the 337 nm radiation from the N_2 laser. (Reprinted with permission from *J. Phys. Chem.* **1994**, *98*, 10419).

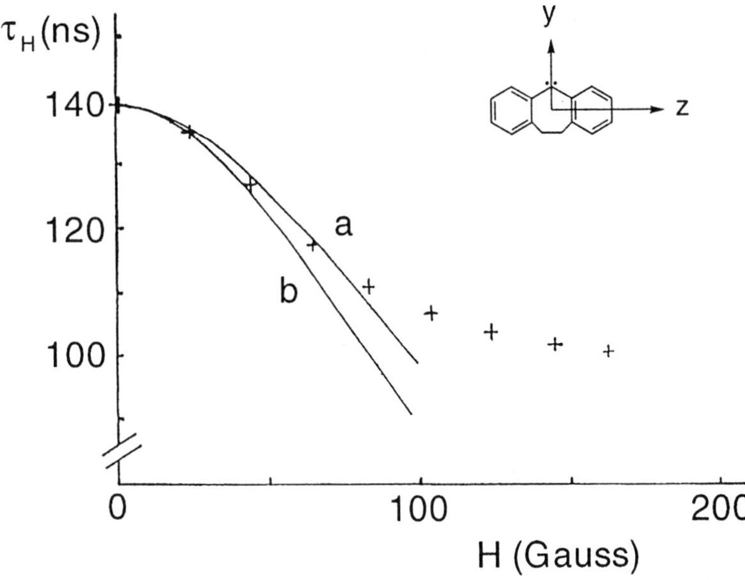

Figure 14. Lifetimes of the slow component in the fluorescence decay of DBC in *n*-hexane at 20 K as a function of a weak magnetic field. The crosses correspond to the observed values. The curves have been calculated for two values of the ZFS parameter D: (a) $D = 0.020$ cm^{-1}, (b) $D = 0.015$ cm^{-1} (Reprinted with permission from *J. Phys. Chem.* **1992,** *96,* 2486).

upon application of the field is due to the mixing of the wave functions of the T_1 sublevels. For randomly oriented carbenes in polycrystalline Shpolskii matrices, the decay law is expected to be multiexponential. In the presence of a weak magnetic field, however, the decay curves are still fitted with a convoluted sum of three exponential functions. The results of such an analysis for DBC, displayed in Figure 14, indicate that the lifetime τ_H of the slow component decreases significantly as the field H increases. This magnetic field effect on τ_H will be used to estimate the ZFS parameters $D(T_1)$ of carbenes in the first excited triplet state T_1.

C. Interpretation

The fluorescence of carbenes corresponds to the spin-allowed transition between T_1 and T_0 states. Since the transition moments are spin-independent, the radiative decay rate constants k^r from the three sublevels are equal. This is not true for the corresponding nonradiative decay rate constants, because the three sublevels, which may belong to different irreducible representations in the molecular group of symmetry, are coupled by spin-orbit coupling (SOC) to different singlet states. For example, in planar *m*-xylylene biradicals of C_{2v} symmetry with y and z axes in the molecular plane, only the Y and Z sublevels can interact with the higher-lying $(\sigma\pi)^*$ singlet states. This SOC induces ISC, which accelerates the depopulation of the Y and Z sublevels only. The X sublevel, which cannot interact with $(\sigma\pi)^*$ singlets, will give rise to the slow component with a total decay rate

$$k^r + k_x^{nr} < k^r + k_{y\,\text{or}\,z}^{nr}$$

As for the m-xylylene biradicals, the nonexponential fluorescence decays of carbenes will be interpreted by different ISC rates from the three triplet sublevels, due to different SOC with the singlet states. This interpretation is confirmed by the significant modification of the fluorescence decays observed in the presence of a magnetic field.

D. Determination of the ZFS Parameter $D(T_1)$ in the T_1 State

The effect of the Zeeman mixing on the radiative and on the nonradiative decay rates of the T_1–T_0 transition has been discussed in detail for m-xylylene biradicals.[36] This has led to the following expression for the effective total decay rate $\langle k_{xH}\rangle$ from the X sublevel (slow component) in the presence of a weak field H ($g\beta H \ll D$)

$$\langle k_{xH}\rangle = k_x + \frac{2}{3}\frac{g^2\beta^2 H^2}{D^2}\left(\frac{k_y + k_z}{2} - k_x\right) \tag{1}$$

where k_x, k_y, and k_z are the total decay rates from X, Y, and Z sublevels in the absence of the field.

For planar (or quasi-planar) carbenes having the X magnetic axis normal to the molecular plane (or some average plane), the slow component (rate k_x) is assumed to be associated with the X sublevel separated by D from the two other spin levels, as in the case of m-xylylene. This assumption is reasonable for DBC, for 2-NPC, and for the quasi-planar DPC conformer emitting the sharp fluorescence, but is not valid for the nonplanar carbenes such as the DPC conformer emitting the broad fluorescence. The ZFS parameter $D(T_1)$ in the T_1 state of planar or quasi-planar carbenes is then estimated from Eq. (1) where every quantity except D is known from the decay measurements.

The dependence of $\tau_H^{cal} = \langle k_{xH}\rangle^{-1}$ as a function of a weak magnetic field has been calculated for different D values by using Eq. (1) and the component rate constants determined from the fluorescence decays. In the case of DBC, the curves obtained for τ_H^{cal} with two tentative D values are displayed in Figure 14. The best fit between the calculated curve and experimental points corresponds to $D(T_1) = 0.020 \pm 0.005$ cm^{-1}. This value and the $D(T_1)$ values obtained in the same way for the other carbenes[24,31] are collected in Table 5 together with the corresponding $D(T_0)$ values determined for the ground T_0 state by EPR spectroscopy. This table also contains the calculated $D(T_0)$ and $D(T_1)$ values of DPC conformers[24] and of 2-NPC.[33] For the quasi-planar DPC, the agreement is reasonable since the calculated value

Table 5. The Observed and Calculated ZFS Parameters $D(T_1)$ and $D(T_0)$ of Carbenes

Carbene	$D(T_1)$ cm^{-1}		$D(T_0)$ cm^{-1}	
	Experimental	*Calculated*	*Experimental*	*Calculated*
DBC	0.020 ± 0.005[20]		0.3932[37]	
Quasi-planar DPC C_{2v} symmetry	0.007[24]	0.0155[24]		0.2073[24]
Nonplanar DPC C_1 symmetry	0.20[24]	0.0100[24]	0.405[9,11]	
Pseudo E/*trans* 2-NPC	0.038[43]	0.0619[33]	0.3898[30] 0.41[31] 0.50[43]	0.3405[33]
Pseudo Z/*cis* 2-NPC	0.011[31]	0.0633[33]	0.4044[30] 0.42[31]	0.3461[33]

$D(T_1) = 0.0155$ cm^{-1} is about twice the experimental value $D(T_1) = 0.007$ cm^{-1}. The situation is less favorable for the pseudo Z/*cis* 2-NPC where the calculated value $D(T_1) = 0.0633$ cm^{-1} is about six times the experimental one $[D(T_1) = 0.011$ cm$^{-1}]$. In contrast, there is a large discrepancy between the computed $D(T_1) = 0.0115$ cm^{-1} and the experimental $D(T_1) = 0.20$ cm^{-1} values for the nonplanar DPC, thus confirming that the method used in the present work to estimate $D(T_1)$ is not appropriate for nonplanar species. Furthermore, this method did not allow the estimation of $D(T_1)$ for the pseudo E/*trans* 2-NPC since the fluorescence decay of this carbene was not influenced by the magnetic field (see Figure 13). These results reveal the limitations of this experimental method.

Examination of Table 5 shows that for all the carbenes studied here, the $D(T_1)$ values are significantly smaller than the $D(T_0)$ values. Since D is proportional to $\langle 1/r^3 \rangle$, r being the average separation between the two unpaired electrons, the observed decrease of D indicates that the average separation r is significantly larger in the T_1 state than in the T_0 state. This is explained by substantial delocalization of the π electron onto the aromatic rings in the first excited state.

E. Discussion

The aromatic carbenes in the ground T_0 state are characterized by a large D value due to the presence of the two unpaired electrons, one in a σ orbital and the other in a π orbital localized on the carbene carbon atom. The electronic structure of the excited T_1 state involves the promotion of the unpaired electron from the π orbital to a π^* orbital. The calculations performed on DPC[24] and on 2-NPC[33] indicate that this π^* orbital has only a minor contribution from the carbenic $2p\pi$ atomic orbital. Thus $\pi \rightarrow \pi^*$ excitation results in a delocalization of the unpaired electron onto the rings and to a decrease of the corresponding spin density on the carbenic center, while the spin density originating from the σ unpaired electron is retained. As a consequence, the average separation between the two unpaired electrons is increased in the excited T_1 state relative to the T_0 ground state, leading to a $D(T_1)$ value much smaller than $D(T_0)$, in agreement with the experimental results listed in Table 5.

Since the distribution of unpaired electrons is very different in the T_0 and T_1 states, the photochemistry of aromatic compounds is expected to be very different in these two states, as suggested by the following results observed with DPC.

F. Specific Chemistry of DPC in the Presence of an Intense Excimer Laser

Aromatic carbenes give rise to a number of chemical reactions involving their lowest singlet or ground triplet states. For example, when two DPC molecules react in their ground triplet state T_0, the resulting product is tetraphenylethylene (TPE).

Ph   Ph   Ph   Ph

Ph   Ph   Ph   Ph

TPE

However, Turro $et\ al.$[38] have found that appreciable amounts of diphenylanthracene (DPA), diphenylphenanthrene (DPP), and fluorene (FL) are formed, in addition to TPE, when the diazo precursor of DPC is decomposed in liquid solution at room temperature with intense excimer laser radiation. The "new" reaction products are not observed when the photolysis is performed with a standard UV lamp.

In liquid solution at room temperature, DPC may have a quasi-planar geometry since the laser-induced fluorescence measured in these conditions has a maximum at 505 to 507 nm,[38] a value that is very close to the maximum at 502.8 nm of the sharp fluorescence origin observed in low-temperature matrices. In the latter case, the small value $D(T_1) = 0.007$ cm^{-1} has been interpreted in terms of the π-electron delocalization onto the DPC rings upon electronic excitation.

It is therefore possible that the formation of the new products is due to *chemical reactions of electronically excited DPC* with this carbene in the ground state T_0 according to Scheme I. Such reactions would be favored in the presence of an intense excimer laser, which creates a high concentration of DPC in the first excited triplet T_1.

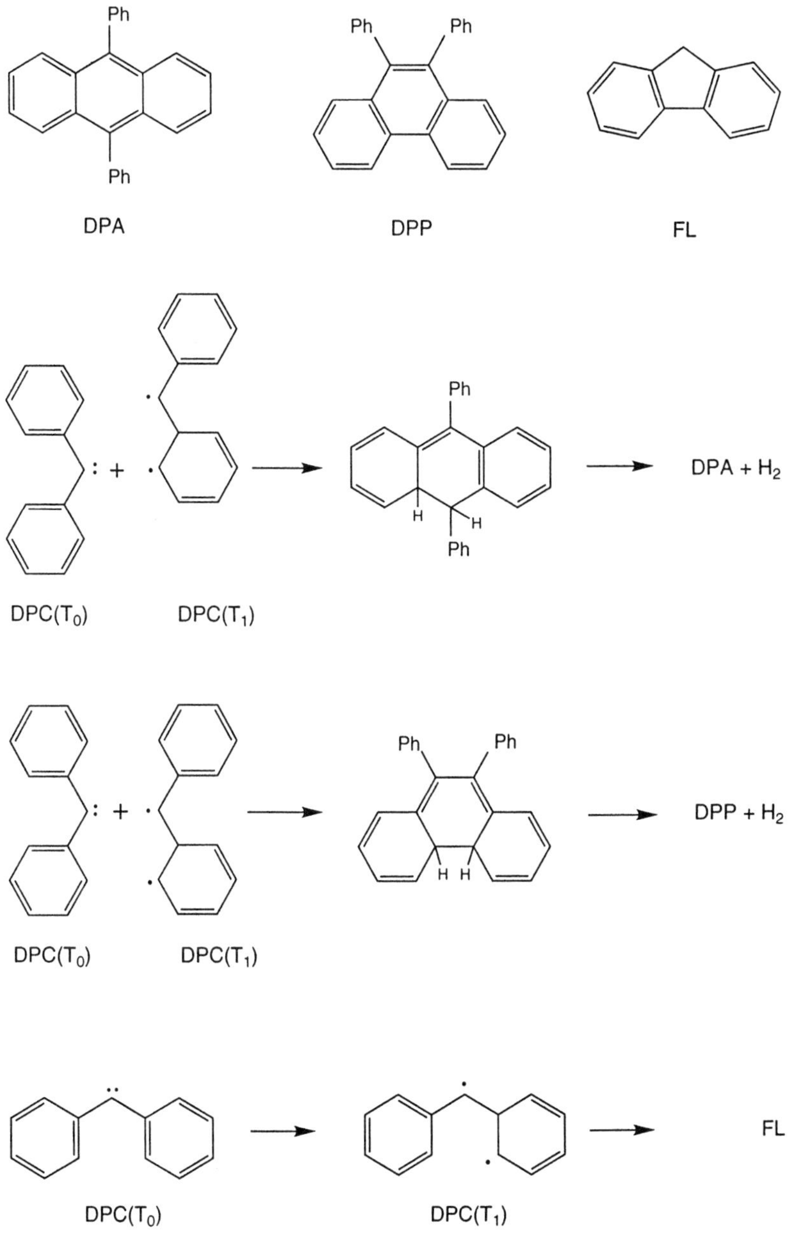

DPA

DPP

FL

DPC(T_0) DPC(T_1)

DPA + H_2

DPC(T_0) DPC(T_1)

DPP + H_2

DPC(T_0) DPC(T_1)

FL

Scheme 1.

G. Conclusion

Magnetic resonance methods used to determine the ZFS parameters D and E require substantial concentrations of molecules in the triplet state. The conventional EPR technique has been successfully used to measure the D and E values of either matrix-isolated carbenes in the ground triplet state[8,11] or matrix-isolated aromatic hydrocarbons having triplet lifetimes of several seconds[39] because the steady-state concentration of triplet species is sufficiently high in these systems. In contrast, for aromatic carbonyl compounds[40] or N-heterocyclics[41] having triplet lifetimes of the order of milliseconds, the D and E parameters have to be determined by the optically detected magnetic resonance (ODMR) method. These two techniques cannot be used, however, for excited species having triplet lifetimes of the order of 10 to 100 ns, since their steady-state concentration is too low. For these reasons, the $D(T_1)$ parameter of aromatic biradicals and carbenes in the first excited triplet state T_1 has been estimated from the magnetic field effect on the fluorescence decays.[36,20] We have seen that this optical method has led to $D(T_1)$ values in reasonable agreement with the theoretical predictions. However, this method has its limitations. In particular, it can only be applied to triplet sublevels that decay with substantially different rates so that the level mixing induced by the magnetic field significantly changes the fluorescence decay. Furthermore this method does not give any information on the ordering of triplet sublevels and, therefore, on the sign of the D parameter.

In 1977, Al'shitz et al.[42] reported the first measurement of the ZFS parameter D of coronene in the first excited triplet state T_1 by using the "phosphorescence line narrowing" method. In this study, the phosphorescence was selectively excited within the origin band of the S_0 (singlet ground state)–T_1 transition of coronene by means of the 514.38 nm narrow line emitted from an argon laser and was detected with a Fabry–Perot interferometer. This important experiment revealed the possibility of directly investigating the triplet zero-field structure from the optical spectra of molecules.

Hole-burning is another high-resolution technique that can provide the ZFS parameters D and E with great precision. In the next section, we will describe a hole-burning study[43] that leads to the determination of the D and E values of carbenes in the first excited triplet state T_1. To our knowledge this is the first hole-burning experiment carried out on a triplet–triplet transition.

IV. SELECTIVE LASER EXCITATION: HOLE-BURNING EXPERIMENTS

The absorption and emission spectra of matrix-isolated molecules or radicals at low temperatures are generally composed of inhomogeneously broadened bands having widths of 0.1 to 10 cm^{-1} in crystalline hosts and 100 to 500 cm^{-1} in glasses. This is due to the distribution of local environments around the absorbing species.

The width of these bands can, however, be considerably reduced by selective laser excitation using either line-narrowing of the luminescence spectrum[44] or hole-burning in the absorption spectrum.[44–47] Both techniques have been used to separate a homogeneous component from the inhomogeneously broadened spectrum and have been extensively described in books[45] as well as in review articles.[44,46,47]

At low temperatures, each spectral band consists of a narrow zero-phonon line (ZPL) and a relatively broad phonon wing (PW), originating from the phototransitions with a simultaneous creation or annihilation of the matrix phonons. Only the systems characterized by a relatively weak electron–phonon coupling and thus a small contribution of PW are suitable for the selective laser experiments. In the other case, an excitation of the so-called "non-resonant" centers can considerably spoil the obtained line pattern. Here we shall limit our description to the systems where the electronic transitions are predominantly concentrated in ZPL.

The homogeneous ZPL has usually a lorentzian shape and a line-width given by $\Gamma = (2\pi T_1)^{-1} + (\pi T_2)^{-1}$, where T_1 is the lifetime of the excited state and T_2 is the pure dephasing time, related to a remaining molecular motion. The second contribution to the line-width vanishes to zero at sufficiently low, cryogenic temperatures and, therefore, the hole-burning experiments are usually performed at temperatures below 2K. A typical range for the line-widths of the ZPL is 1 to 500 MHz. In a real medium like low-temperature glasses or matrices, the incorporated dopant molecules may occupy sites with different local environments. The distribution of local environments induces a manifold of electronic transition energies and leads to a broad, inhomogeneous absorption band of the dopant. Such an inhomogeneous band typically has a gaussian shape with a line-width primary dependent on the host–guest interactions.

A. Selective Laser Excitation of Molecules in Their $S_0 \rightarrow S_1$ Transitions

A narrow-band laser light that has a frequency within the inhomogeneous $S_0 \rightarrow S_1$ 0,0 absorption band of the molecule-matrix system can only be absorbed by the molecules that have their 0,0 homogeneous bands in resonance with the laser frequency. The selectively excited molecules can then emit a sharp luminescence spectrum detected by a high-resolution monochromator and/or a Fabry–Perot interferometer. This is the principle of the fluorescence or phosphorescence line-narrowing method, pioneered by Personov and co-workers.[44] Alternatively, a narrow-band laser light can create a hole within the inhomogeneous absorption band of a photo unstable system. The first experiments of this kind were done by Personov and Rebane groups.[48,50,51] Two different types of persistent hole-burning can be considered. A photochemical hole-burning removes the molecules that absorb the laser light from the inhomogeneous absorption band by means of photochemical reactions (proton transfer, electron transfer, dissociation, etc.). A non-photochemical hole-burning involves a structural rearrangement of the environment in the close neighborhood of the absorbing molecule, and shifts its homogeneous absorption line to some other position within the inhomogeneous band. The line-width of burnt holes is twice the width of the ZPL because in a typical hole-burning experiment the ZPL is used twice, the first time to burn the hole and the second time, at much lower intensity, when scanning the laser over the frequency range in order to monitor the hole shape.

In the case of a $S_0 \rightarrow S_1$ transition, a single hole is burnt at the same frequency as the laser light or a single fluorescence spectrum is emitted from the selectively excited molecules. The situation is different for $S_0 \rightarrow T_1$ or $T_0 \rightarrow T_1$ transitions, for which several holes having different frequencies can simultaneously be burnt with a single laser frequency, or in a luminescence line-narrowing experiment, in which a multiplet is emitted from the selectively excited molecules.

B. Selective Laser Excitation of Molecules in Their $S_0 \rightarrow T_1$ Transitions

A good example of selective laser excitation involving a S_0-T_1 transition is given by the phosphorescence line narrowing experiments of Al'shitz et al. on coronene.[42] The aim of this study was to determine the ZFS parameters in the first excited triplet state T_1. Theory predicts[49] that $E = 0$ for highly symmetrical molecules such as coronene, which has a sixfold symmetry axis. Therefore, the triplet level T_1 of this molecule is only split into two components separated by D. Under selective laser excitation within the inhomogeneous width, two types of centers can be excited with the laser frequency, each type having one sublevel in resonance with the exciting radiation. Each type of center will then emit two phosphorescence lines, but the lines corresponding to the laser frequency should coincide in the two types of centers, as indicated in Figure 15. Consequently, the phosphorescence origin band of coronene will be split into three components with the central band being two times more intense than the two others. The separations of these components give a direct measurement of the $D = 0.01$ cm^{-1} value of coronene. It is interesting to note that the phosphorescence origin is split into three components, while the triplet state T_1 of coronene has only two sublevels.

In the presence of a strong magnetic field, the degeneracy of the spin sublevel due to $E = 0$ is lifted, and the triplet state T_1 of coronene is now split into three Zeeman sublevels.[52] The field should be strong enough to ensure that the Zeeman splitting is considerably greater than the ZFS of the triplet state T_1. Under selective $S_0 \rightarrow T_1$ excitation within the inhomogeneous width, the laser will select three types of centers, each type being excited to one of its Zeeman sublevels. Each type of center will then emit three phosphorescence lines, but the lines corresponding to the laser frequency will coincide in the three types of centers. In this case, the overall phosphorescence origin will be split into five lines of different intensities, as indicated in Figure 16.

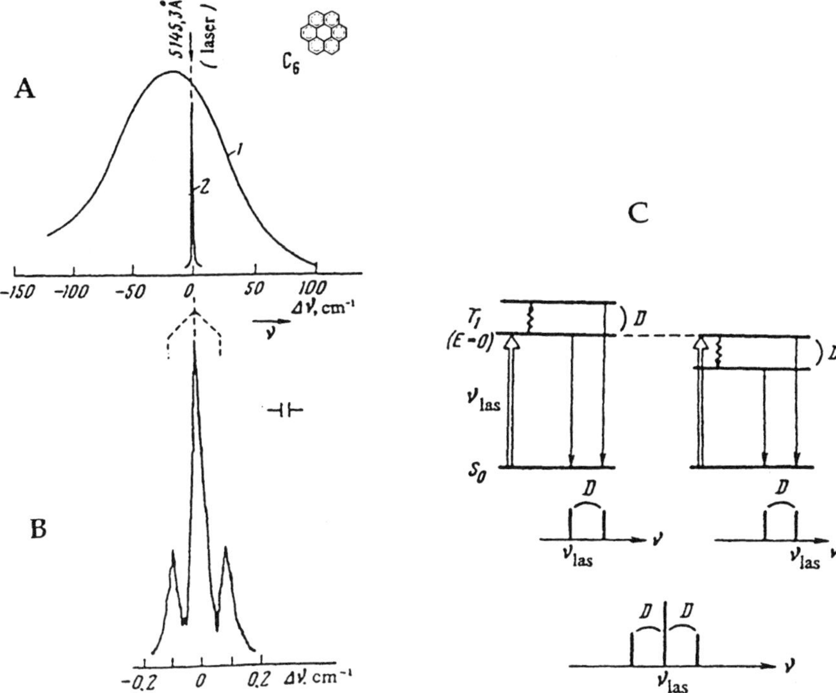

Figure 15. The 0-0 band in the phosphorescence spectrum of coronene in butyl bromide at 4.2 K (A) under ordinary $S_1 \leftarrow S_0$ excitation with $\lambda_{Hg} = 313$ and 365 nm (curve 1) and monochromatic excitation with an argon laser (curve 2). (B) Interferometric recording of the 0-0 line following excitation with an argon laser (with line width ≈ 0.01 cm^{-1}). (C) Energy level diagram of the two types of centers excited by the same laser line ($D \neq 0$, $E = 0$) and the triplet produced in the spectrum. (Reprinted with permission from *JETP Lett.* **1977**, *26*, 586).

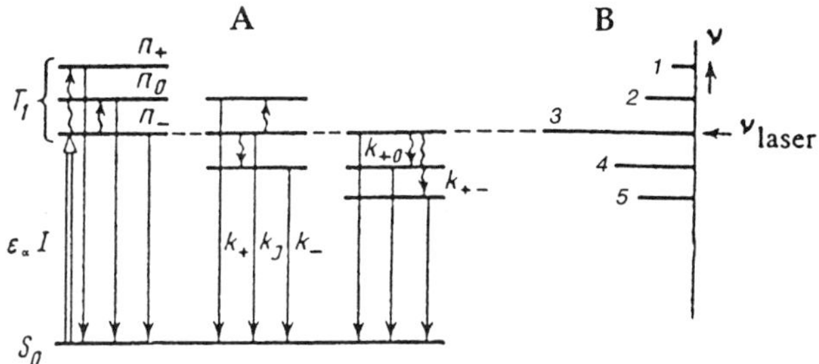

Figure 16. Energy level diagram illustrating the origin of the Zeeman quintet in the phosphorescence spectrum of an isotropic solution under selective $T_1 \leftarrow S_0$ excitation. (A) Zeeman sublevels of the three types of centers excited by the laser radiation. (B) Quintet appearing in the spectrum. (Reprinted with permission from *Sov. Phys. JETP* **1984**, *60*, 428).

C.　Selective Laser Excitation of Carbenes in Their $T_0 \rightarrow T_1$ Transition: The Hole-Burning Studies with 2-NPC

The work that is described in this section represents not only the first hole-burning experiment on a carbene but also, to the best of our knowledge, the first study on a system that has a triplet ground state. There are two main requirements for a successful hole-burning experiment. The electronic transition of the molecule has to be concentrated in a narrow zero-phonon line. The spectral range of this transition has to be covered by an appropriate dye laser. Both requirements are fulfilled with 2-NPC in Shpolskii matrices.

In the following section we will concentrate on the main, photostable 0,0 fluorescence excitation line of 2-NPC, which was attributed to its pseudo E/*trans* conformer. As shown in Figure 17, this line has a maximum at 588.7 nm and a fwhm (full width at half maximum) of 7.5 cm^{-1}. (Recall that the corresponding inhomogeneous fluorescence line has a maximum at 588.8 nm). Using a single-mode dye laser with a 1 to 3 MHz frequency resolution, a complicated pattern of holes is burnt at 1.8 K within the fluorescence excitation origin band, as shown in Figure 18. The holes cover a spectral range of nearly 1 cm^{-1} and they remain for a long time when the sample is maintained at 1.8 K. The holes disappear when the sample is annealed overnight at 130 K and is then cooled down. The typical pattern of burnt holes can be described as follows. The central hole, burnt at the frequency of applied laser light (0 GHz in Figure 18), is symmetrically surrounded by two satellite holes (0.3

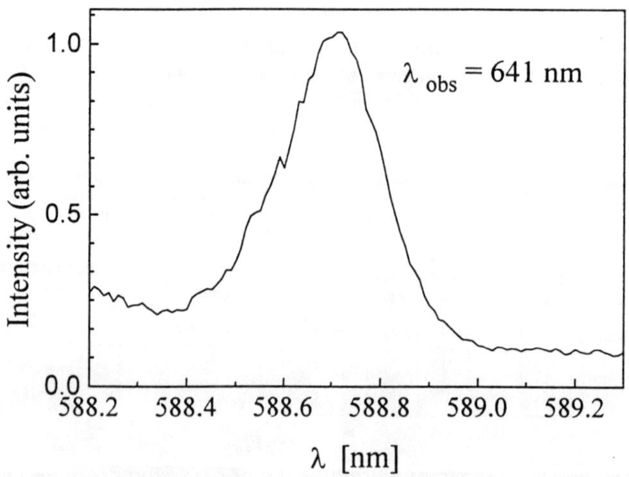

Figure 17.　The inhomogeneous fluorescence excitation origin band at 588.7 nm of the pseudo E/*trans* 2-NPC in *n*-hexane at 1.8 K, obtained with the observation wavelength λ_{obs} = 641 nm. (Reprinted with permission from *Chem. Phys. Lett.* **1995**, *245*, 549).

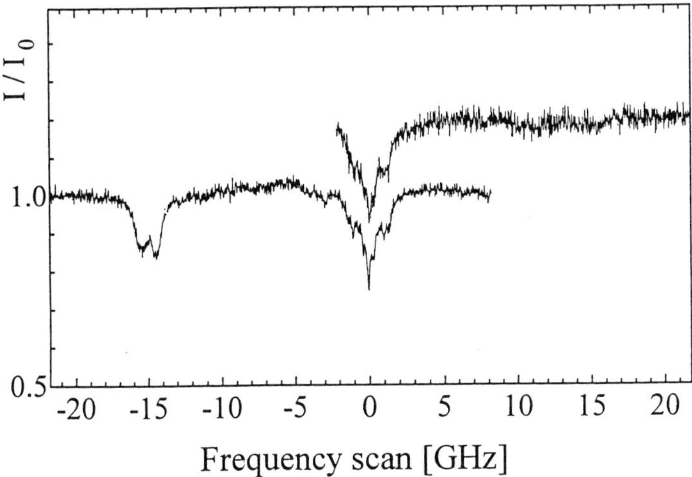

Figure 18. Typical holes burnt within the inhomogeneous excitation origin band of the pseudo E/*trans* 2-NPC at 588.7 nm. The spectral position of holes are given with respect to the frequency of the burning light. The spectrum scanning range (30 GHz) requires separate detection of the high and the low energy sides of the hole spectrum. (Reprinted with permission from *Chem. Phys. Lett.* **1995**, *245*, 549).

GHz away) and two satellite pairs of holes (1.0 and 1.3 GHz away). Another satellite pair of holes, deeper and broader than those mentioned before, is located on the low energy (red) side of the central hole (14.5 and 15.5 GHz away). There are no satellite holes on the high energy side of the central hole.

Interpretation: The appearance of satellite holes in Figure 18 can be explained by taking into account the three sublevels of the T_0 and T_1 states. Let us first consider the simple case of a D_{2h} molecule with ZFS spin axes having the same direction in both the T_0 and T_1 states. In this case, the electronic transitions are only allowed between sublevels involving the same spin functions in the T_0 and T_1 states, as indicated in the energy level diagram shown in Figure 19. The burning laser light will only excite the species that have their $T_0 \rightarrow T_1$ transition in resonance with the laser frequency. This requirement is satisfied by three groups (1, 2, 3) of carbenes having the same energy of their $T_{0z} \rightarrow T_{1z}$, $T_{0y} \rightarrow T_{1y}$, and $T_{0x} \rightarrow T_{1x}$ transitions. The electronic transitions, together with the positions expected for the holes, are displayed in Figure 19. In this predicted pattern, the holes are symmetrically distributed with respect to the central hole, which is not the picture observed in Figure 18. In the latter case, the satellite holes around 15 GHz are detected only on the low energy side of the central hole, not on the high energy side. This indicates that the only group of carbenes (group 1) that is effectively burnt corresponds to the species that absorb the laser light from the lowest energy spin sublevel T_{0z}. The two other groups of carbenes (groups 2 and 3), for which some holes are expected

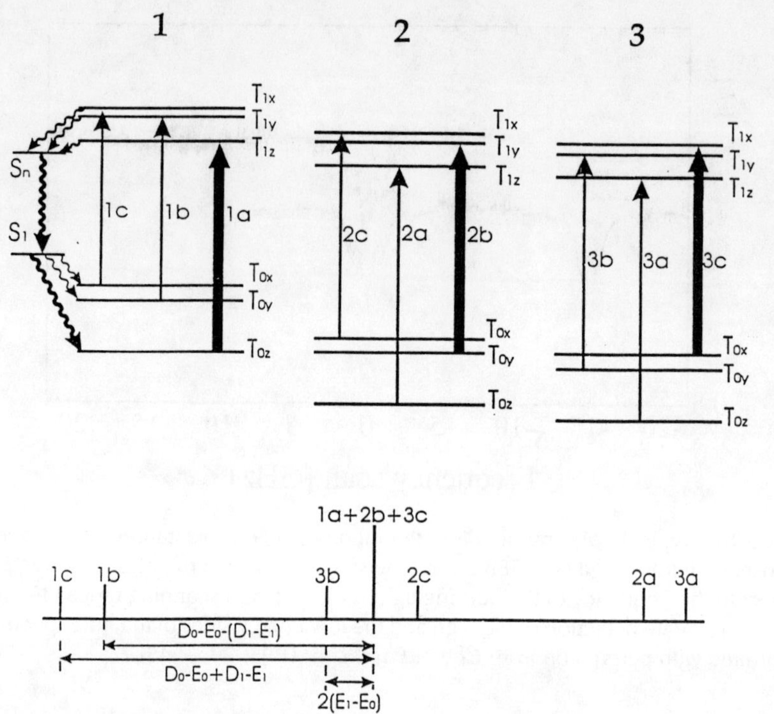

Figure 19. Energy level diagram and electronic transitions of a D_{2h} molecule. The burning transitions, resonant with the laser light, are indicated with thick arrows. The energy separations between sublevels T_z and T_y and between T_y and T_x are, by definition, respectively, $D_0 - E_0$ and $2E_0$ in the T_0 state and $D_1 - E_1$ and $2E_1$ in the T_1 state. The expected spectral pattern of burnt holes and their energy separations are shown at the bottom.

to appear on the high energy side of the central hole, are absent in Figure 18. These observations are interpreted in the following way.

Since the pseudo $E/trans$ conformer of 2-NPC is photochemically stable at 1.8 K, the hole burning should proceed via nonphotochemical processes such as the nonradiative relaxation from the T_1 state. This relaxation energy is dissipated into the surrounding lattice and the carbene/matrix system may jump to another energy minimum, shifting the spectral position of the zero-phonon line and leaving a spectral hole. There are two possible channels for nonradiative relaxation, the internal conversion $T_1 \rightsquigarrow T_0$ and the intersystem crossing (ISC) $T_1 \rightsquigarrow S_4 \rightsquigarrow S_1 \rightsquigarrow T_0$. The ISC relaxation channel is predominant, as indicated[20] by the different nonradiative decay rates from the T_1 sublevels leading to the nonexponential fluorescence decays of carbenes. The most important step of this channel is the final one, the $S_1 \rightsquigarrow T_0$ step, which may selectively populate the spin component T_{0z} only.

Let us mention that the selective population of the spin sublevel T_{0z}, where the z axis passes through the carbene carbon atom and is parallel to the line joining the phenyl centers, has already been demonstrated by the spin echo experiment on DPC.[53] The molecules with $T_{0y} \rightarrow T_{1y}$ and $T_{0x} \rightarrow T_{1x}$ transitions that satisfy the resonance energy requirement (groups 2 and 3) are selectively relaxed by ISC to the T_{0z} component where they are no more in resonance with the laser light. Therefore they remain there during the spin-lattice relaxation time, which is usually long at 1.8 K. The only group of molecules that may effectively supply heat to the surrounding lattice through the ISC process and can thus be burnt is the group of molecules (group 1) having their $T_{0z} \rightarrow T_{1z}$ transition in resonance with the laser light.

2-NPC does not belong to the D_{2h} symmetry group. This carbene has less symmetry elements and the situation is more complicated than that described above. The principal spin axes x, y, and z in the T_0 state may rotate leading to new axes x′, y′, and z′ in the T_1 state. The directions of spin axes follow the redistribution of the electronic density. Under these circumstances, the electronic transitions from each spin sublevel of the T_0 state to each spin sublevel of the T_1 state are allowed, as indicated in the energy level diagram displayed in Figure 20. The three groups of 2-NPC that can contribute to the hole-burning spectrum are those having their $T_{0z} \rightarrow T_{1z'}$, $T_{0z} \rightarrow T_{1y'}$, and $T_{0z} \rightarrow T_{1x'}$ in resonance with the laser frequency. The predicted spectrum of burnt holes is shown at the bottom of Figure 20. The central hole, at the burning frequency, should be symmetrically surrounded by three weaker satellite holes on the low-energy side and three on the high-energy side, separated from the center by $2 \cdot E_1$, $D_1 - E_1$, and $D_1 + E_1$. This substructure around the central hole provides information about the ZFS of the excited triplet state and closely resembles that observed experimentally (see Figure 18). Two other sets of holes, each with the same substructure as that around the burning frequency, should be located on the low energy side, $D_0 - E_0$ and $D_0 + E_0$ away from the center. The energy positions of these holes provide information about the splittings of the triplet ground state. These holes are also present on the low energy side of the experimental hole-burning spectrum. The comparison between the predicted and experimental spectra of holes allows to estimate the ZFS parameters of the T_0 and T_1 states of the pseudo E/*trans* conformer of 2-NPC:

$$D(T_0) = D_0 = 0.50 \text{ cm}^{-1} \pm 0.02 \text{ cm}^{-1},$$

$$E(T_0) = E_0 = 0.017 \text{ cm}^{-1} \pm 0.003 \text{ cm}^{-1},$$

$$D(T_1) = D_1 = 0.038 \text{ cm}^{-1} \pm 0.003 \text{ cm}^{-1},$$

$$E(T_1) = E_1 = 0.005 \text{ cm}^{-1} \pm 0.001 \text{ cm}^{-1}.$$

The above data may be compared with the corresponding parameters shown in Table 5.

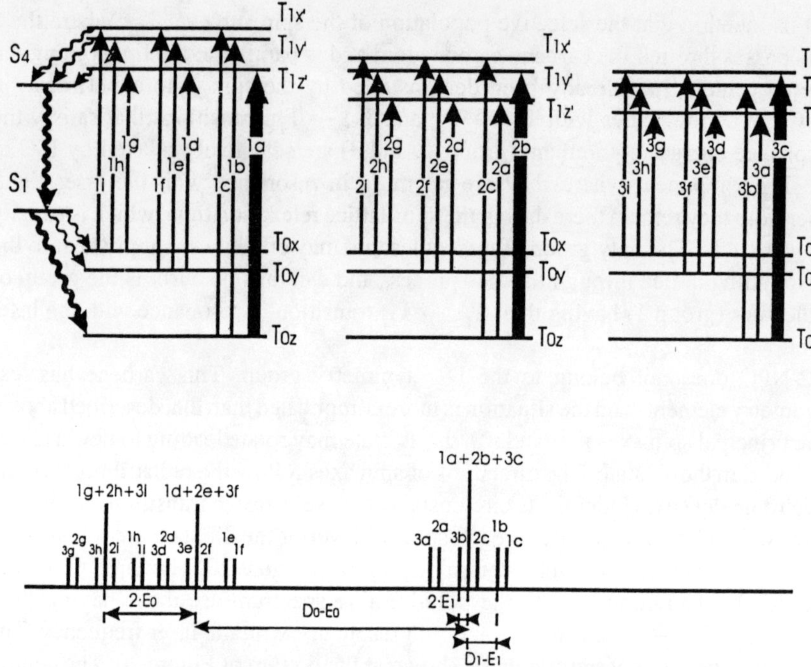

Figure 20. Energy level diagram and electronic transitions of 2-NPC. The burning transitions, resonant with the laser light, are indicated with thick arrows. The energy separations between sublevels T_z and T_y and between T_y and T_x are, by definition, respectively, D_0–E_0 and $2E_0$ in the T_0 state, and D_1–E_1 and $2E_1$ in the T_1 state. The expected spectral pattern of burnt holes and their energy separations are shown at the bottom. The ordering of the spin sublevels in the ground state is the same as in Ref. 53. (Reprinted with permission from *Chem. Phys. Lett.* **1995**, *245*, 549).

It was also possible to burn holes on the photosensitive 0,0 fluorescence excitation line with a maximum at 600.1 nm, attributed to the pseudo Z/*cis* conformer of 2-NPC.[43,54] The presence of unusually deep holes, as well as holes that seem to not be related to the energy separations of the triplet ground T_0 and excited states T_1, is unexpected. Although the observation may be related to the light induced conformational transformation from the pseudo Z/*cis* to the pseudo E/*trans* form,[20] a convincing interpretation of the spectra needs further elaboration.

In summary, the first hole-burning experiment on 2-NPC has led to the precise determination of the ZFS parameters in the triplet ground T_0 and excited T_1 states of the photostable conformer. It seems to also provide some information about a conformational transformation. We think that our hole-burning work opens a new field of investigations and that this technique will be successfully applied in the future to many other systems with triplet ground states.

REFERENCES

1. Hine, J. *J. Am. Chem. Soc.* **1950**, *72*, 2438.

2. Hine, J. *Divalent Carbon*; Ronald Press: New York, 1964; Ch. 1.

3. Dumas, J.B. *Ann. Chem. Phys. [2]* **1835**, *28*.

4. Geuther, A. *Liebigs. Ann. Chem.* **1862**, *123*, 121.

5. Doering, W.v.E.; Hoffmann, A.K. *J. Am. Chem. Soc.* **1954**, *76*, 6162.

6. Herzberg, G.; Shoosmith, J. *Nature* (London) **1959**, *183*, 1801; Herzberg, G. *Proc. Roy. Soc.* (London) **1961**, *A262*, 291; Herzberg, G.; Johns, H.W.C. *Proc. Roy. Soc.* (London) **1966**, *A295*, 107.

7. For a review of matrix isolation spectroscopy of carbenes see Sander, W.; Bucher, G.; Wierlacher, S. *Chem. Rev.* **1993**, *93*, 1583.

8. Murray, R.W.; Trozzolo, A.M.; Wasserman, E.; Yager, W.A. *J. Am. Chem. Soc.* **1962**, *84*, 3213.

9. Brandon, R.W.; Closs, G.L.; Hutchison Jr, C.A. *J. Chem. Phys.* **1962**, *37*, 1878.

10. Closs, G.L. in: *Carbenes*, Vol. II, Moss, R.A.; Jones, M., Jr., Eds.; Wiley: New York, 1975, p 159.

11. Trozzolo, A.M.; Wasserman, E. in: *Carbenes,* Vol. II, Moss, R.A.; Jones, M., Jr., Eds.; Wiley: New York, 1975, p 185; Wasserman, E.; Hutton, R.S. *Acc. Chem. Res.* **1977**, *10*, 27; Trozzolo, A.M. *Acc. Chem. Res.* **1968**, *1*, 329.

12. Doetschman, D.C.; Hutchison, Jr., C.A. *J. Chem. Phys.* **1972**, *56*, 3964.

13. Trozzolo, A.M.; Gibbons, W.A. *J. Am. Chem. Soc.* **1967**, *89*, 239.

14. Closs, G.L. in: *Chemically Induced Magnetic Polarization*, Lepley A.R.; Closs, G.L., Eds.; Wiley: New York, 1973, 95; Roth ,H.D. *Acc. Chem. Res.* **1977**, *10*, 83.

15. *Kinetics and Spectroscopy of Carbenes and Biradicals*, Platz, M.S., Ed.; Plenum: New York, 1990.

16. Shpolskii, E.V.; Ilina, A.A.; Klimova, L.A. *Dokl Acad. Nauk SSSR* **1952**, *87*, 935; Shpolskii, E.V. *Sov. Phys. Usp* (Engl. Transl.) **1960**, *3*, 372; Shpolskii, E.V. *Sov. Phys. Usp* (Engl. Transl.) **1962**, *5*, 522; Shpolskii, E.V. *Sov. Phys. Usp* (Engl. Transl.) **1963**, *6*, 411; Shpolskii, E.V.; Bolotnikova, T.N. *Pure Appl. Chem.* **1974**, *37*, 183.

17. Richards, J.L.; Rice, S.A. *J. Chem. Phys.* **1971**, *54*, 2014.

18. Ware, W.R.; Sullivan, P.J. *J. Chem. Phys.* **1968**, *49*, 1445; Ono, Y.; Ware, W.R. *J. Phys. Chem.* **1983**, *87*, 4426.

19. Anderson, R.J.M.; Kohler, B.E.; Stevenson, J.M. *J. Chem. Phys.* **1979**, *71*, 1559.

20. Després, A.; Lejeune, V.; Migirdicyan, E.; Platz, M.S. *J. Phys. Chem.* **1992**, *96*, 2486.

21. Saebo, S.; Boggs, J.E. *Theochem.* **1982**, *4*, 365.

22. Burkett, U.; Allinger, N.L. *Molecular Mechanics,* ACS Monograph 177; American Chemical Society: Washington, DC, 1982.

23. Haider, K.W.; Platz, M.S.; Després, A.; Lejeune, V.; Migirdicyan, E. *J. Phys. Chem.* **1990**, *94*, 142.

24. Després, A.; Lejeune, V.; Migirdicyan, E.; Admasu, A.; Platz, M.S.; Berthier, G.; Parisel, O.; Flament, J.P.; Baraldi, I.; Momicchioli, F. *J. Phys. Chem.* **1993**, *97*, 13358.

25. Anderson, R.J.M.; Kohler, B.E. *J. Chem. Phys.* **1976**, *65*, 2451.

26. Momicchioli, F.; Baraldi, I.; Bruni, M.C. *Chem. Phys.* **1982**, *70*, 161.

27. Trozzolo, A.M.; Wasserman, E.; Yager, W.A. *J. Am. Chem. Soc.* **1965**, *87*, 129.

28. Hutton, R.S.; Roth, H.D. *J. Am. Chem. Soc.* **1982**, *104*, 7395.

29. Roth, H.D.; Hutton, R.S. *Tetrahedron* **1985**, *41*, 1567.

30. Maloney, V.; Platz, M.S. *J. Phys. Org. Chem.* **1990**, *3*, 135.

31. Kozankiewicz, B.; Després, A.; Lejeune, V.; Migirdicyan, E.; Olson, D.; Michalak, J.; Platz, M.S. *J. Phys. Chem.* **1994**, *98*, 10419.

32. Trozzolo, A.M.; Wasserman, E.; Yager, W.A. *J. Chim. Phys.* **1964**, 1663.

33. Parisel, O.; Berthier, G.; Migirdicyan, E. *Can. J. Chem.* **1995,** *73*, 1869.

34. Roth, H.D.; Platz, M.S. *J. Phys. Org. Chem.* **1996**, *9*, 252.

35. Lejeune, V.; Després, A.; Migirdicyan, E.; Siebrand, W. *J. Phys. Chem.* **1991**, *95*, 7585.

36. Lejeune, V.; Després, A.; Migirdicyan, E. *J. Phys. Chem.* **1990**, *94*, 8861.

37. Moritani, I.; Murahashi, S.I.; Nishino, M.; Yamamoto, Y.; Itoh, K.; Mataga, N. *J. Am. Chem. Soc.* **1966**, *89*, 1259.

38. Turro, N.J.; Aikawa, M.; Butcher Jr, J.A.; Griffin, G.W. *J. Am. Chem. Soc.* **1980**, *102*, 5127.

39. Hutchison Jr, C.A.; Mangum, B.W. *J. Chem Phys.* **1961**, *34*, 908.

40. Cheng, T.H.; Hirota, N. *J. Chem. Phys.* **1972**, *56*, 5019; *Mol. Phys.* **1974**, *27*, 281.

41. Schmidt, J.; Veeman, W.S.; Van der Waals, J.H. *Chem. Phys. Letters* **1969**, *4*, 341.

42. Al'shitz, E.I.; Personov, R.I.; Kharlamov, B.M. *JETP Letters* **1977**, *26*, 586.

43. Kozankiewicz, B.; Bernard, J.; Migirdicyan, E.; Orrit, M.; Platz, M.S. *Chem. Phys. Letters* **1995**, *245*, 549.

44. Personov, R.I. in: *Spectroscopy and Excitation Dynamics of Condensed Molecular Systems*; Agranovich, V.M.; Hochstrasser, R.M., Eds.; North-Holland: Amsterdam, 1983, p 555.

45. Moerner, W.E. *Persistent Spectral Hole-burning, Science and Applications*; Eds; Springer Verlag: Berlin, 1988.

46. Völker, S. *Ann. Rev. Phys. Chem.* **1989**, *40*, 499.

47. Orrit, M.; Bernard, J.; Personov, R.I. *J. Phys. Chem.* **1993**, *97*, 10256.

48. Personov, R.I.; Al'shitz, E.I.; Bykovskaya, L.A. *Opt. Commun.* **1972**, *6*, 169.

49. El Sayed, M.A. *Acc. Chem. Res.* **1971**, *4*, 23.

50. Kharlamov, B.M.; Personov, R.I.; Bykovskaya, L.A. *Opt. Commun.* **1974**, *12*, 191.

51. Gorokhovski, A.A.; Kaarli, R.K.; Rebane, L.A. *Zh Eksp. Teor. Fiz. Pisma Red.* **1974**, *20*, 474 (in Russian).

52. Kharlamov, B.M.; Al'shitz, E.I.; Personov, R.I. *Sov. Phys. JETP* **1984**, *60*, 428.

53. Doetschman, D.C.; Botter, B.J.; Schmidt, J.; Van der Waals, J.H. *Chem. Phys. Lett.* **1976**, *38*, 18.

54. Kozankiewicz, B.; Bernard, J.; Migirdicyan, E.; Orrit, M.; Platz, M.S. *Mol. Cryst. Liq. Cryst.* **1996**, *283*, 191.

ISSUES AND CHALLENGES IN THE CHEMISTRY OF ALKYLCARBENES

Matthew S. Platz

Advances in Carbene Chemistry
Volume 2, pages 133–174
Copyright © 1998 by JAI Press Inc.
All rights of reproduction in any form reserved.
ISBN: 1-55938-837-4

I. INTRODUCTION

A renaissance in the study of alkyl- and dialkylcarbenes is in progress. The
emergence of nano, pico, and femtosecond time-resolved experimental methods
and powerful computational methods have given modern carbene chemists new
tools with which to study old, but still fundamental issues.

These powerful new tools have enabled the community of carbene chemists to
observe simple species of fundamental interest directly and to measure the rates[1]
at which they undergo unique and beautiful rearrangement reactions.[2] Modern
physical organic chemists will spend the next few years determining how carbene
structure, temperature, and solvent polarity affect the rates of rearrangement
reactions. Recent progress concerning these issues will be addressed in the accompanying chapter in this volume by Toscano.

Contemporary experimental techniques can also be used to study other, equally
fundamental issues such as how, and even if, an alkyl- or dialkylcarbene is produced
from a given "precursor"; and this topic which is the primary focus of this review
has a rich history spanning several decades. This essay does not intend to compre-
hensively review every contribution to this field. Rather, the author hopes to
highlight a few experiments that have most influenced his thinking. Apologies are
given at the outset to those scientists whose important contributions to this field are
not discussed in detail.

Alkyl- and dialkylcarbenes have been studied for decades of course. Brinton and
Vollman considered the gas-phase photochemistry of diazoethane (**1**) in 1951.[3]

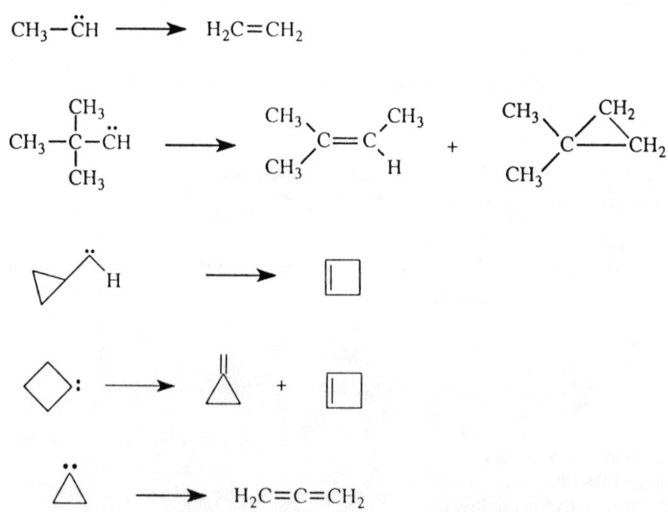

Scheme 1.

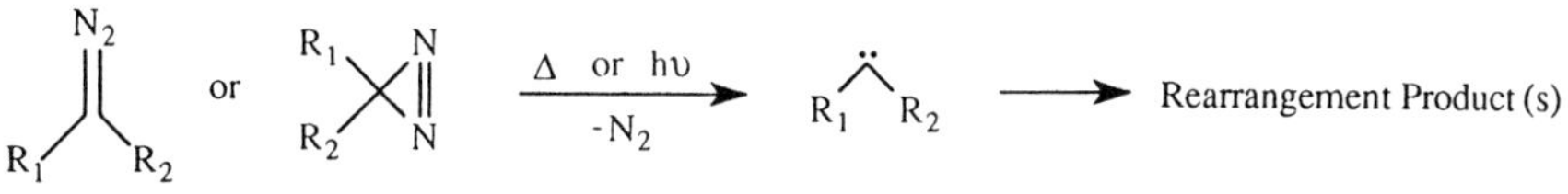

Scheme 2.

They discovered that photolysis of **1** produces nitrogen and ethylene in good yield and postulated the intermediacy of a reactive intermediate, methylcarbene (MC), or ethylidene.

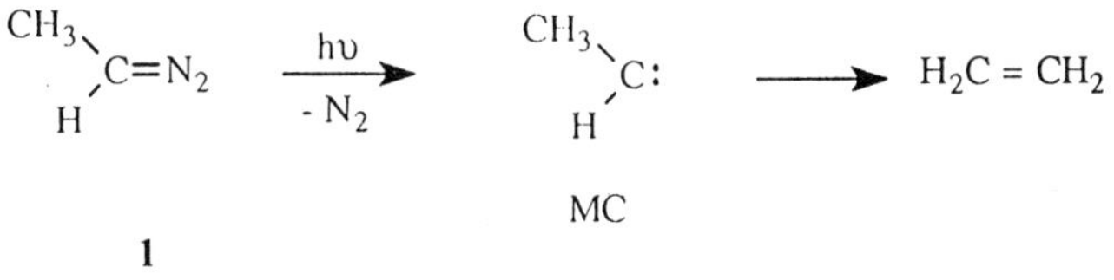

These workers announced that their work in the area would be discontinued due to the depressing frequency of violent explosions realized during the attempted purification of **1**. The difficulty of handling alkyl diazo compounds and perhaps the experiences reported by Brinton and Vollman meant there were only a few subsequent studies of this type. More than ten years would pass before additional research on the mechanistic photochemistry of diazoethane and other photochemical sources of methylcarbene would be reported.[4]

The development of simple and practical syntheses of diazirines by Paulsen and by Schmitz and Ohme[5] led to renewed interest in alkylcarbenes in the 1960s. Seminal papers were published by Mansoor and Stevens,[6] and by Frey and coworkers[7] on the photolysis and pyrolysis of diazirines and of the salts of tosylhydrazones which produce diazo compounds. During that period, analysis of reaction mixtures via gas–liquid chromatography was the most powerful tool available to the carbene chemist. Many fascinating and exotic rearrangements, attributed to alkylcarbenes, were discovered, a few of which are shown in Scheme 1.[2] It became customary to write the mechanism shown in Scheme 2 and this depiction still finds its way into standard textbooks. This simple mechanism is economical but was known to be wrong, or at least seriously incomplete, over 30 years ago for several systems.[6,7]

II. MIGRATION OF HYDROGEN

Pyrolysis of tosylhydrazone salt **2** and pyrolysis and photolysis of diazirine **3** should all produce ethylmethylcarbene (EMC). As the enthalpic barriers to rearrangement of EMC are probably rather small (5 kcal/mol), the $\Delta\Delta H^{\ddagger}$ to the formation of

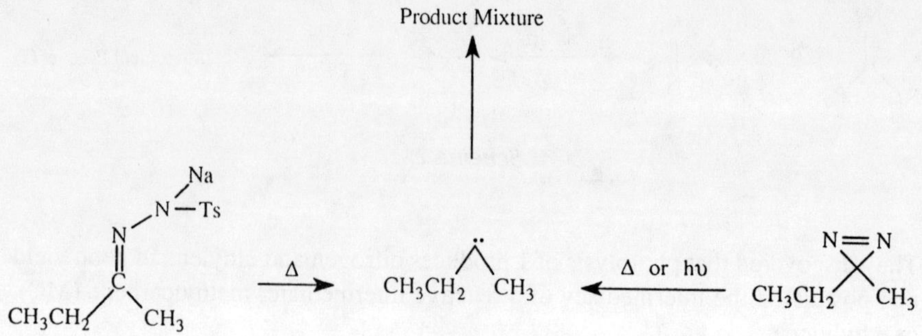

Product Mixture

Scheme 3.

products should be close to zero. Thus, EMC should rearrange to a characteristic mixture of alkenes which should be rather insensitive to temperature. These expectations were not realized. Pyrolysis of **2** and **3** gives the same mixture of products (Scheme 3), but the compounds isolated after photolysis of **3** are produced in markedly different proportions (Table 1). Clearly heat and light do not produce

Table 1. The Mixture of Products Formed upon Decomposition of Various Precursors of Ethymethylcarbene[6–8]

Precursor	Conditions	(structure 1)	(structure 2)	(structure 3)	(structure 4)	(structure 5)
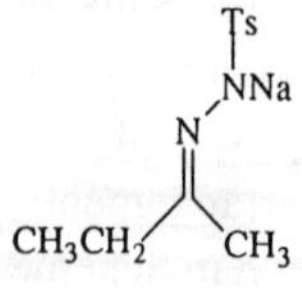 **2**	Δ	28	67	5	0.5	0.0
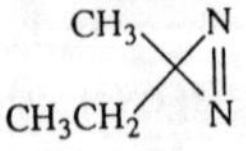 **3**	hν	34.7	38.0	23.2	3.7	0.3
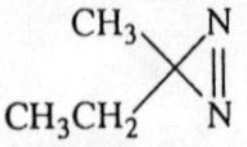 **3**	Δ	29.5	66.6	3.3	0.5	0.0

the same intermediate from the nitrogenous precursors.[6–8] It was clearly recognized by the original investigators that there must be two product-forming routes, one of which was taken to involve the relaxed carbene for the sake of simplicity. The second route must involve a different intermediate such as a vibrationally or electronically excited state of the precursor or of the carbene.

Several workers felt[6–8] that the pyrolysis route afforded pure relaxed carbene, because it produced the simpler product mixture, and that photolysis generated two intermediates that formed products in distinctly different ratios. However, at that point, the data were equally consistent with clean photochemistry and complex thermochemistry or with complications in both decomposition modes.

Consider the products observed when attempts are made to generate *tert*-butylcarbene.[7] Photolysis of **4** and its linear diazo isomer (entries 1 and 8, Table 2) and pyrolyses (entries 3–7, Table 2) of **4** and **5** produce an alkene/cyclopropane mixture that depends on the method of decomposition of the precursor (Table 2).[9] Please note that these differences are in the condensed phase. This behavior indicates, in this case at least, that *vibrationally* excited precursor or carbene is not responsible for the differences in product ratios as a function of decomposition mode.

Shechter and Chang[10] made the remarkable discovery (Table 2, entry 2) that triplet photosensitized decomposition of *tert*-butyldiazirine and direct photolysis of the tosylhydrazone salt (Table 2, entry 8) in solution and pyrolysis gave the same mixture of products as pyrolysis of the diazirine. This outcome implied an *electronically* excited singlet state of the diazirine or of *tert*-butylcarbene as the origin of the unusual product distribution. This provides more evidence that the migratory patterns discovered in the *pyrolysis* of the nitrogenous precursors of *tert*-butylcarbene were the result of rearrangements of the free carbene and not formed via nitrogen containing intermediates.

Additional evidence in this regard was provided by independent generation of *tert*-butylcarbene with a non-nitrogenous precursor (Table 2, entries 9–12).[11–12] Decomposition of a geminal diodide precursor (Table 2, entries 10–12) where the intervention of a carbenoid[12] is unlikely gave the same distribution of products as the pyrolysis of diazirines,[4] tosylhydrazone salts[5] and triplet sensitized photolysis of *tert*-butyldiazirine.[4]

However, in at least one case there is evidence that the thermal generation of a carbene from a diazo precursor is complicated by the presence of a second product-forming pathway (Table 3).[13,14] Producing "ethylphenylcarbene" by the multistep phenylcarbene rearrangement[15] (Table 3, entries 5–8) surely gives a nitrogen-free intermediate. Nevertheless the β-methylstyrenes formed in this method have small but real differences in their *cis/trans* composition from that obtained on pyrolysis of phenyldiazopropane (Table 3, entries 2, 3, 10–12). Generation of the carbene by reaction of propiophenone with atomic carbon (a non-nitrogenous precursor) gives the same mixture of products observed when "ethylphenylcarbene" is produced by the phenylcarbene rearrangement. In this

Table 2. The Ratio of Products Formed on Gas and Solution Phase Decomposition of *tert*-Butylcarbene Precursors[7,9–12]

Precursors	Conditions	⋛═/◁⋉	Entry
4 $(CH_3)_3C-C-H$ (diazirine)	$h\nu$	49/51	1
4 $(CH_3)_3C-C-H$ (diazirine)	$h\nu$ (triplet sens, 25°)	11.6/88.4	2
4 $(CH_3)_3C-C-H$ (diazirine)	$\Delta 145°$	12.4/87.6	3
	$\Delta 160°$	8/92	4
	$\Delta 180°$	11.6/88.4	5
5 $(CH_3)_3C-C-H$ (N–N–Ts, Na)	$\Delta 130°$	9.9/86.8	6
		7/92	7
$(CH_3)_3C-C-H$ (tosylhydrazone salt)	$h\nu$ (THF, 25°)	46.5/46.7	8
6 $(CH_3)_3CCHI_2$ $+CH_3Li$	40° (ether)	4/96	9
	21° (gas-phase)	9.8/90.2	10
	100° (gas-phase)	17.7/82.3	11
	146° (gas-phase)	23.7/76.3	12
$(CH_3)_3CH_2Cl$ + PhNa	100° (decane)	3.5/87	13

system neither pyrolysis nor photolysis (entry 4) seems to produce the desired free carbene intermediate cleanly.

Many workers, beginning with Frey,[7] have presented evidence for the involvement of precursor excited states in what can be written as formal carbene rearrangements. We will propose a specific mechanism for such a process in a later section.

Table 3. Product Mixtures from Decomposition of Various Precursors of "Ethylphenylcarbene"[13]

Precursor	Conditions	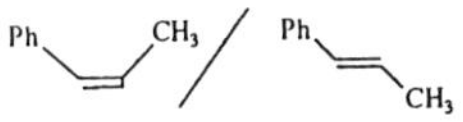 $\dfrac{Ph\!-\!CH\!=\!CH\!-\!CH_3}{Ph\!-\!CH\!=\!CH\!-\!CH_3\ (CH_3)}$	Entry
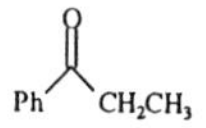 **7** (Ph–CO–CH₂CH₃)	Atomic Carbon −196°	36/64	1
8 (Ph–C(=N₂)–CH₂CH₃)	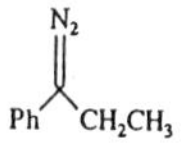 500°	24/76	2
	320°	28/72	3
8 (Ph–C(=N₂)–CH₂CH₃)	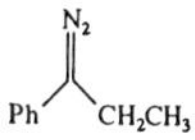 25° (hν)	23/77	4
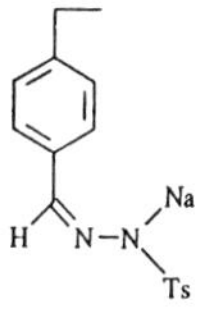 (tosylhydrazone sodium salt)	500°	37/63	5
	350°	26/74	6
	500° (acid)	6/94	7
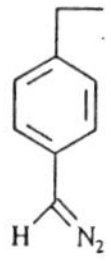 (diazo, H–C=N₂)	500°	37/63	8
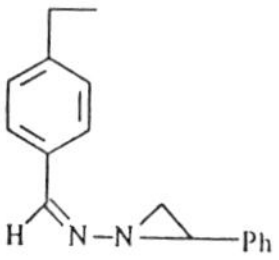 (N–N–aziridinyl–Ph)	320°	36/64	9
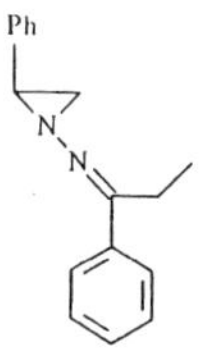 (aziridinyl imine, Ph)	250°	21/79	10
	350°	25/75	11
	500°	27/73	12

The examples of rearrangement involving a second pathway mentioned so far involved 1,2 or 1,3 migration of hydrogen. However, extremely compelling examples of rearrangements in precursor excited states involve the migration of carbon.

III. REARRANGEMENTS OF CARBON IN PRECURSOR EXCITED STATES

In the opinion of this author, the most spectacular demonstration of a 1,2 shift of carbon in concert with nitrogen extrusion in a dialkyldiazirine is provided by the study of the homocubylcarbene system.[16] Photolysis of **9** (Scheme 4) in pentane containing concentrated deuterated alcohol led to the isolation of ether **13**. One cannot postulate that homocubylcarbene **12** rearranges more rapidly to alkene **11** than it can be trapped, because this intermediate **12** is easily captured by deuterated alcohol when generated from hydrocarbon precursor **10** under the same conditions. The same mixture of ethers **13** and **14** is produced upon individual photolysis of either **9** or **10** in the presence of dilute alcohol. This last result indicates that **11** and **12** can equilibrate.[16] Theory and experiment agree that the equilibrium constant in this system is close to unity. It is perhaps rather generous to refer to the strained intermediate **11** as an alkene. Calculations of Borden and Hrovat[16d] indicate that it is perhaps better thought of as a biradical, since the two "alkene" carbons are pyramidal and there is little overlap of the orbitals in question.

Thus, carbene **12** is not the first species generated upon photolysis of diazirine **9**. Light-induced extrusion of nitrogen is accompanied by a rearrangement to the alkene. Labeling studies by Eaton and White[17] demonstrated that the phenyl analogues of **11** and **12** interconvert by a C–C bond shift and not by a 1,2 migration of hydrogen. Thus, the C–C bond shift must proceed in a nitrogen containing species upon photolysis of **9**.

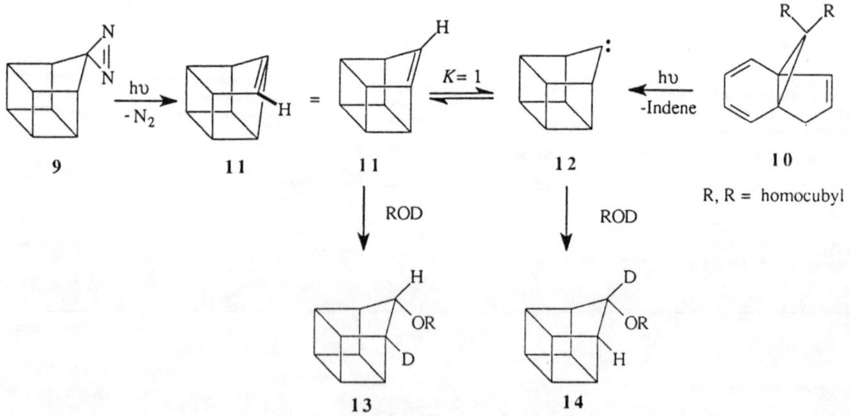

Scheme 4.

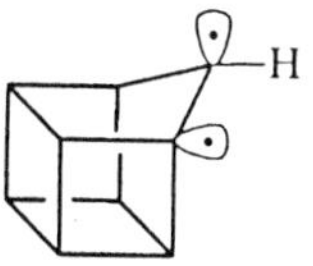

1 1

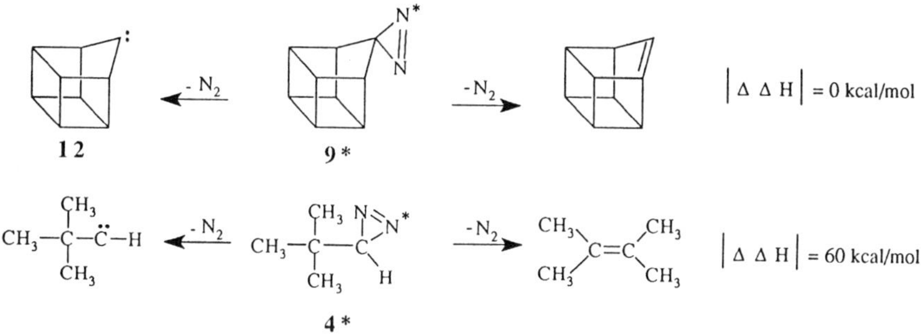

1 2

9 *

$\left| \Delta \Delta H \right|$ = 0 kcal/mol

$$CH_3 - \underset{CH_3}{\overset{CH_3}{C}} - \ddot{C} - H \xleftarrow{-N_2} CH_3 - \underset{CH_3}{\overset{CH_3}{C}} - \underset{H}{\overset{N}{C}} \xrightarrow{-N_2} \underset{CH_3}{\overset{CH_3}{C}} = \underset{CH_3}{\overset{CH_3}{C}}$$

$\left| \Delta \Delta H \right|$ = 60 kcal/mol

4 *

Perhaps there is a compelling stereoelectronic factor that drives the homocubyl rearrangement and makes this system unique. Alternatively one can argue that this system is atypical because the great strain energy present in the homocubyl system predisposes it to rearrangement. However, the alkene product of rearrangement (e.g. **11**) is just as strained as carbene **12**. Recall that the carbene–alkene equilibrium constant is close to unity. Unlike a simple analogue (e.g. ethylmethyldiazirine), there is no large thermodynamic push to rearrange the diazirine excited state in the homocubyl system.

By way of comparison, there is a tremendous ($\approx$ 60 kcal/mol) thermodynamic driving force for rearrangement in concert with nitrogen extrusion relative to carbene formation in a simple system such as *tert*-butyldiazirine. Thus, this author believes that simple diazirine excited states will be as prone to rearrangement of carbon and hydrogen as homocubyldiazirine **9**, and perhaps even more so.

IV. DIAZOCARBONYL COMPOUNDS

There are other important examples of carbon migration in precursor excited states which predate the work of Chen and Jones.

Roth observed that upon direct photolysis of diazoacetone **15** in chloroform no CIDNP signals are observed.[18] Formation of a singlet carbene that can abstract halogen atoms from solvent could have produced radical pairs and subsequent polarization. One can explain this negative result by postulating that Wolff rearrangement (WR) of the singlet carbene is faster than reaction with chloroform (Scheme 5):

$$(\text{e.g. } k_{WR} > k_{CHCl_3}[CHCl_3])$$

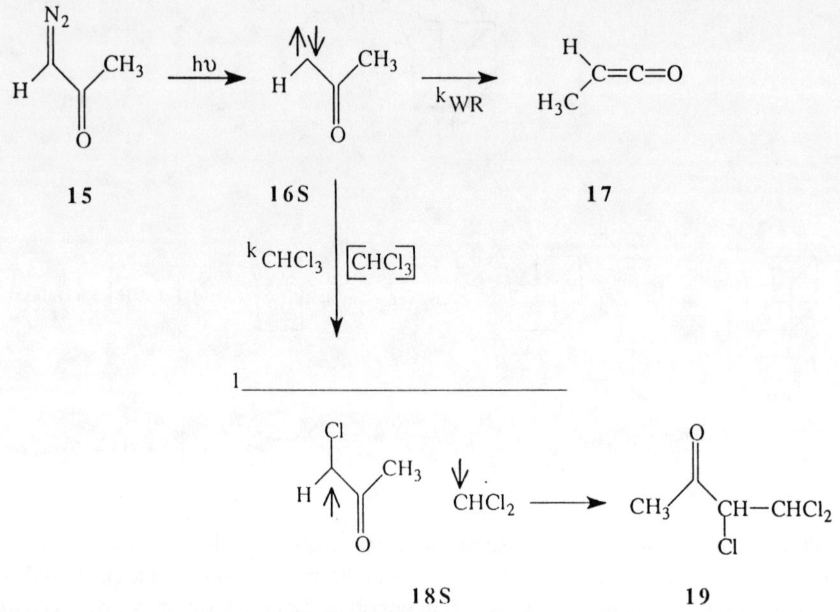

Scheme 5.

However, *singlet* radical pair polarization is observed upon *triplet* photosensitized photolysis in chloroform (Scheme 6). Triplet sensitized photolysis produces the triplet carbene which can access the singlet carbene (or the singlet surface), followed by abstraction of a halogen atom from solvent. The singlet radical pair so produced collapses to form polarized product **19**. This demonstrates that the singlet carbene does not rearrange to ketene so fast as to preclude bimolecular chemistry, a view supported by recent high level calculations.[19] The Roth experiment demonstrates that the ketene produced upon direct photolysis of the precursor is not formed from the singlet carbene. Apparently free singlet carbene **16S** is not formed upon direct photolysis of **15**. The photochemical Wolff rearrangement must proceed in the excited state of the precursor. However, it is important to note that this does not mean the Wolff rearrangement cannot proceed in the free carbene.

It is also important to recall the studies of Kaplan et al., which indicate that the photochemistry of diazoketones is a sensitive function of the conformation of the precursor.[20] A *syn* disposition of the carbonyl and diazo group as shown in Scheme 7 with formyldiazomethane favors carbon migration in the diazo excited state and Wolff rearrangement; whereas the *anti* arrangement favors carbene formation. The *syn* conformation is generally most favored in simple diazo ketenes[20] which means that little trappable carbene is produced with these precursors.

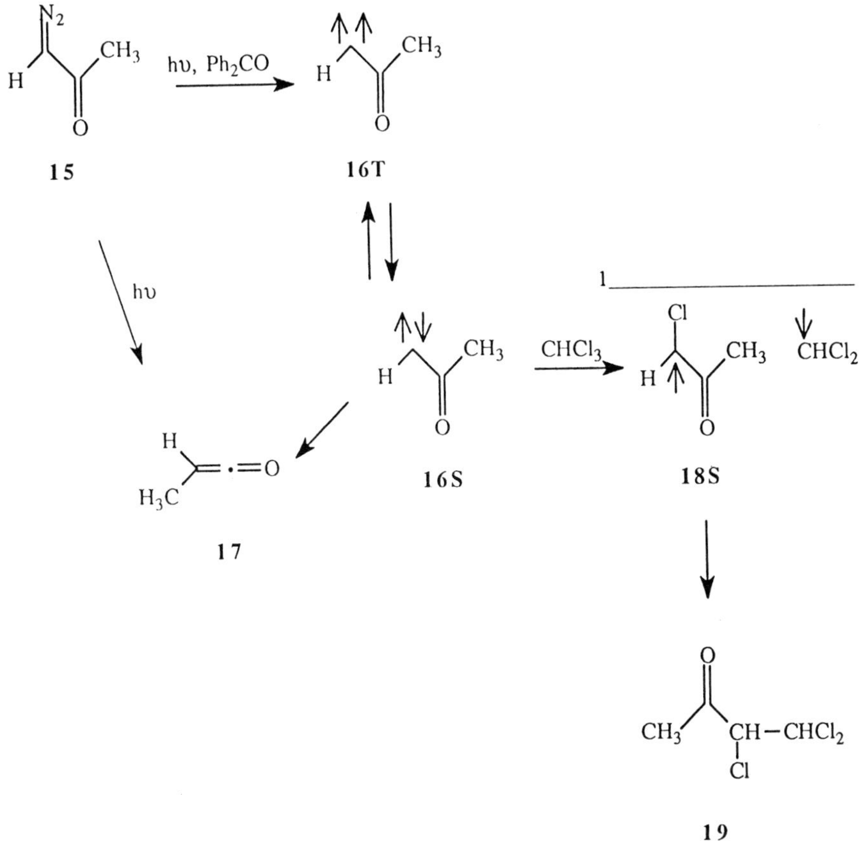

Scheme 6.

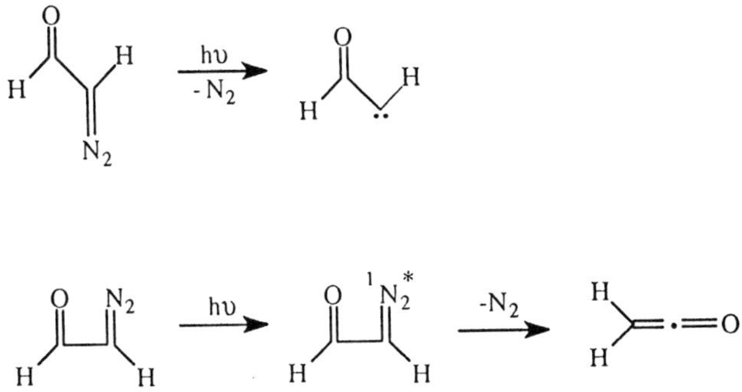

Scheme 7.

V. N,N-DIETHYLDIAZO AMIDES

Product studies reported by several groups demonstrate that diazo carbonyl excited states can "masquerade" as carbenes and undergo both carbon and hydrogen migrations.[21] The work of Strausz and Tomioka et al. are representative[22] and particularly insightful. Many related studies have also appeared.[23]

Photolysis of **20** in cyclohexane produces two lactams **21** and **22** and a product of reduction **23**. In the presence of methanol, two additional products, **26** and **27**, are formed. A traditional mechanistic scheme involving carbene **24** can explain the data presented to this point is shown in Scheme 8.

If this mechanism is correct, then increasing concentrations of methanol, a good carbene trap, should lead to increasing yields of **26** and decreased yields of **21–23**, and of ketene **25** (as measured by the yield of **27**). These predictions are not realized (Table 4) with **21** and **27**, and realized only in part with **22**. Lactam **21** is formed in equal yield in cyclohexane and in neat methanol, an excellent carbene trap. This finding demonstrates that lactam **21** is not formed from a free carbene intermediate. Lactam **22** is formed by two pathways, one of which involves a trappable intermediate.

Table 4. Products Formed on Photolysis of **20** under Various Conditions[22]

% CH$_3$OH	Additive	21	22	26	27
0	—	36.5	45.8	0	4.4[a]
2	—	37.6	35.3	10.1	11.9
100	—	36.6	16.4	33.0	11.4
100	Ph$_2$CO	2.0	8.3	26.8	<1.0
100	Piperylene	36.8	15.3	30.2	11.1

Note: a) CH$_3$OH added immediately following photolysis.

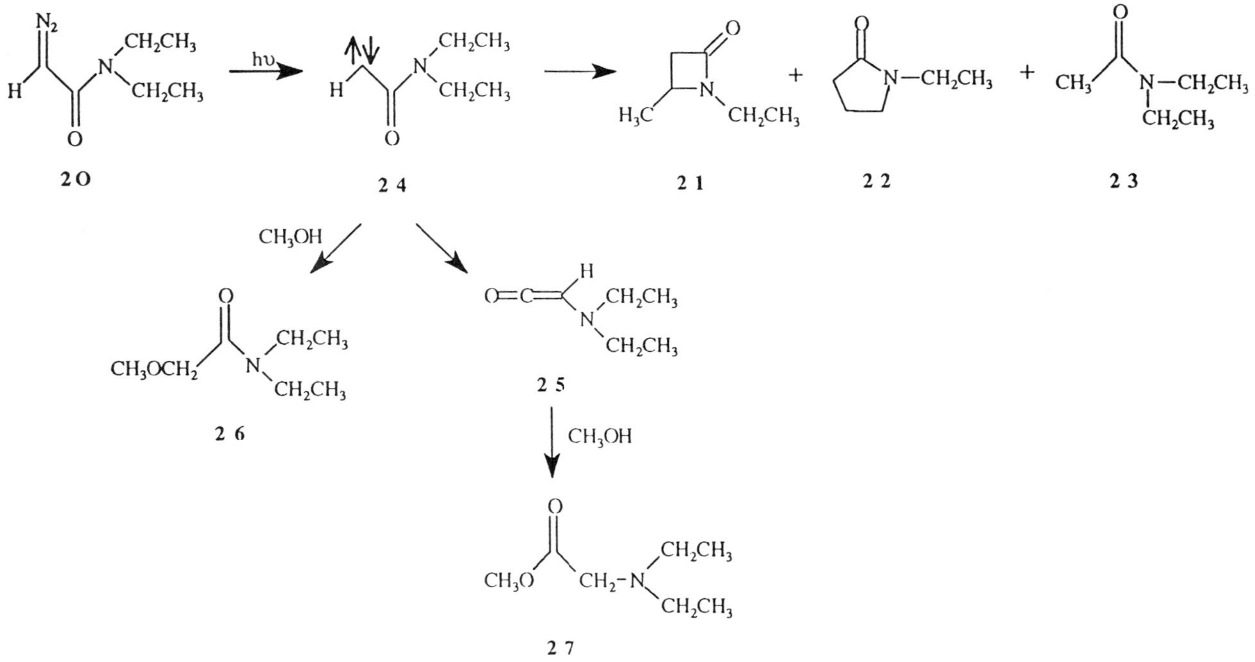

Scheme 8.

The yield of ketene **25** (as measured by the yield of **27**) does not vary as the concentration of methanol increases from 2 to 100%. Again this result demonstrates that the photochemical Wolff rearrangement does not involve a free singlet carbene intermediate. In fact, when one forms the free singlet carbene (via the triplet carbene) by triplet sensitization, in neat methanol, it greatly prefers to react with solvent rather than undergo Wolff rearrangement ($k_{CH_3OH}[CH_3OH] \gg k_{WR}$). This result is in accord with recent calculations which indicate that there is a substantial barrier to Wolff rearrangement of singlet carbonylcarbenes.[19] *Ab initio* theory also clearly predicts that singlet carbonylcarbenes are nonplanar species.[19]

Thus, the combined data indicate that the results of the direct photolysis are best written as shown in Scheme 9.

Only the singlet excited state of the precursor $^1\mathbf{20}^*$ must be involved in product formation as piperylene, a good quencher of triplet excited states, has no effect on the products formed. The following conclusion seems unmistakable: the diazo excited state must undergo 1,2 migration of the diethylamino group and 1,4 migration of hydrogen to give products typically associated with carbene rearrangements (Scheme 10).

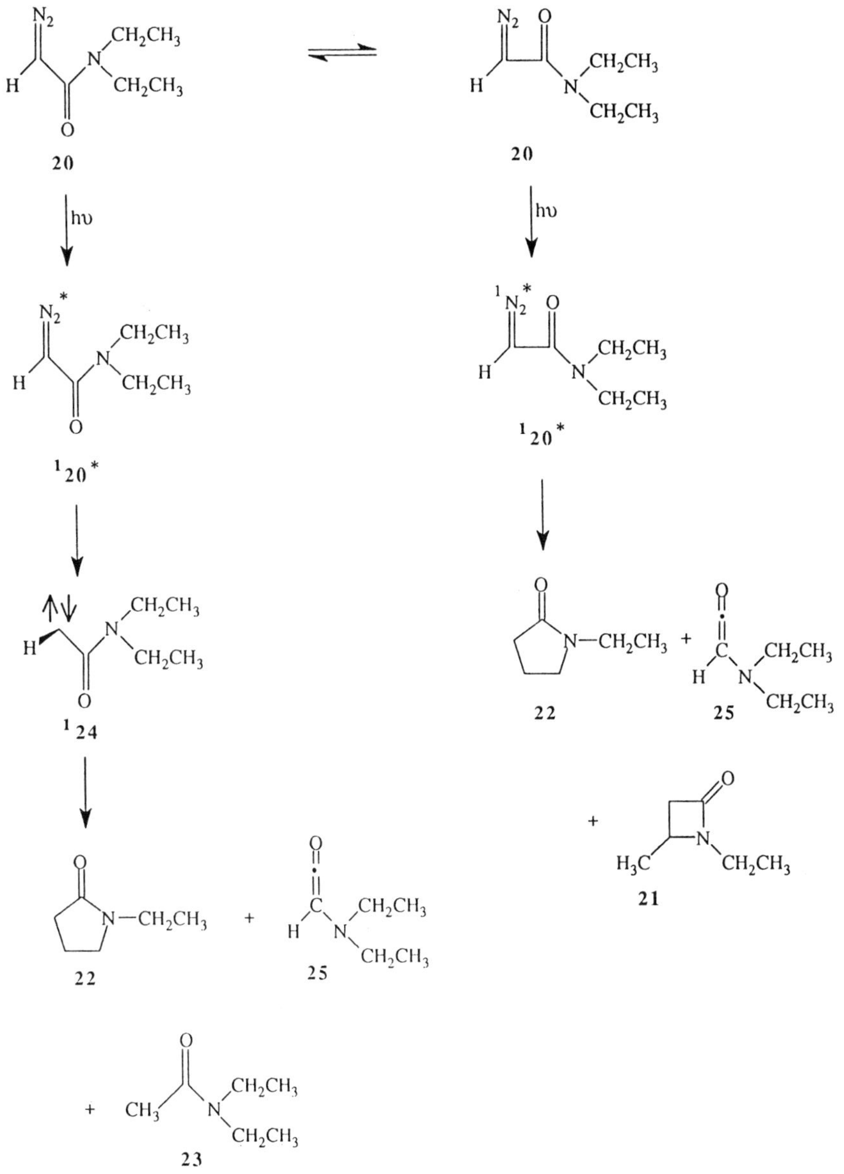

Scheme 9.

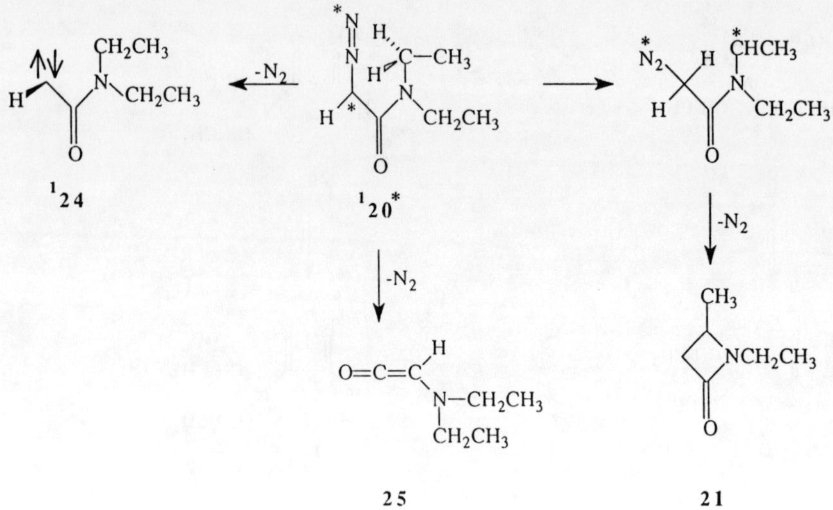

$* = \bullet, +, -$

Scheme 10.

Despite the fact that matrix EPR spectroscopy has demonstrated that many carbonyl-substituted carbenes have triplet ground states,[24] there is little pressure from product studies of simple systems to view them as product-forming intermediates. Part of this is due to the fact that diazocarbonyl excited states clearly form carbenes only inefficiently.[2,21–25] However, calculations indicate that singlet carbonyl-substituted carbenes are nonplanar, whereas the lower energy triplet carbenes are planar.[19] Thus, singlet-to-triplet intersystem crossing has a poor "Franck–Condon" factor, and singlet formate ester and amide carbenes may react with solvent or rearrange faster than they relax to triplet ground states. Alternatively the triplet carbenes may reform the corresponding singlet carbenes presumably because their reactions with solvent are much slower than those of the singlet carbene.

VI. DIAZOMALONATE

Jones et al. found that photolysis of diazomalonate **28** in neat *cis* alkene gives cyclopropane adducts with 90% stereospecificity.[26] Upon slight dilution of the alkene with hexafluorobenzene (presumably an inert additive) the stereospecificity actually slightly increased. It seems clear that the intermediate that adds specifically to the alkene is ^{1}**29**, but this species cannot be the first product-forming intermediate. The first trappable intermediate produced on photolysis of **28** must be the singlet excited state, ^{1}**28*** (see Scheme 11). Thus, the excited state intermediate in this case does not react stereospecifically with alkenes.

Study of this system also provides evidence that the diazo excited state can lead to formal CH insertion products. Direct photolysis of diazomalonate in alkanes gives products of formal CH insertion.[26] These products are not formed upon triplet sensitized photolysis of diazomalonate even though methanol trapping studies indicate that the triplet carbonylcarbene can access the singlet surface.[26]

Toscano et al. have studied this system by laser flash photolysis (LFP) techniques.[27a–d] The carbene was studied by trapping with pyridine to form a UV-Vis active ylide **30** (Scheme 11). The lifetime of the carbene was found to be the same in both cyclohexane and cyclohexane-d$_{12}$, which indicated that the carbene was not decaying in alkane solvent by reaction with the medium but was most likely consumed by Wolff rearrangement.[27a–d] To this author, the results with diazo compounds **20** and **28** indicate that Wolff rearrangement of ester-substituted carbenes is much faster than that of amide substituted carbenes.

In further support of this viewpoint, Toscano et al.[27a–d] found that the yield of carbene produced in freon solvent (where formal CH insertion with solvent is not possible) is much greater than that produced in cyclohexane, per laser pulse. Thus, complications due to precursor chemistry are not restricted to intramolecular processes (Scheme 11).

This conclusion was, in fact, first realized by Wulfman, et al.[27e] This group studied the photochemistry of methyldiazomalonate in cyclohexene. Three products were identified: the expected cyclopropane and the allylic and homoallylic products of C–H bond insertion. The ratio of cyclopropane product to allylic C–H insertion adduct was the same at 0 and 54 °C. However, the ratio of cyclopropane product to homoallylic CH insertion product doubles upon lowering the temperature from 54 to 0 °C. These authors concluded that the cyclopropane and allylic CH insertion product derive from a common intermediate (dicarbomethoxycarbene), but that the product of homo allylic C–H bond insertion is formed from a second intermediate which we associate with the excited state of the diazo compound. This explains why the yield of carbene–ylide determined by Toscano et al.[27a–d] depends upon the solvent (fluorocarbon > deuterocarbon > hydrocarbon) and why formal CH insertion adducts of the carbene are formed[26,27] even though the carbene lifetime shows no solvent isotope effect. The simplest explanation is that much of

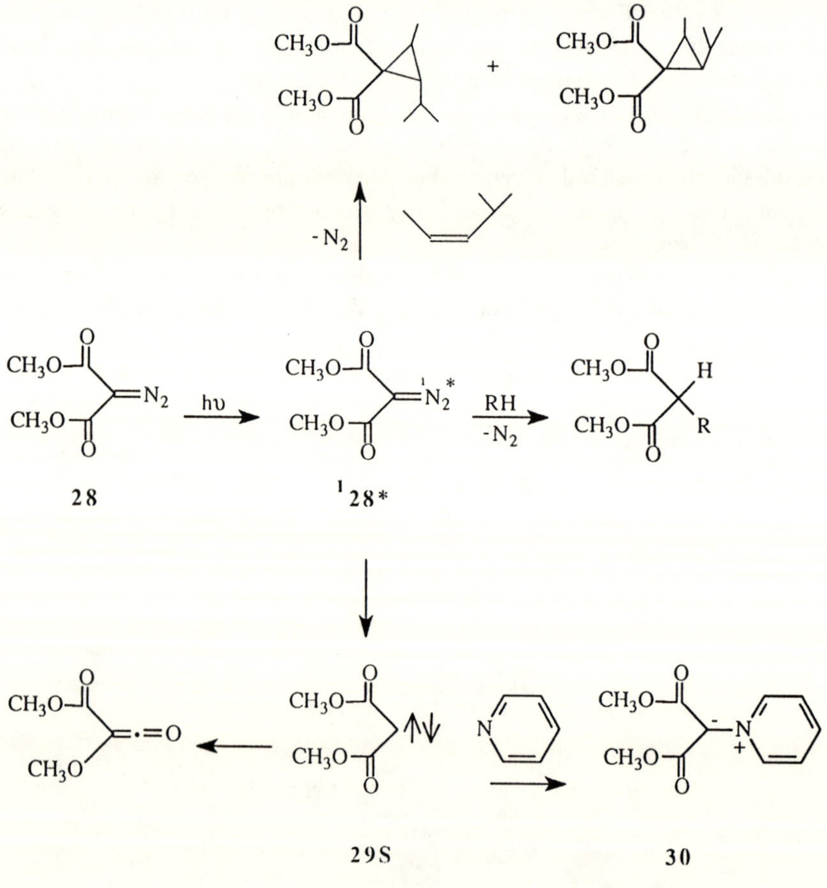

Scheme 11.

the C–H insertion adducts observed by Jones et al.[26] derive from the excited state of methyldiazomalonate rather than from dicarbomethoxycarbene.

VII. CHLOROCYCLOPROPYLCARBENES

The Moss group has studied cycloalkylhalocarbenes in great detail.[28] Their work with chlorodimethylcyclopropylcarbene is particularly revealing. Because of the chlorine substituent, this carbene will certainly have a singlet ground state. Photolysis of **31** gives a mixture of rearrangement products **32–33** and fragmentation products **34–35**.When **31** is photolyzed in pentane in the presence of the alkene trimethylethylene (≤ 9 M) these same products were accompanied by the new ones **36** through **39** (Scheme 12). If the simple mechanistic scheme shown in Scheme 13 was accurate, then a plot of the ratio of yields of **(36–39)/(32–35)** versus [alkene] should be linear with slope $k_{al}/(k_R + k_F)$. However this plot (Figure 1) is most definitely curved, which led Moss et al. to consider a more complex mechanism (Scheme 14).[28]

Two routes to the formation of cyclobutenes **32** and **33** must exist. One involves a trappable carbene **40**; the other was attributed to the excited state of the precursor. It is interesting to note that the trappable carbene gives **32/33** in a ratio of 3.8:1, but that the second route originating from ^{1}**31*** gives **32/33** in a ratio of 5.2:1.

Photolysis of **31** also produces chloroacetylene and isobutylene, two products of fragmentation in approximately 20% yield. The proportion of these products is not

	31	**32**	**33**	**34**	**35**
CDCl$_3$:		59.2%	19.7%	(21.1%)	
Pentane:		67.4%	14.0%	(18.6%)	
Pentane:		32.4%	8.52%	(19.0%)	

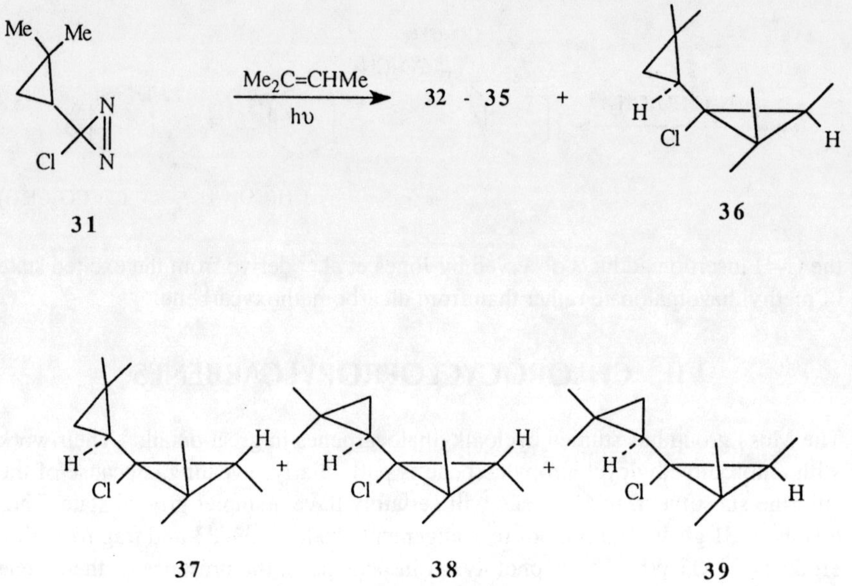

31

36

Scheme 12.

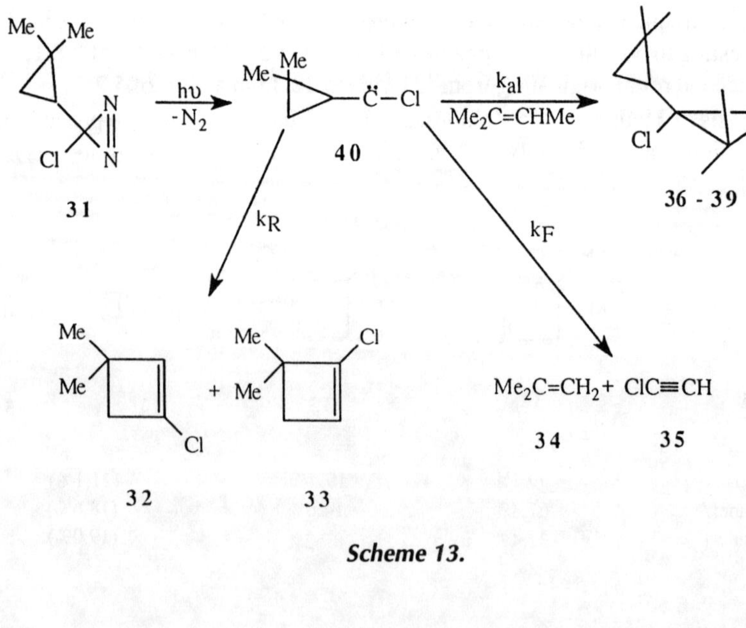

37   **38**   **39**

31   **40**   **36 - 39**

k_R k_F

32   **33**   **34**   **35**

Scheme 13.

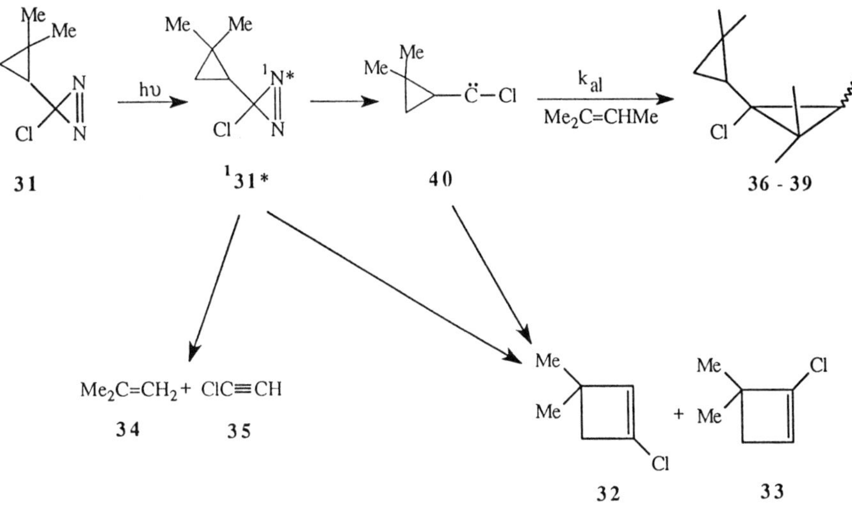

Scheme 14.

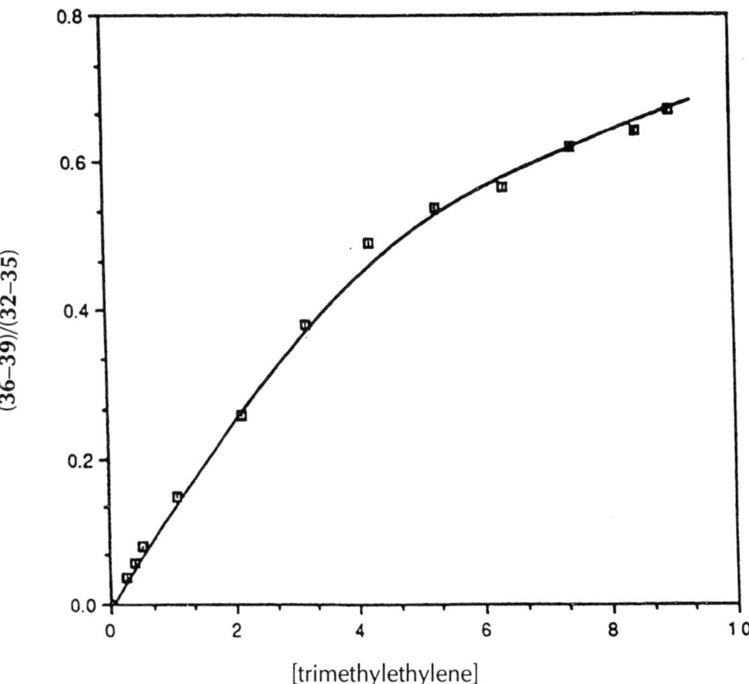

Figure 1. A plot of **(36–39)/(32–35)** versus trimethylethylene (reprinted with permission from reference 28).

reduced in the presence of 9 M alkene at 25 °C, although substantial yields of cyclopropane are formed under these conditions. Moss et al. therefore concluded that fragmentation proceeds not via carbene **40** at 25 °C but through the diazirine excited state [1]**31***.

Laser flash photolysis studies by the same group allowed the measurement of the absolute lifetime of carbene **40** in pentane and its absolute rate constant of reaction with trimethylethylene. These studies revealed that 9 M alkene is sufficient to trap all carbene generated upon photolysis. On this basis one can dissect the decomposition of the photolysis at 25 °C into percentages of products derived directly from the diazirine excited state [1]**31*** (17–20%) and the carbene **40**, as shown in Scheme 15.

It is interesting to note that pyrolysis of **31** gives a different picture. Upon pyrolysis of **31** in pentane (60–80 °C), cyclobutene products (**32** and **33**) again are observed. However the yield of fragmentation products is now very much diminished by the presence of alkene trap. This is further evidence that the fragmentation products can be formed by a carbene reaction at elevated temperatures.

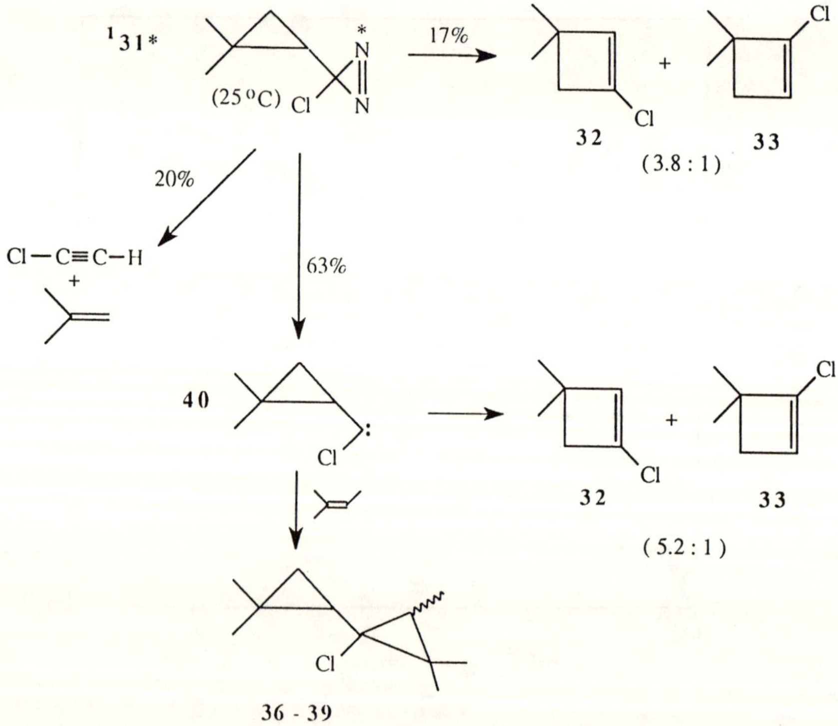

Scheme 15.

VIII. *trans-tert*-BUTYLCYCLOPROPYLDIAZIRINE

I cannot resist the temptation to illustrate this picture further with a recent result by Haiyong Huang in our laboratory. Huang has studied the decomposition of diazirine **41**.[29] Photolysis (350 nm, –25 to +25 °C) and pyrolysis (100 °C) of **41** in freon largely forms cyclobutene **42** (Scheme 16, Tables 5 and 6). A complex mixture of halogenated products and azine, detected by GC-MS, is also formed.

Adduct **44** is formed on decomposition (Δ,hυ) of the diazirine in the presence of tetramethylethylene (TME). However, the presence of TME during the photolysis (350 nm, 0 °C) of **41** has no effect on the yield of cyclobutene (Table 5). The presence of TME during the pyrolysis of **41** reduces the yield of cyclobutene from 64 to 57% at 25 °C even though a 37% yield of adduct is realized (Table 6). Scheme 16 is incorrect; the cyclobutene and the adduct are not formed from a common intermediate. The cyclopropylcarbene is not the source of the cyclobutene formed in solution upon pyrolysis or photolysis of the precursor. We believe this product is formed by ring opening of the precursor as shown in Scheme 17.

I believe that there is nothing "wrong" with the cyclopropylcarbene rearrangement in solution. It is not observed under our experimental conditions because in our opinion the carbene reaction with solvent is much faster than ring expansion. This viewpoint is in accord with theory which predicts that ring expansion must surmount a barrier of 5 kcal/mol.[30]

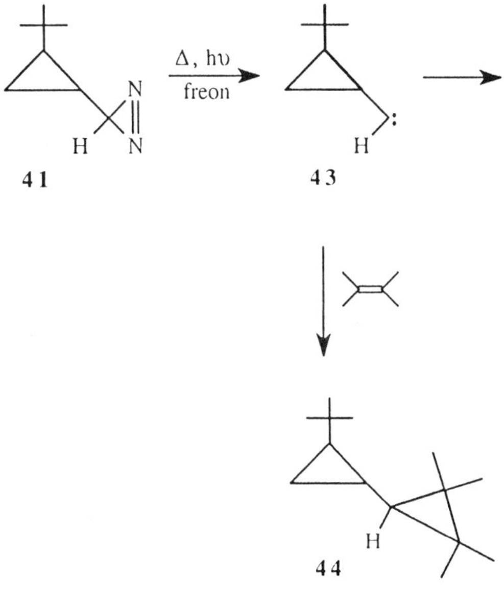

Scheme 16.

Table 5. Absolute Product Yields in the Photolysis (350nm) of **41** in CF$_2$ClCFCl$_2$ (–25 °C and 25 °C)[29]

	–25 °C			0 °C		25 °C	
TME(M)	%42	%44	%42	%44	%42	%44
0.017	44	13	46	10	49	7
0.034	47	19	44	14	46	10
0.067	45	22	46	20	43	15
0.134	46	26	48	24	47	22
0.269	45	30	45	27	43	24
0.420	46	32	47	28	46	27

We note that photolysis of **41** does lead to some fragmentation.[29] The yield of fragmentation is independent of [alkene] upon photolysis at 25 °C but drops markedly in the presence of alkene upon pyrolysis. Following Moss,[28] we believe that fragmentation is a diazirine excited state process at 25 °C but can arise from both precursor and carbene at higher temperature.

Results in our laboratory are in good agreement with a related study of the groups of Jones and Shevlin with cyclopropylmethylcarbene.[31]

The cyclopropylcarbene rearrangement was discovered by Friedman and Shechter in 1960 upon pyrolysis of a tosylhydrazone salt to give the mixture shown.[32]

Generation of cyclopropylmethylcarbene in like manner by Kirmse and Smith leads mainly to 1-methylcyclobutene along with vinylcyclopropane as a minor

Table 6. Absolute Prodct Yields in the Pyrolysis of **41** in CF$_2$ClCFCl2 (100 °C)[29]

TME(M)	% 42	% 44
0.0	64	0
0.036	64	4.9
0.067	65	8.7
0.135	64	14
0.261	64	20
0.506	60	26
0.948	57	32
1.94	57	37

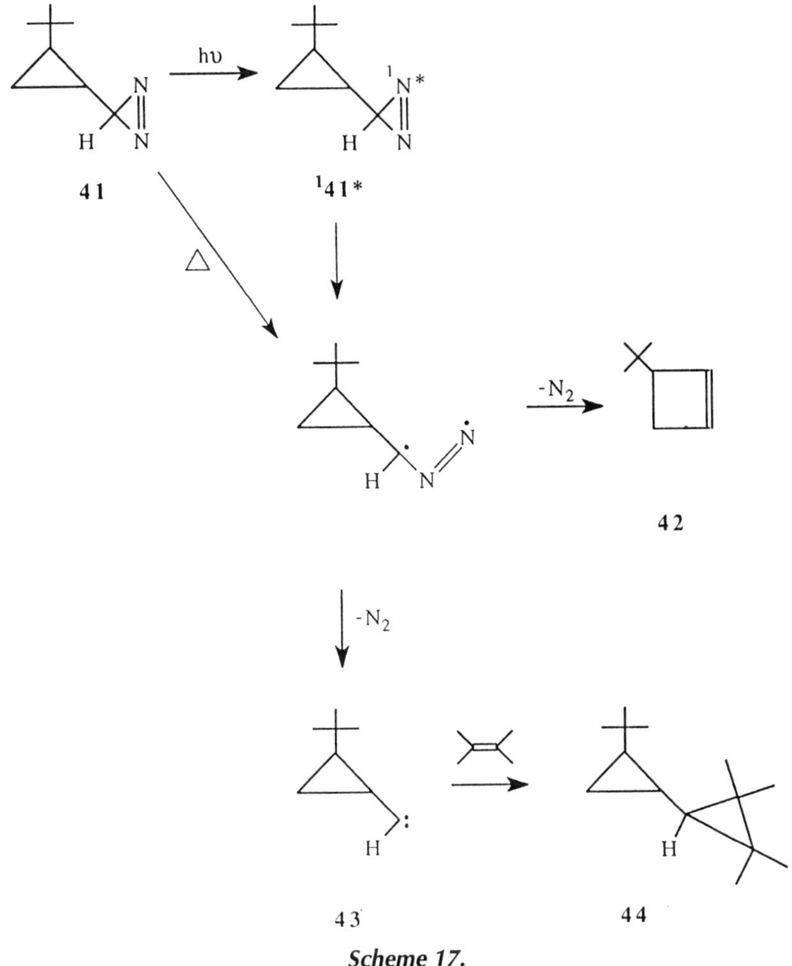

Scheme 17.

product.[33] However, Jones and Shevlin find that non-nitrogenous carbene precursors **45** and **46** lead to vinylcyclopropane as the major product.[31] Again, one can argue that the nitrogenous precursors suffer rearrangement and can form 1-methylcyolobutene by a pathway that bypasses the carbene. Presumably the

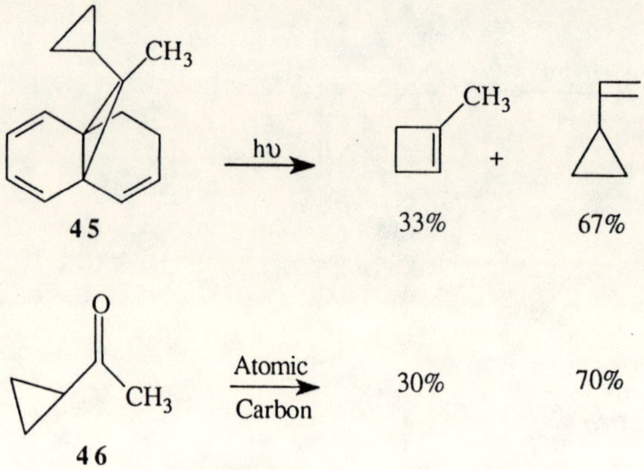

non-nitrogenous precursors **45** and **46** fail to suffer rearrangements in their excited
states as do their nitrogen containing counterparts.

IX. ISSUES OF CONFORMATION OF CYCLOPROPYLCARBENES

One can attempt to explain the data of Huang[29] and Jones and Shevlin[31] without
recourse to the intervention of precursor excited state. Cyclopropylcarbene can
exist in both a *syn* and an *anti* conformation. The *anti* form (A) is disposed to
rearrangement to cyclobutene, whereas the *syn* form (B) is inclined to fragment.[30]
Theory finds that the barrier to interconversion of these two conformers is quite
large. Thus it is unlikely that they will interconvert at a rate that is competitive with
the decay of the carbene.[30]

It is certainly true that the *anti* form must overcome a larger steric blockade in
its reaction with TME than the *syn* form. Thus, one can explain Huang's data by

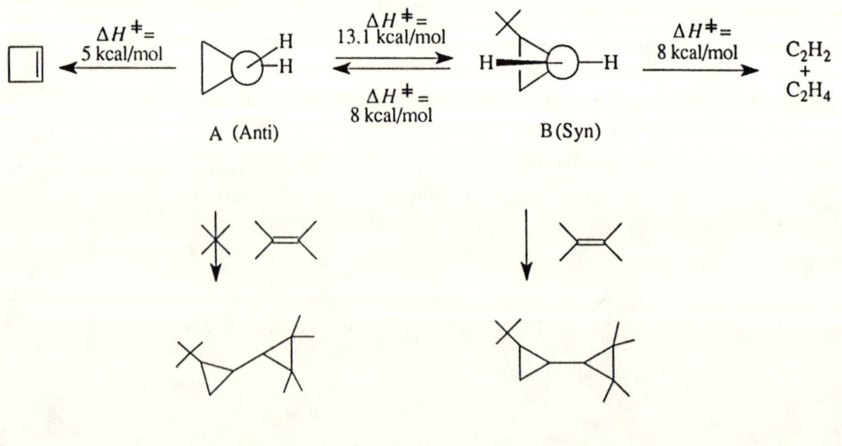

postulating that the *anti* form *always* rearranges and never reacts with TME for steric reasons. However, this explanation seems unlikely because of the findings of Moss et al.[28] In the chlorocyclopropyl system the yield of cyclobutene is depressed by roughly 50% by the presence of trimethylethylene. Thus Moss's data require that either *anti* chlorocyclopropylcarbene can react with alkene or that the *syn* conformation can rearrange to cyclobutene. As the latter process appears to be geometrically unlikely, one must concede that chlorocyclopropylcarbene is not so sterically blocked so as to preclude its reaction with trimethylethylene. Thus, by analogy, it seems unlikely that *anti-trans-tert*-butylcyclopropylcarbene is so sterically encumbered that it can not react with TME. The conformational argument also fails to explain why the ratio of **32/33**, produced on photolysis of **31**, changes in the presence of trimethylethylene (Schemes 14 and 15).

X. INTERMOLECULAR REACTIONS OF PRECURSOR EXCITED STATES

The complications we have discussed to this point have been restricted largely to intramolecular rearrangements with the exception of dicarbomethoxycarbene **29**.[26,27] Chambers and Jones, however, have seen the same type of effects in an intermolecular process.[34]

Table 7. The Spectrum of Products Realized upon Decomposition of **47**

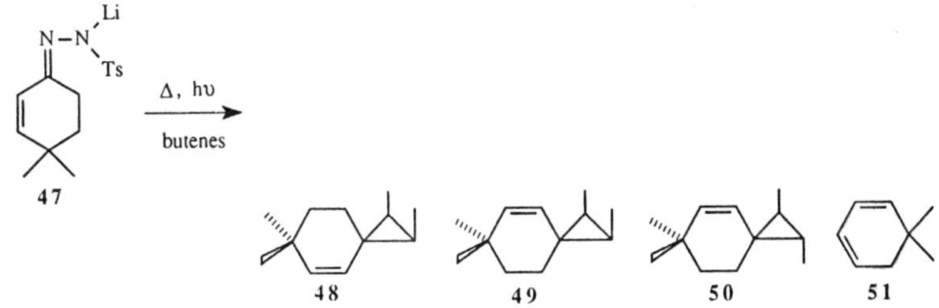

	Conditions	48	49	50	51
	hν, PH₂CO	29	22	14	35
	125°	34	23	13	30
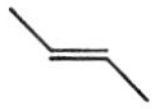	hν	0	0	35	65
	hν	23	25	0	52

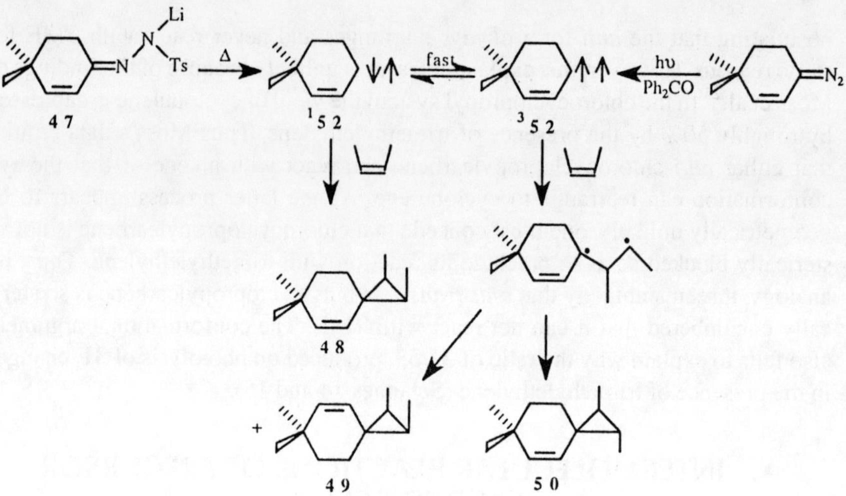

Scheme 18.

Once again, pyrolysis and triplet photosensitized photolysis of **47** in alkane solvent gives nearly the same mixture of products (Table 7). In this case, the products comprise adducts with the alkene (**48–50**) in addition to simple products like **51** deriving from intramolecular rearrangement. However, *direct* photolysis of **47** in *cis*-2-butene gives a totally different spectrum of products than does pyrolysis or triplet sensitized photolysis.

Chambers and Jones[34] postulated that pyrolysis and triplet photosensitized decomposition produced rapidly equilibrating singlet and triplet carbenes which react to form three cyclopropane adducts, in accord with the Skell–Woodworth[35] rules (Scheme 18).

Apparently, direct photolysis of **47** fails to give a spin-equilibrated carbene **52**. The reactive intermediate produced instead must react stereospecifically with *cis*-2-butene. Chambers and Jones associated this intermediate with a carbene excited state, but with hindsight it can also be attributed to an excited state of the diazo compound.

XI. BENZYLCHLOROCARBENE AND CARBENE–ALKENE COMPLEXES

Tomioka and Liu have obtained fascinating results with benzylchlorodiazirine **53**.[36] Photolysis of **53** in cyclohexane at 10 °C produced the two expected 2-chlorostyrenes (**54**, **E** and **Z**) (Scheme 19). Upon photolysis of **53** in the presence of an alkene, the yield of styrenes dropped, as expected, and the yield of cyclopropane adducts (e.g. **55**) increased.

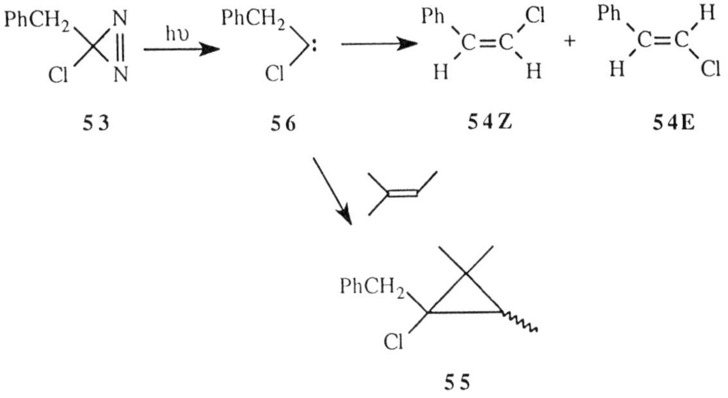

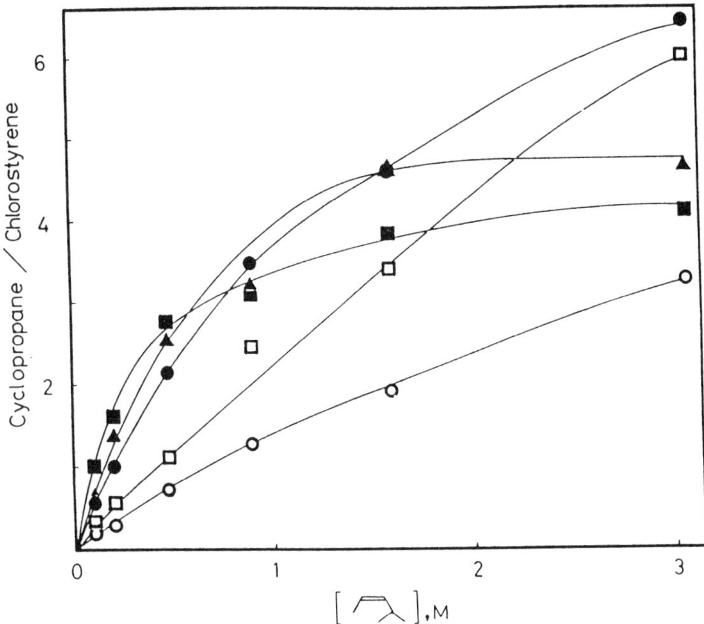

Scheme 19.

If the simple mechanism of Scheme 19 were operative, the ratio **55/(54E** plus **Z)** would increase smoothly with alkene concentration. Surprisingly this was not the case (see Figure 2). The data require a mechanism that is more complex than that shown above and must allow a second route to the formation of chlorostyrene. The

Figure 2. A plot of **55/(54E + 54Z)** versus trimethylethylene (reprinted with permission from reference 36a).

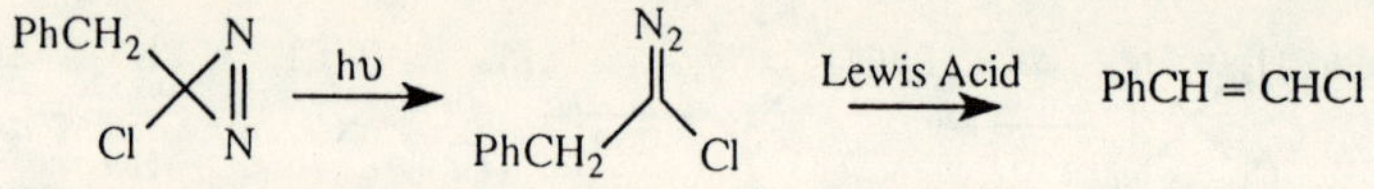

second intermediate could be a diazirine excited state ^{1}**53*** or an unstable diazo isomer **57** which can decompose by ionic mechanisms. Alternatively, Tomioka and Liu associated the second intermediate with carbene-alkene complex **58** (Scheme 20).

Additional evidence for a second intermediate was soon forthcoming. Liu et al. found that the *E/Z* ratio of chlorostyrenes **54** varied with alkene concentration. Again these observations were attributed to the intervention of a carbene–alkene complex. To explain the data the reasonable assumption was made that **58** forms styrenes in a different *E* and *Z* ratio than does the free carbene.[36]

Complexes of this type were postulated previously by Moss and Turro[37] to explain the negative activation energies observed in the reaction of chlorophenyl-carbene with alkenes. The kinetics required that any complex so formed must be

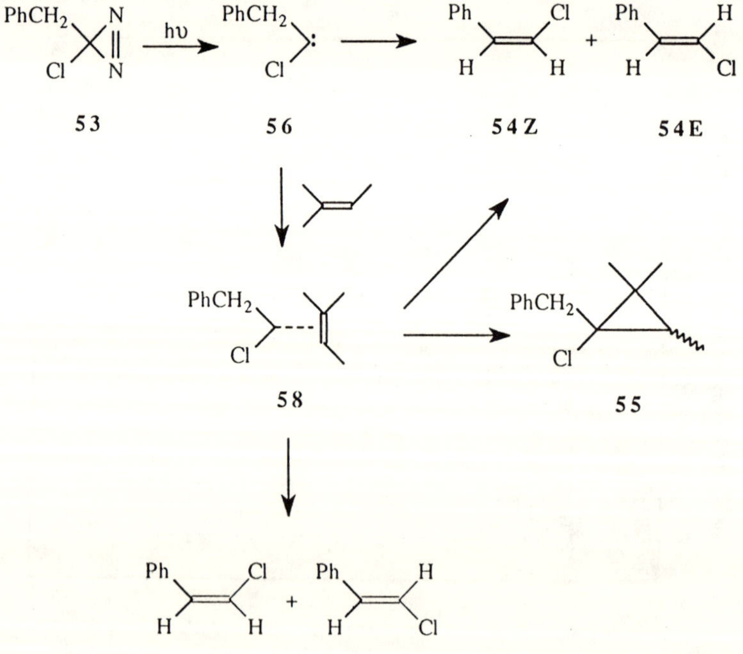

Scheme 20.

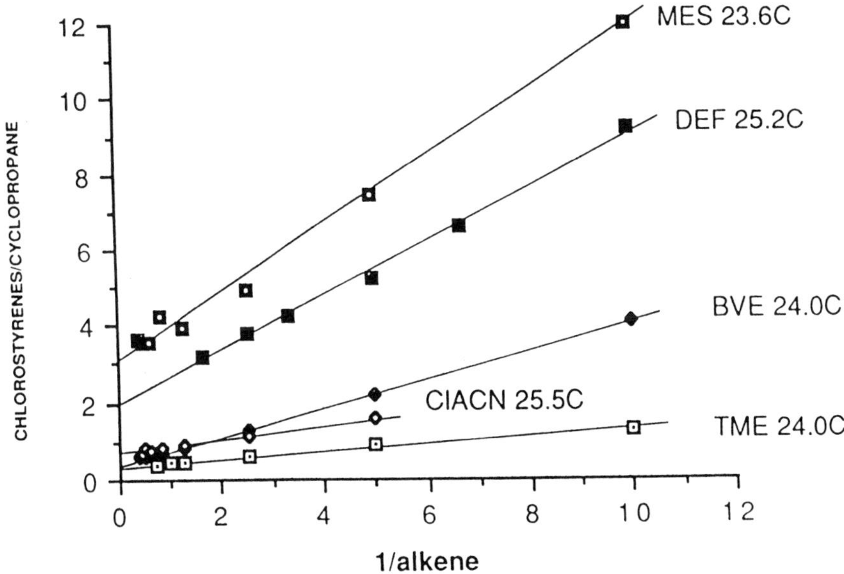

formed reversibly. The complex would be held together by charge transfer from the π electrons of the alkene to the empty orbital of the carbene as postulated earlier by Skell and Cholod[38] and by Hoffmann et al.[30c]

As predicted by this mechanism, the ratio of chlorostyrene/cyclopropane is alkene dependent (Figure 3).[39] This behavior will not be the case if the second product-forming intermediate is the diazirine excited state. However, if the second intermediate is the diazo compound, the results could vary with alkene and the diazo compounds' ability to undergo cycloaddition reactions with alkene.

Figure 3. Plots of chlorostyrenes/cyclopropanes versus alkene, with various alkenes (reprinted with permission from Reference 39).

Laser flash photolysis studies strongly support the need for a second pathway to beta chlorostyrene product. Liu and Bonneau have measured the absolute rate constants of rearrangements of benzylchlorocarbene **56** and of its reactions with alkenes.[39] According to Scheme 19, one should be able to suppress all intramolecular reactions of **56** in the presence of moderate concentrations of alkene. Yet this is not the case in the presence of alkenes, methanol and pyridine as traps.[40] Thus, one must posit complexes with these traps as well to explain the data.

Liu et al. have obtained curved Arrhenius plots for the rearrangement reactions of **56** and (**56-d$_2$**) in hydrocarbon solvents.[41] Curvature is not observed in chloroform. This led to the postulation of a carbene–alkane complex **59**. However, photolysis of diazirine precursor in isooctane over an 80 °C temperature range leads to no detectable CH insertion product. Thus, unlike the carbene–alkene complex, the putative carbene alkane complex **59** never leads to an insertion product but exclusively suffers rearrangement of hydrogen. In summary, Liu proposes that **56** forms a wide variety of complexes and that these intermediates have far-reaching consequences for product distributions and carbene kinetics.[36,39,41]

The alert reader will note that the cycloaddition data of Moss with chlorocyclopropylcarbene,[28] the data of Huang[29] with *tert*-butylcyclopropylcarbene, and the data of Chambers and Jones[34] all can be explained in terms of carbene–alkene complexes. Complexes, however, cannot explain the early data with ethylmethylcarbene and *tert*-butylcarbene.[6–10]

This author prefers the diazirine excited state mechanism to those involving carbene–alkene complexes. Although such complexes have often been postulated, *ab initio* theory finds no evidence for their existence.[42]

The second factor to influence my thinking is economy. Photolysis of diazirines and diazo compounds necessarily produces electronically excited states. Thus it is not *necessary* to invoke a carbene–alkene intermediate. Furthermore, the idea of rearrangements in diazirine excited states provides a unified mechanistic picture that satisfactorily explains the older gas-phase product studies data and the more recent solution-phase work. There are of course no alkenes present during the photolysis of ethylmethyldiazirine and *tert*-butyldiazirine and the idea of carbene–alkene complexation cannot explain the results obtained in these systems.

Sadly, nature is under no obligation to be economical. The more elaborate mechanism involving complexes may still be operative or the second product-forming

59

$$R = CH_2Ph$$

60   **61**

mechanism may vary with the structure of the precursor. There is no definitive evidence available at this time which disqualifies either mechanism.

This may soon change, however. What is needed are independent, non-nitrogenous, precursors of carbenes such as those reported by Johnson[43] (e.g. **60**) and by Jones[44] (e.g. **61**). If these precursors show the same alkene-dependent effects as do the diazirines then, the idea of carbene–alkene complexes is indisputable. If the effects disappear or change dramatically with precursor, however, then this concept is disqualified for at least this particular carbene.

XII. THE MECHANISM OF EXCITED STATE REARRANGEMENTS

Several times in this review, rearrangements in diazirine excited states have been posited but few details of this process have been put forth. One can fill this void with the help of theory.[45]

Theorists and spectroscopists agree that the characteristic UV absorption of diazirines above 300 nm is due to an n to π^* transition. This absorption is associated with a weak fluorescence.[46,47] At ca. 3 kcal/mol above the first excited state (S_1) lies the n to σ^*, S_2 state.

Fragmentation of S_1 (n to π^*) leads to excited states of both methylene and nitrogen. It is the S_2 state (n to σ^*) that correlates with the ground states of methylene and nitrogen and with isomerization to diazomethane. Thus, there must be an n to σ^*/n to π^* conical intersection as pointed out by Jug.[45b] As stated more recently by Yamamoto et al.: "For a direct photochemical pathway that starts on the n to π^* state and proceeds onto the n to σ^* state, the reaction path must curve so that the crossing becomes avoided and a transition state occurs."[45c] The conical intersection leading to the n to σ^* state resembles a diradical type of structure **62** (Scheme 21).

This picture explains the known temperature dependence of the fluorescence.[47] The fluorescence intensities of dialkyldiazirines increase as the temperature decreases. This result demonstrates that the fluorescence of S_1 competes against a thermally activated decay process. The apparent "activation energy" of this process is 2.8 kcal/mol, the energy required to reach the S_1, S_2 surface crossing.[45–47]

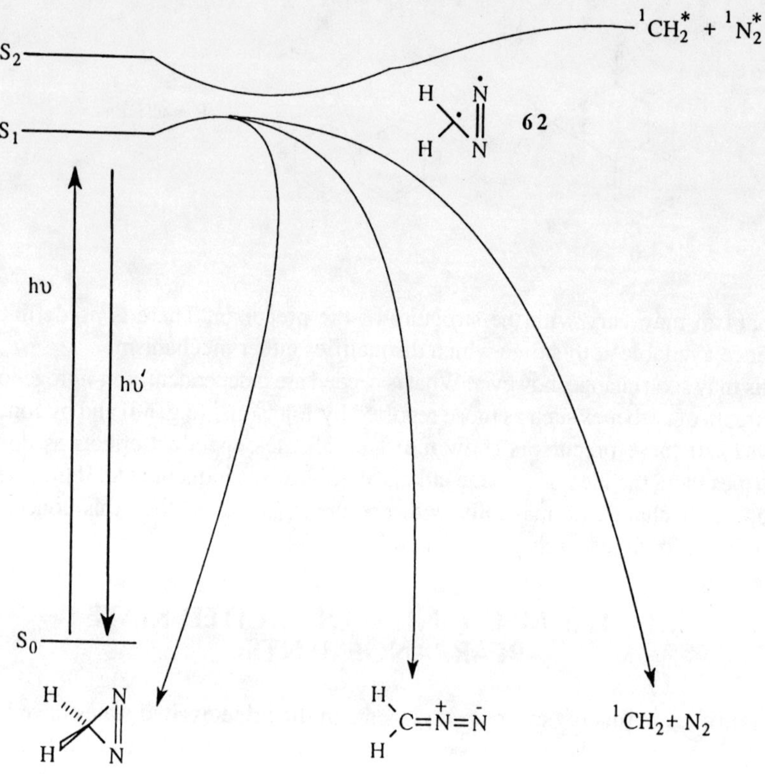

Scheme 21.

Ring opening of a diazirine to a diradical is hardly a revolutionary idea:

Azo compounds can fragment one C–N at a time,[48]

and cyclopropenes can ring open to form vinylcarbenes,[49]

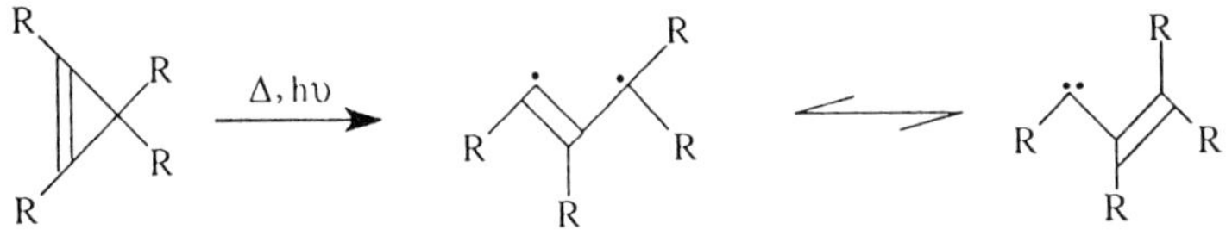

in addition to other processes.[50]

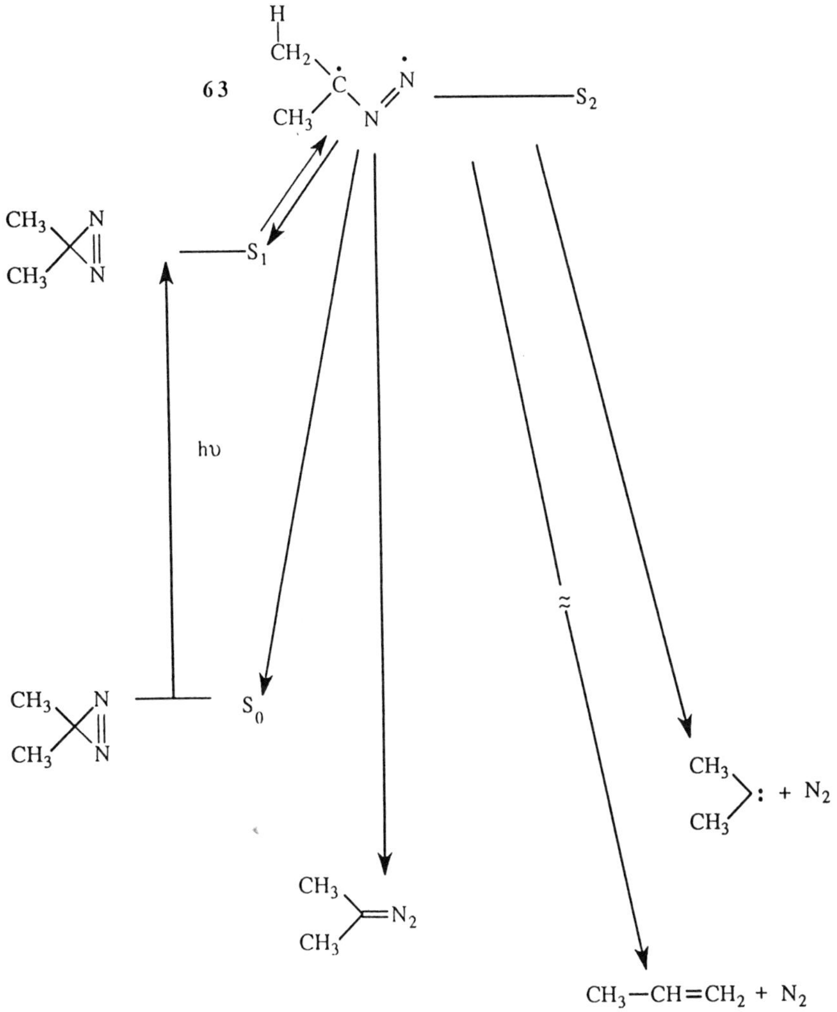

Scheme 22.

Our laboratory posits that the diradicals can suffer rearrangement in competition with the formation of diazirine and diazo compounds, and cleavage to carbene plus nitrogen.

Hydrogen migration to form ethylene in concert with nitrogen extrusion is certainly the most exothermic decay route available to diradical **63** (Scheme 22), derived from dimethyldiazirine or 2-diazopropane. If hydrogen shift is competitive with other, well-known decay routes, then the diradical can be the "second intermediate" needed to explain both the gas-phase and solution-phase data. It is important to note that the same diradical can be formed thermally or by activation of diazo precursors.

This concerted decay route needs to be studied by *ab initio* theory to provide it with more justification than the present convenience of economy. Hopefully this will soon be forthcoming by masters of the art.

Finally, we note that these carbon nitrogen diradicals can also explain the carbon migrations postulated in the case of homocubyldiazirine,[16]

and of cyclopropyldiazirines,[28,29]

as well as the fragmentation observed in this system. Diradicals may also intervene in Wolff rearrangements in the excited states of diazo compounds.[18,20–24]

XIII. PHOTOPHYSICAL AND PHOTOCHEMICAL EVIDENCE FOR EXCITED STATE REARRANGEMENT

Spectroscopic evidence has been advanced in favor of the excited state rearrangement mechanism by our laboratory, but we must admit that it is not overwhelming.[46,47]

According to our view, the fluorescence of diazirines is weak because it competes against other efficient excited state processes. Reducing the efficiency of the competitive processes by variation of diazirine structure should increase the fluorescence intensity. The observed order of fluorescence intensity shown is consistent with this expectation.[51] Adamantyldiazirine and parent diazirine, which cannot rearrange hydrogen in concert with nitrogen extrusion, fluoresce more brightly than does dimethyldiazirine which can rearrange.

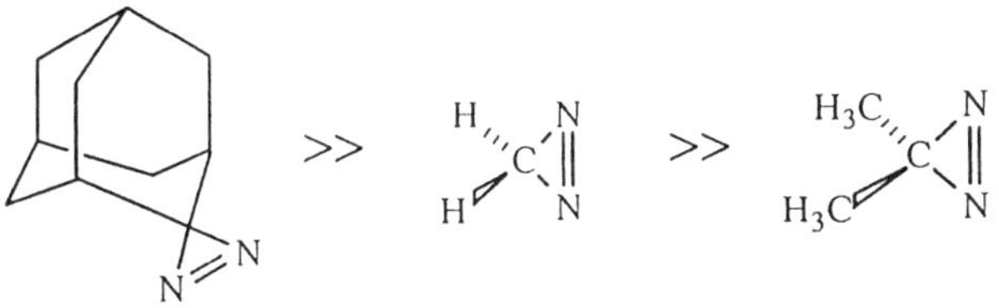

The alert reader will realize that fluorescence proceeds from S_1, but that rearrangement transpires in a diradical formed at or just after the S_1/S_2 surface crossing. This interpretation requires that the diradical must be able to reform the S_1 state in competition with hydrogen migration.

The same explanation is required to explain isotope effects on the fluorescence of dimethyldiazirine (DMD) and its perdeuterated isotopomer (DMD-d_6). Modarelli et al. have reported that DMD-d_6 is $55 \pm 20\%$ more fluorescent than is DMD at ambient temperature.[47] Furthermore, DMD-d_6 produces 50% more trappable carbene than does DMD, per laser pulse. Thus suppression of the excited state rearrangement by isotopic substitution improves the yield of the competitive processes (e.g. fluorescence and cleavage to carbene). Recently this work has been repeated and verified by Ford using more highly purified samples and a more sensitive fluorometer.[51]

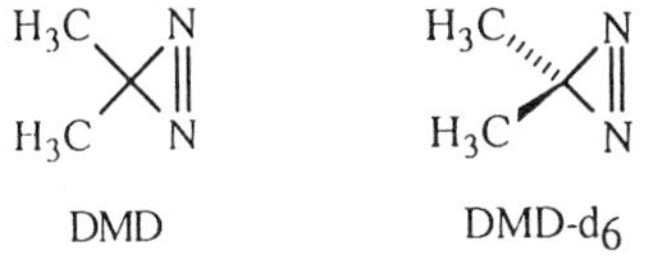

It is also possible to explain the isotope effects as a consequence of enhanced internal conversion (IC) in DMD-d_6 relative to DMD. However, Ford has performed the same type of experiments with diazirine and diazirine-d_2. Surely the isotope effects on IC will be greater in the simpler system. However, large isotope effects on the yield of fluorescence or the yield of carbene are not observed with diazirine and diazirine-d$_2$.[51]

XIV. KINETICS AND MECHANISMS OF CARBENE REARRANGEMENTS

In this chapter I have been primarily concerned with how carbenes are formed from precursors and whether they react as free species or as complexes. It would be wrong not to consider the kinetics and mechanisms of their rearrangements as well, an important thrust in the area of contemporary carbene research. The accompanying chapter of Toscano in this volume will elaborate on this subject in detail.

Let me close by saying that it is a very exciting time to be a carbene chemist. Alkyl and dialkylcarbenes were once thought to be so short-lived so as to be impossible to study. This is not the case. The lifetimes of all of the following species have now been measured in solution.[1,52] Some acylcarbene lifetimes have been deduced as well.[27]

Physical organic chemists know what to do next. In the next few years, we will gleefully measure isotope, temperature, solvent, and structural effects on the rate of carbene rearrangements. Dicyclopropylcarbene[52] and adamantanylidene[53] (Ad) have already been characterized by matrix UV-vis and IR spectroscopy. Undoubtedly this will be followed by more studies of dialkyl carbenes in matrices. One can no longer underestimate the power of theory. *Ab initio* theory can accurately predict the IR spectrum of a large carbene such as adamantanylidene. Thus one expects theory and spectroscopy to work together in the next few years to reveal the electronic ground states and geometries of simple carbene intermediates of fundamental interest.

Provocative experimental results are already in hand. It is clear that the 1,2 H shift is accelerated by polar solvents indicating that charge develops in the transition state.[1] Placement of alkyl groups adjacent to the migrating C–H bond accelerates rearrangement.[1] The 1,2 H shift is characterized by large negative entropies of activation, a result that is not predicted by theory.[1,54] Does this mean that quantum mechanical tunnelling and unusual dynamical effects are involved? The peculiar isotope effects that have been reported by Goodman also point in this direction.[55]

Thus, the realization that small carbenes of fundamental interest are accessible to direct experimental study means that there will be substantial focus on such systems in the next few years. Future progress will require the efforts of synthetic chemists, to prepare a new generation of precursors, and of kineticists, spectroscopists, and theorists. It seems inevitable that we are poised on the brink of major new discoveries and insights into the chemistry of alkyl- and dialkylcarbenes.

ACKNOWLEDGMENTS

The author gratefully acknowledges the hard work of his co-workers cited in the references and their enthusiasm for carbene chemistry. I am indebted to Professors Jones, Kirmse, Moss and Nickon for a critical reading of the manuscript and absolve them of all responsibility for remaining errors. Support of this work by the NSF is gratefully acknowledged.

REFERENCES

1. Platz, M.S.; Modarelli, D.A.; Morgan, S.; White, W.R.; Mullins, M.; Çelebi, S.; Toscano, J.P.; Huang, H. *Prog. React. Kinetics* **1994**, *19*, 93.
2. (a) Kirmse, W. *Carbene Chemistry*; Academic Press: New York, 1971. (b) Baron, W.J.; DeCamp, M.R.; Hendrick, M.E.; Jones, M., Jr.; Levin, R.; Huang, H.; Sohn, M.B. *Carbenes;* Jones, M., Jr.; Moss, R.A., Eds.; Wiley: New York, 1973, Vol. I, p 1.
3. Brinton, R.K.; Volman, D.H. *J. Chem. Phys.* **1951**, *19*, 1394.
4. a) Frey, H.M. *J. Chem. Soc.* **1962**, 2293. b) Kibby, C.L.; Kistiakowsky, G.B. *J. Phys. Chem.* **1966**, *70*, 126. c) Frey, H.M.; Stevens, I.D.R. *Proc. Chem. Soc.* **1962**, 79. d) Frey, H.M.; Stevens, I.D.R. *J. Chem. Soc.* **1962**, 3865.
5. a) Paulsen, S.R. *Angew. Chem.* **1960**, *72*, 781. b) Schmitz, E.; Ohme, R. *Tetrahedron Lett.* **1961**, 612. c) Schmitz, E. *Diazirines*; Liu, M.T.H., Ed.; CRC Press: Boca Raton, FL, 1987, Vol. I.
6. a) Mansoor, A.M.; Stevens, I.D.R. *Tetrahedron Lett.* **1966**, 1733. b) Frey, H.M.; Stevens, I.D.R. *J. Am. Chem. Soc.* **1965**, 3101. c) Frey, H.M. *J. Am. Chem. Soc.* **1962**, *84*, 2647.
7. Frey, H.M. *Adv. Photochem.* **1964**, *4*, 225.
8. Friedman, L.; Shechter, H. *J. Am. Chem. Soc.* **1959**, *81*, 5512.
9. Frey, H.M.; Stevens, I.D.R. *J. Chem. Soc.* **1965**, 3101.
10. Chang, K.-T.; Shechter, H. *J. Am. Chem. Soc.* **1979**, *101*, 5082.

11. a) Kirmse, W.; Wedel, W.v. *Ann. Chem.* **1963**, *666*, 1. b) Friedman, L.; Berge, J.G. *J. Am. Chem. Soc.* **1961**, *83*, 500. c) Goldstein, M.J.; Dolbier, W.R., Jr. *J. Am. Chem. Soc.* **1965**, *87*, 2293.

12. Fukushima, M.; Jones, M., Jr.; Brinker, U.H. *Tetrahedron Lett.* **1982**, *23*, 2212.

13. Fox, J.M.; Scacheri, J.E.G.; Jones, K.G.L.; Jones, M., Jr.; Shevlin, P.B.; Armstrong, B.; Szytrbicka, R. *Tetrahedron Lett.* **1992**, *33*, 5021.

14. a) Closs, G.L.; Coyle, J.J. *J. Org. Chem.* **1966**, *31*, 2759. b) Moritani, I.; Yamamoto, J.; Murahashi, S.-I. *Tetrahedron Lett.* **1968**, 5697. c) Moss, R.A.; Joyce, M.A. *J. Am. Chem. Soc.* **1977**, *99*, 1209. d) Abelt, C.J.; Pleier, J.M. *J. Am. Chem. Soc.* **1989**, *111*, 1795. e) Yamamoto, Y.; Murahashi, S.-I; Moritani, I. *Tetrahedron* **1975**, *31*, 2663. f) Overberger, C.G.; Anselme, J.-P. *J. Org. Chem.* **1964**, *29*, 1188.

15. a) Vander Stouw, G.G., Ph.D. Thesis, The Ohio State University, **1964**. b) Joines, R.C.; Turner, A.B.; Jones, W.M. *J. Am. Chem. Soc.* **1969**, *91*, 7754. c) Gasper, P.P.; Hsu, J.-P.; Chari, S.; Jones, M., Jr. *Tetrahedron* **1985**, *41*, 1479. d) Schissel, P.C.; Kent, M.E.; McAdee, D.J.; Hedaya, E. *J. Am. Chem. Soc.*, **1970**, *92*, 2147. e) Matzinger, S.; Bally, T.; Patterson, E.V.; McMahon, R.J. *J. Am. Chem. Soc.* **1996**, *118*, 1535.

16. a) Chen, N.; Jones, M., Jr.; White, N.R.; Platz, M.S. *J. Am. Chem. Soc.* **1991**, *113*, 4981. b) Eaton, P.E.; Appell, R.B. *J. Am. Chem. Soc.* **1990**, *112*, 4055. c) Eaton, P.E.; Hoffman, K.L. *J. Am. Chem. Soc.* **1987**, *109*, 5285. d) Hrovat, D.A.; Borden, N.T. *J. Am. Chem. Soc.* **1992**, *114*, 2719.

17. Eaton, P.E.; White, A.J. *J. Org. Chem.*, **1990**, *55*, 1321.

18. Roth, H.D. *Accts. Chem. Res.* **1977**, *10*, 85.

19. a) Scott, A.P.; Nobes, R.H.; Schaefer, H.F., III; Radom, L. *J. Am. Chem. Soc.* **1994**, *116*, 10159. b) Vacek, G.; Galbraith, J.M.; Yamaguchi, Y.; Schaefer, H.F., III; Nobes, R.H.; Scott, A.P.; Radom, L. *J. Phys. Chem.* **1994**, *98*, 8660. Please see these papers for the history of calculations pertaining to the Wolff rearrangement.

20. a) Kaplan, F.; Meloy, G.K. *J. Am. Chem. Soc.* **1966**, *88*, 950. b) Kaplan, F.; Mitchell, M.L. *Tetrahedron Lett.* **1979**, 759. c) Arnold, Z.J. *Chem. Soc., Chem. Commun.* **1967**, 299.

21. a) Moss, R.A.; Jones, M., Jr. *Reactive Intermediates*; Jones, M., Jr.; Moss, R.A., Eds.; Wiley: New York, 1981, p 61. b) Moss, R.A.; Jones, M., Jr. *Reactive Intermediates*; Jones, M., Jr.; Moss, R.A., Eds.; Wiley: New York, 1981, p 95.

22. a) De Minh, T.; Strausz, O.P.; Gunning, H.E. *J. Am. Chem. Soc.* **1969**, *91*, 1261. (b) Tomioka, H.; Okuno, H.; Izawa, Y. *J. Org. Chem.* **1980**, *45*, 5278.

23. a) Rando, R.R. *J. Am. Chem. Soc.* **1970**, *92*, 6706. b) Wulfman, D.S.; Poling, B.; McDaniel, R.S., Jr. *Tetrahedron Lett.* **1975**, 4519.

24. a) Hutton, R.S.; Roth, H.D. *J. Am. Chem. Soc.* **1978**, *100*, 4324. b) Murai, H.; Torres, M.; Strausz, O.P. *J. Am. Chem. Soc.* **1979**, *101*, 3976; **1980**, *102*, 1421; **1980**, *102*, 5104; **1980**, *102*, 7390; **1981**, *103*, 6422. c) Trozzolo, A.M. *Acc. Chem. Res.* **1968**, *1*, 329. d) Torres, M.; Safarik, I.; Murai, H.; Strausz, O.P. *Rev. Chem. Intermed.* **1986**, *7*, 243. e) Trozzolo, A.M.; Wasserman, E. *Carbenes*; Moss, R.A.; Jones, M., Jr., Eds.; Wiley: New York, 1975, Vol. 2, p 185. f) McMahon, R.J.; Chapman, O.L.; Hayes, R.A.; Hess, T.C.; Krimmer, H.-P. *J. Am. Chem. Soc.* **1985**, *107*, 7597. g) Hayes, R.A.; Hess, T.C.; McMahon, R.J.; Chapman, O.L. *J. Am. Chem. Soc.* **1983**, *95*, 7786. h) Schultz, R.; Schweig, A. *Angew. Chem. Int. Ed. Engl.* **1984**, *23*, 509. i) Murata, S.; Ohtawa, Y.; Tomioka, H. *Chem. Lett.* **1989**, 853. j) Torres, M.; Clement, A.; Strausz, O.P. *Rev. Chem. Intermed.* **1990**, *13*, 1. k) Bucher, G.; Sander, W. *J. Org. Chem.* **1992**, *57*, 1346. l) Roth, H.D. *J. Am. Chem. Soc.* **1972**, *94*, 1761. m) Roth, H.D. *J. Am. Chem. Soc.* **1971**, *93*, 1527. n) Roth, H.D. *J. Am. Chem. Soc.* **1971**, *93*, 4935. o) Roth, H.D. *J. Am. Chem. Soc.* **1972**, *94*, 1400. p) Roth, H.D.; *J. Am. Chem. Soc.* **1977**, *99*, 85.

25. a) Marchand, A.P.; Brockway, N.M. *Chem. Rev.* **1974**, *74*, 431. b) Meier, H.; Zeller, K.-P. *Angew Chem., Int. Ed. Engl.* **1975**, *14*, 32. c) Torres, M.; Lown, E.M.; Gunning, H.E.; Strausz, O.P. *Pure Appl. Chem.* **1980**, *52*, 1623. d) Zeller, K.-P. *Tetrahedron Lett.* **1977**, 707. e) Zeller, K.-P.; Meier, H.; Kolshor, H.; Müller, E. *Chem. Ber.* **1972**, *105*, 1875. f) Timm, U.; Zeller, K.-P.; Meier, H. *Tetrahedron*, **1977**, *33*, 453.

26. a) Jones, M., Jr.; Ando, W.; Hendrick, M. E.; Kulczycki, A., Jr.; Howley, P.M.; Hummel, K.F.; Malament, D.S. *J. Am. Chem. Soc.* **1972**, *94*, 7469. b) Kammula, S.C.; Tracer, H.L.; Shevlin, P.B.; Jones, M., Jr. *Org. Chem.* **1977**, *42*, 293.

27. a) Toscano, J.P.; Platz, M.S.; Nikolaev, V.; Popik, V. *J. Am. Chem. Soc.* **1994**, *116*, 8146. b) Wang, J.-L.; Toscano, J.P.; Platz, M.S.; Nikolaev, V.; Popik, V. *J. Am. Chem. Soc.* **1995**, *117*, 5477. c) Toscano, J.P.; Platz, M.S.; Nikolaev, V. *J. Am. Chem. Soc.* **1995**, *117*, 4712. d) Toscano, J.P.; Platz, M.S.; Nikolaev, V.; Cao, Y.; Zimmt, M.B. **1996**, *118*, 3527. e) Wulfman, D.S.; Poling, B.; McDaniel, R.S. ,Jr. *Tetrahedron Lett.* **1975**, 4519.

28. Ho, G.-J.; Krogh-Jespersen, K.; Moss, R.A.; Shen, S.; Sheridan, R.S.; Subramanian, R. *J. Am. Chem. Soc.* **1989**, *111*, 6875. Moss, R.A.; Ho, G.-J.; Shen, S.; Krogh-Jespersen, K. *J. Am. Chem. Soc.* **1990**, *112*, 1638.

29. Huang, H.; Platz, M.S. *Angew Chem., Int. Ed. Engl.* submitted.

30. a) Shevlin, P.B.; McKee, M.L. *J. Am. Chem. Soc.* **1989**, *111*, 519. b) Armstrong, B.M.; McKee, M.L.; Shevlin, P.B. *J. Am. Chem. Soc.* **1995**, *117*, 3689. c) Hoffmann, R.; Zeiss, G.D.; VanDine, G.W. *J. Am. Chem. Soc.* **1968**, *90*, 1485. d) Schoeller, W. W. *J. Org. Chem.* **1980**, *45*, 2161.

31. Private communication from Professors Jones and Shevlin.

32. a) Friedman, L.; Shechter, H. *J. Am. Chem. Soc.* **1960**, *82*, 1002. b) Smith, J.A.; Shechter, H.; Bayless, J.; Friedman, L. *J. Am. Chem. Soc.* **1965**, *87*, 659. c) Kaufman, G.M.; Smith, J.A.; Vander Stouw, G.G.; Huang, H.; Shechter, H. *J. Am. Chem. Soc.* **1965**, *87*, 935.

33. a) Kirmse, W.; von Bülow, B.G.; Schepp, H. *Liebigs Ann. Chem.* **1966**, *41*, 691. (b) Smith, J.A. *Dissertation Abstracts* **1965**, *xxv*, 4.

34. Chambers, G.R.; Jones, M., Jr. *J. Am. Chem. Soc.* **1980**, *102*, 4516.

35. Skell, P.S.; Woodworth, R.C. *J. Am. Chem. Soc.* **1956**, *78*, 4496, 3409, 5430.

36. a) Tomioka, H.; Hayashi, N.; Izawa, Y.; Liu, M.T.H.; *J. Am. Chem. Soc.* **1984**, *106*, 454. b) Liu, M.T.H.; Soundararajan, N.; Paike, N.; Subramanian, R. *J. Org. Chem.* **1987**, *52*, 4223. c) Liu, M.T.H.; Chishti, N.H.; Tencer, M.; Tomroka, H.; Izawa, Y. *Tetrahedron* **1984**, *40*, 887. d) Liu, M.T.H.; Suresh, R. *J. Org. Chem.* **1989**, *54*, 486. e) Liu, M.T.H.; Bonneau, R.; Wierlacher, S.; Sander, W. *J. Photochem. Photobio. A: Chem.* **1994**, *84*, 133. f) Wierlacher, S.; Sander, W.; Liu, M.T.H. *J. Am. Chem. Soc.* **1993**, *115*, 8943. g) Bonneau, R.; Liu, M.T.H. *J. Photochem. Photobio. A: Chem.* **1992**, *68*, 97.

37. Moss, R.A.; Turro, N.J. *Kinetics and Spectroscopy of Carbenes and Biradicals*, Platz, M.S., Ed., Plenum: New York, 1990, p 213.

38. Skell, P.S.; Cholod. *J. Am. Chem. Soc.* **1969**, *91*, 7131.

39. Liu, M.T.H.; Bonneau, R. *J. Am. Chem. Soc.* **1990**, *112*, 3915.

40. White, W.R., III; Platz, M.S. *J. Org. Chem.* **1992**, *17*, 2841.

41. Liu, M.T.H.; Bonneau, R.; Wierlacher, S.; Sander, W. *J. Photochem. Photobiol. A; Chem.* **1994**, *84*, 133.

42. Houk, K.N.; Rondan, N.G.; Mareda, *Tetrahedron* **1985**, *41*, 1555.

43. Chateauneuf, J.E.; Johnson, R.P.; Kirchoff, M. *J. Am. Chem. Soc.* **1990**, *112*, 3217.

44. a) Hartwig, J.F.; Jones, M., Jr.; Moss, R.A.; Lawrynowist, W. *Tetrahedron Lett.* **1986**, *27*, 5907. b) Glick, H.; Likhotvorik, I.R.; Jones, M., Jr. *Tetrahedron Lett.* **1995**, *36*, 5715.

45. a) Bigot, B.; Ponec, R.; Sevin, A.; Devaquet, A. *J. Am. Chem. Soc.* **1978**, *100*, 6573. b) Müller-Remmers, P.L.; Jug, K. *J. Am. Chem. Soc.* **1985**, *107*, 7275. c) Yamamoto, N.; Bernardi, F.; Bottoni, A.; Olivucci, M.; Robb, M.A.; Wilsey, S. *J. Am. Chem. Soc.* **1978**, *116*, 2064.

46. Buterbaugh, J.; Toscano, J.P.; Gustafson, T.; Platz, M. S., unpublished research at The Ohio State University.

47. Modarelli, D.A.; Morgan, S.; Platz, M.S. *J. Am. Chem. Soc.* **1992**, *114*, 7034–7041.

48. Engel, P.S. *Chem. Rev.* **1980**, *80*, 99.

49. a) Gajewski, J.J. *Hydrocarbon Thermal Isomerizations;* Wiley: New York, 1981, pp 22–25. b) York, E.J.; Dittmar, W.; Stevenson, J.R.; Bergman, R.G. *J. Am. Chem. Soc.* **1972**, *94*, 2882; **1973**, *95*, 5680.

50. Likhotvorik, I.R.; Brown, D.W.; Jones, M., Jr. *J. Am. Chem. Soc.* **1994**, *116*, 6173.
51. Ford, F.; Platz, M.S., unpublished research at The Ohio State University.
52. Modarelli, D.A.; Platz, M.S.; Sheridan, R.S.; Amman, J.R. *J. Am. Chem. Soc.* **1993**, *115*, 10, 440.
53. Bally, T.; Matzinger, S.; Truttmann, L.; Platz, M.S.; Morgan, S. *Angew Chem. Int. Ed. Engl.* **1994**, *33*, 1964.
54. a) Matzinger, S.; Fulscher, M.P. *J. Phys. Chem.* **1995**, *99*, 10747. b) Richards, C.A., Jr.; Kim, S.-J.; Yamaguchi, Y.; Schaefer, H.F., III *J. Am. Chem. Soc.* **1995**, *117*, 10104. c) Khodabadeh, S.; Carter, E.A. *J. Phys. Chem.* **1993**, *97*, 4360. d) Ma, B.; Schaefer, H.F., III *J. Am. Chem. Soc.* **1994**, *116*, 3539. e) Evanseck, J.D.; Houk, K.N. *J. Phys. Chem.* **1990**, *94*, 5518. f) Gallo, M.M.; Schaefer, H.F., III *J. Phys. Chem.* **1992**, *96*, 1515. g) Schaefer, H.F., III *Acc. Chem. Res.* **1979**, *12*, 28.
55. a) Dix, E.J.; Herman, M.S.; Goodman, J.L. *J. Am. Chem. Soc.* **1993**, *115*, 10424. b) Storer, J.W.; Houk, K.N. *J. Am. Chem. Soc.* **1993**, *115*, 10426.

PERSISTENT TRIPLET CARBENES

Hideo Tomioka

Advances in Carbene Chemistry
Volume 2, pages 175–214
Copyright © 1998 by JAI Press Inc.
All rights of reproduction in any form reserved.
ISBN: 1-55938-837-4

I. INTRODUCTION

Radical chemistry has had a unique and interesting history. During the late 1800s, most chemists felt that free radicals could not exist. This extreme view was upset in 1900 when Moses Gomberg announced the discovery of triphenylmethyl, the first moderately stable free radical.[1] Naturally, Gomberg's interpretation of his product as free triphenylmethyl was not generally accepted at first. Moreover, most organic chemists had little general interest in radical chemistry in these early years. This changed in the 1930s, when the idea of small free radicals as intermediates in high-temperature gas-phase reactions had been considerably developed and free radicals had been recognized as an important species in ordinary liquid-phase organic chemistry. Therefore, the "modern" history of radical chemistry is usually considered to have begun in the 1930s. However, Gomberg is still regarded as "Moses who had led chemists from the wilderness."[2] Despite Gomberg's statement that he wished to reserve research in this field for himself, in the next few years, papers on triphenylmethyl appeared from at least a dozen laboratories. In a modern context, triphenylmethyl combined the novelty of something like buckyballs with the controversial nature of something like polywater.[3] Thus, Gomberg's radicals have constantly and uniquely developed since their discovery and have still received a good deal of attention because of their stability and ambient reactivity. Furthermore, they are now attractive as potential units for organic ferromagnet.[4]

While the pioneering work of Staudinger in the 1910s on the decomposition of diazo compounds and ketenes contributed much to the recognition of carbenes as a new reactive species, carbene chemistry started by their being generally regarded as diradicals, since radicals were the intermediates *en vogue* in the following decades.[5] Therefore, it might not be very surprising that carbene chemistry has developed without the upsetting discovery equivalent to Gomberg's radicals. However, it is rather surprising to note that stable carbenes have attracted little attention until quite recently when two groups have announced the discovery of these species, i.e., phosphinocarbene as a distillable red oil[6] in 1988 and imidazol-2-ylidene as a crystal with a high melting point[7] in 1991. These carbenes are, however, stabilized by substituents bearing a heteroatom connected to the carbenic atom, and thus have a singlet ground state. Moreover, their electronic configurations are still a topic of debate.[8] Since triplets resemble radicals in terms of reactivity and electronic properties, one may naturally consider that stable triplet carbenes must be the counterparts of Gomberg's radicals. It is then extremely surprising to note that when we started to work in this field in 1990, no papers concerning intentional attempts to generate stable carbenes in triplet states under normal conditions has appeared in the literature, except one by Zimmerman and Paskovich in 1964.[9]

Motivated by this rather naive overview of the historical background, we have initiated efforts to stabilize triplet carbenes to the extent that they are able to survive under normal conditions. However, as one may easily expect from their highly reactive nature with only two modifiable substituents, this turned out to be a very challenging project. For instance, the lifetime of triplet diphenylcarbene in solution at room temperature is two orders of magnitude smaller than that of diphenylmethyl and is even shorter-lived than methyl radical. At present, we are still not able to obtain a triplet carbene that can survive, even in a dilute solution for, say, a period of one hour. This, therefore, is a progress report and not the final summation of our efforts to isolate triplet carbenes.

II. RELATIONSHIP BETWEEN STRUCTURE AND GROUND STATE MULTIPLICITIES

In order to design and generate persistent triplet carbenes, we need to know the relationship between structure and stability of both carbene multiplicities. The carbene carbon is linked to two adjacent groups by covalent bonds, and possesses two nonbonding electrons that may have antiparallel spins (singlet state) or parallel spins (triplet state). If the carbene unit were linear, it should have two degenerate p orbitals, and Hund's first rule would predict a triplet ground state. In contrast, if the carbene unit is not linear, the two orbitals will become different. The orbital perpendicular to the plane defined by the three atoms is designated as p, while that parallel to this plane is called σ. The σ orbital will acquire s character and thereby become stabilized, while the p remains largely unchanged. Actually, most carbenes are not linear and the ground state multiplicity depends upon the relative energy of the singlet and triplet states. The four lowest energy configurations of a carbene have an electronic configuration described as $\sigma^1 p^1$, σ^2, or p^2. The electron spins in the $\sigma^1 p^1$ configuration may be paired, a singlet, or parallel to form a triplet, while the σ^2 and p^2 configurations must be electron-paired singlets. Thus, the triplet state has a $\sigma^1 p^1$ ($^3B_1$) configuration, while the σ^2 ($^1A_1$) is generally thought to be the lowest energy configuration for the singlet (Figure 1).

In a singlet state σ^2 or p^2 carbene, the electron–electron coulomb repulsion would be severe, since two electrons are constrained to the same small molecular orbital (MO). On the other hand, the triplet configuration is stabilized by relief of the coulomb repulsion and "exchange repulsion," although one pays a price for separation of the electrons into different MOs. Thus, the magnitude of the energy

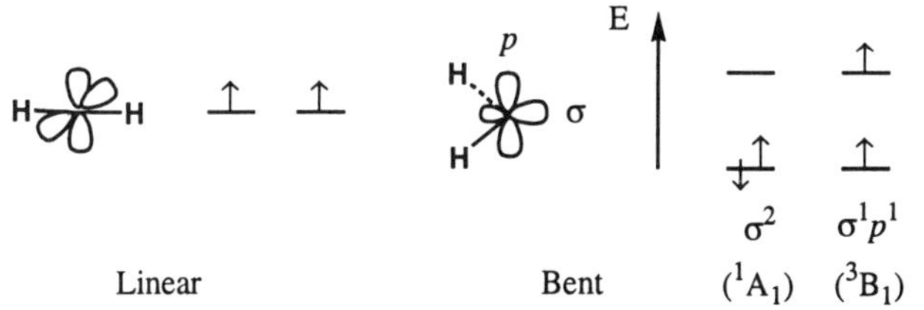

Figure 1. Linear methylene with two degenerate p orbitals and bent methylene with σ and p orbitals.

difference between the triplet and singlet states (the singlet–triplet splitting, ΔG_{ST}) is roughly equal to the electron–electron repulsion minus the energy required to promote an electron from the σ to the p-nonbonding orbital. In other words, as the energy separation between σ and p states increases, the promotion energy becomes large enough to overcome the repulsion energy, while if the spacing is small, the species will still have a triplet ground state.[10] The small difference between the energies of S_o and T_1 may easily be overturned by the effects of substituents on the carbene center. The factors that influence the spacing can be analyzed in terms of electronic and steric effects.

A. Electronic Effects

Because of more favorable overlap, the interaction of the carbon $2p$ orbital with the p or π orbitals of the substituent(s) is expected to dominate. The σ (sp^n) orbital, which lies in the nodal plane of the p or π orbital of the substituent, will not interact, except more weakly with the σ orbitals of the substituent. Its energy is thus mostly unperturbed by the substituent(s).

According to Fleming, substituents interacting with a π system can be classified into three classes: X (π-electron donors such as -NR$_2$, -OR, -SR, -F, -Cl, -Br, and -I), Z: (π-electron acceptors such as -COR, -SOR, -SO$_2$R, -NO, and -NO$_2$), and C: (conjugating such as alkenes, alkynes, or aryl groups).[11]

As shown in Figure 2(a), substituent of type X, which has a p orbital or other suitable doubly occupied orbitals that will interact with the π bond, raises the $2p$ orbital of the carbene, thereby increasing the separation of the $2p$ and sp^n (σ) orbitals. The ground state of an X-substituted carbene becomes a singlet. Many carbenes in this class are known. The most familiar ones are the halocarbenes. In :CHF, for instance, S_o is lower than T_1 by 14.7 kcal/mol.[12] In :CF$_2$, the separation is 56.6 kcal/mol (see also Table 2).[13]

On the other hand, Z and C substituents having a p or π^* orbital and evenly spaced π and π^* orbitals, respectively, either lower the $2p$-sp^n gap or leave it about the same as shown in Figures 2(b) and 2(c). In either case, the ground state for these carbenes is expected to be T_1, although the magnitude of ΔG_{ST} may vary. It has been demonstrated by EPR studies that most aryl and diarylcarbenes have triplet ground states.[10,14]

B. Steric Effects

The magnitude of ΔG_{ST} is expected to be sensitive to the carbene–carbon bond angle. A linear carbene has two degenerate p orbitals, which is calculated to provide the maximum value of ΔG_{ST}. Bending of the R–C–R bond angle of the carbene removes the orbital degeneracy and reduces ΔG_{ST}. As the carbene–carbon bond angle is further contracted, the σ-orbital picks up more s-character and conse-

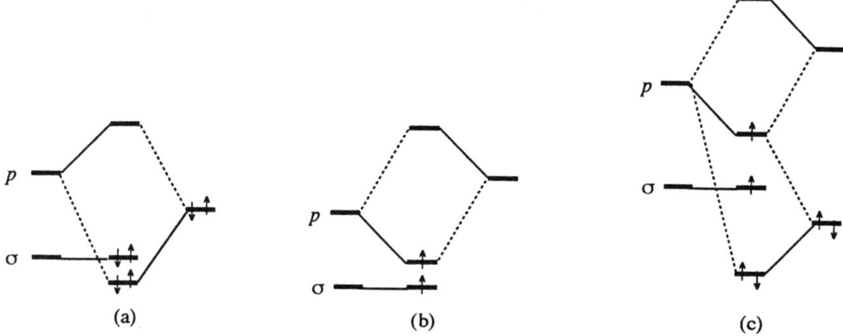

Figure 2. A carbene center interacting with (a) an X substituent, (b) a Z substituent, and (c) a C substituent.

quently moves even lower in energy. The smaller the bond angle, the more energy it takes to promote an electron from the σ to the p-orbital and the smaller ΔG_{ST} becomes.

This is shown more quantitatively by calculations for methylene. The calculations predict that the energy of singlet methylene will drop below that of the triplet state for carbenes with bond angles less than about 90°. On the other hand, theory also suggests that opening of the central angle strongly destabilizes the singlet state, but requires very little additional energy for the triplet, thus making ΔG_{ST} larger.[15] In accordance with this prediction, cyclopropenylidene has been shown to have a singlet ground state,[16] while diadamantylidene has a triplet ground state.[17] Although the bond angle effect is not always easily separated from other electronic effects, the ground state multiplicities of these two carbenes are principally determined by the central bond angle.

III. STRUCTURE–REACTIVITY RELATIONSHIPS

The persistence of a carbene depends on its environment. For instance, the lifetime of a triplet carbene is dramatically decreased by minor impurities in the surrounding medium, e.g., a trace of oxygen. Therefore, the persistence of a carbene should be quantitatively described under specified experimental conditions by the rate constant for the bimolecular or unimolecular process by which the carbene decays. Although laser-flash photolysis has taken its place among the standard tools that can be brought to bear on the study of carbenes in recent years,[18] few systematic data from the studies along the lines suggested above are presently available. Moreover, the persistence of a carbene is complicated by the presence of two electronic states that typically differ greatly in reactivity and lie close in energy. Most carbenes react from the singlet state regardless of the ground state multiplicities. Singlet carbenes can show both electrophilic and nucleophilic characteristics in their reactions, especially with alkenes.[19] Predominant reactions of ground-state triplet carbenes having a small ΔG_{ST} are often not observed with some substrates, even when they are generated selectively by triplet-sensitized photolysis. This is interpreted to mean that the rate of the singlet reaction is much faster than that of the triplet. The reaction from the triplet state can only become dominant when efficient triplet quenchers such as oxygen or excellent hydrogen donors are employed as substrates.

Thus, in this section we try to give an overview of how those electronic and steric effects are operating to control the reactivity of each multiplicity of the carbenes based on the kinetic data thus far reported.

A. Singlet Carbenes

The effects of structure on the reactivity of arylhalocarbenes (ArCX) toward alkenes have been systematically studied by measuring absolute rate constants (Table 1). It has been shown that the order of absolute XCPh reactivity, determined by the magnitude of k_q, is Br > Cl > F not only toward electron-rich alkenes (e.g., $Me_2C{=}CMe_2$)[20a] but also toward electron-poor alkenes (e.g., $CH_2{=}CClCN$).[20b] The observed selectivity ordering (F > Cl > Br) is that anticipated for halogen lone pair resonance stabilization of an adjacent, vacant, $2p$ orbital of a singlet carbene. Reactivity/selectivity relations are also studied in the additions of $p\text{-}X\text{-}C_6H_4CCl$ to alkenes and the order of ArCCl reactivities depends on the carbene's *para* aryl substituents X : MeO < Me < H < Cl < CF_3.[20c] As in the case of substituent variations directly at the carbenic center, resonance donation by the aryl substituent decreases carbene reactivity. Simply interpreted, one might argue that the donor-substituted ArCX are thermodynamically more stabilized.[20,21]

Table 1. Effect of Substituents on Rate Constant (k_q) for Addition of Haloarylcarbenes to Alkenes[20]

		k_q (M⁻¹s⁻¹)	
$Ar\overset{\cdot\cdot}{\frown}X$		$Me_2C{=}CMe_2$	$CH_2{=}CClCN$
Ph	F	1.6×10^8 [a]	1.2×10^8 [b]
Ph	Cl	2.8×10^8 [a]	2.1×10^8 [b]
Ph	Br	3.8×10^8 [a]	3.3×10^8 [b]
$p\text{-MeOC}_6H_4$	Cl	1.4×10^7 [c]	—
$p\text{-MeC}_6H_4$	Cl	1.2×10^8 [c]	—
$p\text{-ClC}_6H_4$	Cl	3.3×10^8 [c]	—
$p\text{-CF}_3C_6H_4$	Cl	1.5×10^9 [c]	—

Notes: [a]Cox et al.[20a]
[b]Moss et al.[20b]
[c]Moss et al.[20c]

Thermodynamic parameters such as heat of formation (ΔH_f), carbene stabilization energy (ΔE_{stab}), and ΔG_{ST} have been estimated by theoretical calculations for a series of singlet ground state carbenes. It is thus quite intriguing to correlate these data with the experimental "lifetime" data (Table 2). It should be noted that the theoretical values depend highly on the level of the calculation, while the "lifetime" is sensitive to the environment as well as decay pathways. Nevertheless, the data summarized in Table 2 exhibit surprisingly good correlations, and again completely support the idea that singlet carbene is thermodynamically more stabilized as substituent π donation increases. The lifetime of dimethoxycarbene, which is one of the most stabilized carbenes, is, for instance, two orders of magnitude greater than that of dichlorocarbene, but it does not survive over 10 ms. In this light, imidazol-2-ylidenes (IM) are exceptionally stable; 1,3-dimethylimadazol-2-ylidene (IM:R=Me, R′=H) is shown to be stable in solution at room temperature and the tetramethyl analogue (IM:R=R′=Me) is isolated as stable crystals that melt above 100 °C.[7e] A variety of spectroscopic measurements as well as high-level quantum mechanical calculations have shown that the π electrons of the carbon–carbon double bond are localized on those two atoms, and each nitrogen lone pair is delocalized on its nitrogen, suggesting that π-electron delocalization around the ring is not an important feature. The unusual stability of the imidazolylidene is

Table 2. Thermodynamic Data and Lifetimes of Several Singlet Carbenes

Carbene	$\Delta G_{ST}{}^{a(Ref.)}$ (kcal/mol)	$\Delta H_f{}^{b}$ (kcal/mol)	$\Delta E_{stab}{}^{c}$ (kcal/mol)	τ^{d}(Ref.)	Solv	Decay Kinetic
:CCl$_2$	−13.5(22)		27	10 μs(25)	Cyclohexane	
:CF$_2$	−56.6(13)	−46	63	~2 ms(26)	Gas	2
PhĊOMe				15–30 μs(27)	3-Metylpentane	1
ClĊOMe		−22	60			
FĊOMe		−53	74	~1 ms(23)	Pentane	2
(MeO)$_2$C:	−77(23)	−61	80	~2 ms(28)	Pentane	2
MeOĊNMe$_2$			93	—		
(IM)	−79.4(7d)			∞(7)		

Notes: [a]Estimated by *ab initio* calculations.

[b]Heat of formation determined either empirically or semiempirically, Du et al.[23]

[c]Calculated by the isodesmic reaction ($CH_2 + CH_3X + CH_3Y \rightarrow CXY + 2CH_4$) for carbene CXY, Rondan et al.[24]

[d]Lifetime in solution at room temperature.

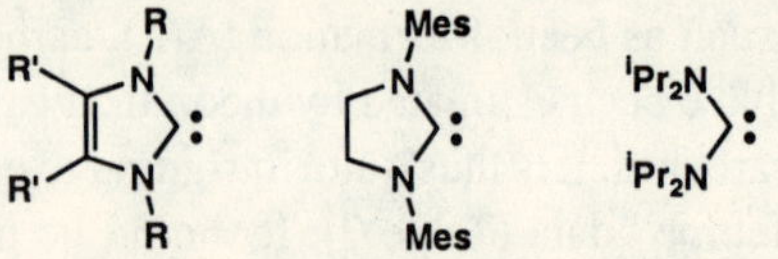

$$\text{IM} \qquad \text{IMH}_2 \qquad \text{IPC} \qquad \text{PC}$$

largely attributed to the accumulated electron density on the nitrogen centers as well as the $C_4=C_5$ double bond that surrounds the carbene center, protecting the imidazolylidene from nucleophilic additions and kinetically stabilizing the molecule.[7g] More recently, however, imidazolin-2-ylidene (IMH_2)[29] and even bis(diisopropylamino)carbene (IPC)[30] have been isolated as stable crystals. This indicates that unsaturation in the imidazole ring is not required to produce a stable nitrogen-substituted carbene. Moreover, it has been shown that there is as much double-bond character to the C–N bonds, especially in bis(diisopropylamino)carbene, as those in the corresponding amidinium ion. Phosphinocarbene (PC) has also been isolated as a distillable oil and shown to react either as a carbene or a phosphaacetylene.[6a] More recent theoretical and X-ray crystallographic studies suggest that the molecule is best described as a phosphaacetylene.[6e] Thus, it is reasonable to suppose that most carbenes that are stabilized as singlets by substituents bearing heteroatoms are at least partially ylids, and the ylidic character depends on the donating ability of the substituents.

B. Triplet Carbenes

Electronic effects on triplet carbene reactivities are not straightforward compared to those on the singlets. Effects of a wide range of p-substituents on the reactivities of diarylcarbenes have been studied (Table 3).[31] Singlet reactions (CH_3OH quenching) are shown to be sensitive to substituent effects, while there is little effect on characteristic triplet reactions (H-atom abstraction). Presumably due to the radical nature of the triplet, carbene substitution has only a minimal effect on their decay lifetimes and only for those cases where considerable delocalization into the substituent is possible (e.g., R=Ph) is stability of the triplet enhanced. On the other hand, the reactions with methanol show a more distinct substituent dependence, where carbenes bearing electron-releasing groups exhibit accelerated rates, while electron-withdrawing groups retard the reaction.

The insertion reaction of carbenes with alcohols to form ethers is usually considered to occur from the singlet states of these species.[10,32] However, time-resolved experiments have shown that the ground triplet state of diphenylcarbene is efficiently quenched by methanol. Pre-equilibrium and surface-crossing mechanisms have been proposed for this reaction. According to the pre-equilibrium mechanism,[33] triplet to singlet interconversion (k_{TS}) occurs prior to O–H insertion of the singlet carbene (k_S), Eq. (1). On the other hand, Griller et al. suggested that triplet carbenes can react

Table 3. Effect of *Para*-Substituents on Reactions of Diphenyl Carbenes (4-RC$_6$H$_4$ÇC$_6$H$_4$R′-4′)[31]

R	R′	τ (μs)[a]	k_q (10^{6}M^{-1}s^{-1})[b]	E_a(kcal/mol)[c]	log A(M^{-1}s^{-1})[c]
H	H	1.5	12	1.45	8.35
Br	Br	1.6	2.0	2.02	8.33
Me	Me	1.2	104	1.52	9.51
Me	CN	3.8	6.2	6.55	11.30
H	CN	5.4	0.87	6.13	10.46
H	CO$_2$Me	5.0	1.3	6.19	10.60
H	Ph	7.8	7.9	—	—

Notes: [a]Reciprocal of the rate of growth of the corresponding benzhydryl radicals in cyclohexane.
[b]Bimolecular rate constants for the reaction with methanol at 300 K in cyclohexane.
[c]Arrhenius parameters for methanol quenching.

$$^3R_2C: \underset{k_{ST}}{\overset{k_{TS}}{\rightleftharpoons}} \, ^1R_2C: \underset{R'OH}{\overset{k_S}{\longrightarrow}} R_2CHOR' \tag{1}$$

directly with alcohols, with surface crossing after the carbene has begun to interact with the O–H bond.[34] According to Eq. (1), the observed rate constant for the reaction of a triplet carbene with alcohol is

$$k_q = \frac{k_{TS}k_S}{k_{ST} + k_S\,[ROH]} \tag{2}$$

At relatively low concentrations of alcohol, with $k_{ST} > k_S$ [ROH], k_q reduces to

$$k_q = k_S K, \text{ where } K = \frac{k_{TS}}{k_{ST}} = \frac{[S]}{[T]} \tag{3}$$

Equation (3) correlates k_q with the magnitude of the singlet–triplet energy gap, $\Delta G_{ST} + RT\ln K$. Electron-donating substituents generally reduce the magnitude of ΔG_{ST} while electron-accepting substituents should enhance it; k_q in Table 3 seems to vary accordingly. However, based on the assumption that the singlet DPC could be stabilized by either a single electron-donating or a single electron-withdrawing group interacting with the empty or filled orbital,[35] respectively, of the singlet carbene, along with the significant difference in substituent effects on the Arrhenius parameters between the first three carbenes listed in the table and the last three, Platz et al. propose that they are reacting by a different mechanism. Thus, the data for the first three are consistent with the surface-crossing mechanism, while the data for the last three agree with the classical pre-equilibrium mechanism.[31]

The two phenyl rings of triplet diphenylcarbene do not lie in the same plane. This is most likely true for singlet diphenylcarbene as well. Thus, interpretation of effects of ring substituents on the reactivities of diphenylcarbene is accompanied with some complexity associated with this flexibility. In order to get more quantitative insight into the effects of ring-substituents on aromatic carbenes, the reactivities of a series of cyclic aromatic carbenes incorporated into a presumably planar ring of five or six atoms have been investigated.[36] The carbene bond angle of those examples containing a five-membered ring, for instance, is expected to be constant and the diversity of chemical behavior must thus be associated primarily with electronic changes. Physical and chemical properties of these aromatic carbenes are summarized in Table 4, which indicates that the diversity is substantial. It is clear that electron donating and withdrawing groups have an opposite influence on the magnitude of ΔG_{ST}, so that the substituents control the chemical properties of arylcarbenes.

For instance, with mesitylbora-anthrylidene (BA), which is at one extreme among those studied, k_{ST} is estimated to be $> 10^{10}$ s^{-1} based on the observations that ^{1}BA reacts with ^{i}PrOH with a rate constant five times slower than the diffusion limit. Its lifetime is too short (< 100 ps) to react with a low concentration of α-methylstyrene, while an upper limit of k_{TS} is estimated to be $\leq 2 \times 10^6$ s^{-1} from the measured rate constant for reaction of ^{3}BA with ^{i}PrOH ($3.4 \pm 0.9 \times 10^6$ M^{-1} s^{-1}) in conjunction with the absence of ether formation from the triplet-sensitized irradiation. These values give $K_{eq} \geq 4000$ and thus $\Delta G_{ST} \geq 5.2$ kcal/mol. Most of the bimolecular reactions of the triplet carbene are faster than the formation of the singlet from the triplet. Thus, intersystem crossing from ^{1}BA to ^{3}BA is irreversible, and if ^{3}BA is formed first, then there will be no reaction at all of ^{1}BA. Some reaction of ^{1}BA occurs when the rate of this process is fast in comparison with that of intersystem crossing (k_{ST}).

On the other hand, another extreme carbene 9-xanthylidene (XA) does not react measurably with oxygen; the lifetime of ^{1}XA is the same both in O$_2$-saturated and deoxygenated cyclohexane. K_{eq} in this case is determined from k_{O_2} [O$_2$] divided by the rate difference between the rate constant for reaction of ^{1}XA in oxygen saturated solution and that in the absence of oxygen, where the triplet carbene is presumed to react irreversibly with O$_2$ with a rate constant close to the diffusion limit ($k_{O_2} \approx k_{diff}$). Since no significant difference in these rate constants can be detected, the estimate of K_{eq} is a limit set in part by the experimental uncertainty in k_{obs}. This value $K_{eq} \geq 1 \times 10^{-4}$ corresponds to $\Delta G_{ST} \geq -5$ kcal/mol, indicating that the singlet carbene is lower in energy than the triplet. Thus, conversion of ^{1}XA to its higher energy triplet does not occur to a significant extent before the singlet carbene is consumed.

Other aromatic carbenes fall into intermediate positions in this range. For these cases, reactions characteristic of the two spin states are observed since significant concentrations of both can be present simultaneously in the equilibrium mixture.

Table 4. Substituent Effects on Properties of Diarylcarbenes[36]

Carbene	$K_{eq}{}^a$	$\Delta G_{ST}{}^b$ (kcal/mol)	Lifetime^c (Ref.)	Solvent
BA	≥4000	≥5.2	250 μs (37a)	PhH
DCFL	1000	4.1	~10 ns (37b)	MeCN
DPC	200	3.2	1.5 μs (31)	$c\text{-}C_6H_{12}$
FL	20	1.9	1.7 ns (37c)	MeCN
BFL	5	1.0	260 ps (37d)	$c\text{-}C_6H_{12}$
DMFL	~0.03	~ −2	51 ns (37e)	PhH
XA	≤0.0002	≤−5	50 μs (37f)	C_5H_{12}

Notes: $^aK_{eq} = k_{ST}/k_{TS}$ where k_{TS} and k_{ST} are the rate of intersystem crossing which could be measured by picosecond time-resolved laser spectroscopic measurements.
^{b}Determined by $\Delta G_{ST} = -RT \ln K_{eq}$.
^{c}Lifetime (τ) or half-life ($t_{1/2}$) in solution at room temperature.

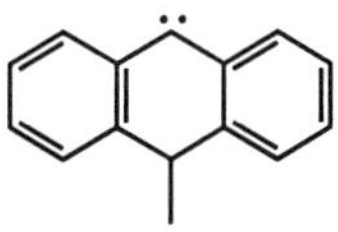

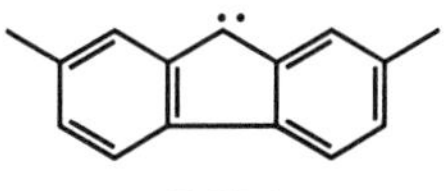

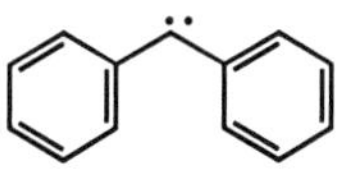

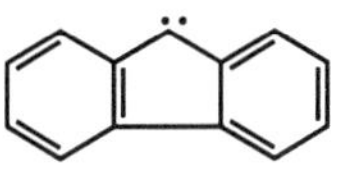

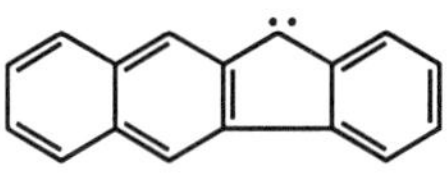

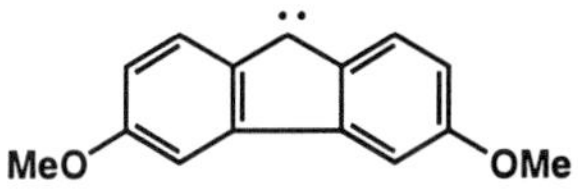

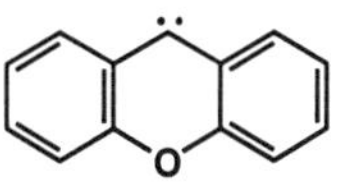

The basic concepts of molecular orbital theory (see Figure 2) can also be applied to explain the change in the magnitude of ΔG_{ST} with the change in structure for these aromatic carbene systems. For example, for BA, the aromatic LUMO (lowest unoccupied molecular orbital) is significantly lowered by the presence of the vacant aromatic orbital of the boron and is thus mixed with the carbene p orbital. This interaction lowers the energy of this non-bonding orbital, thus resulting in an increase in ΔG_{ST}. An analogous explanation is applied for analysis of XA, where the occupied aromatic orbital is raised by the electron-donating ability of oxygen and is mixed with the carbene p orbital. In this case, the splitting of the orbitals in the mixed state is increased sufficiently beyond that of a prototypic carbene to make ΔG_{ST} negative. Exactly the same prediction is made from a valence-bond viewpoint: electron-donor groups, the oxygen in XA or the methoxy-substituent of 3,6-dimethoxyfluorenylidene (DMFL), stabilize the electrophilic singlet carbene more than they do the radical-like triplet, while electron-withdrawing groups, the boron atom in BA, destabilize the singlet and lead to a greater ΔG_{ST}.

While the relationship between ΔG_{ST} and electronic effects is well characterized in these aromatic carbenes, their persistence has not been explicitly examined in terms of ΔG_{ST}. Lifetimes of these carbene are listed in Table 4. Again, the lifetimes shown here are not determined under specified experimental conditions and hence cannot be examined quantitatively. Nevertheless, inspection of the data suggests that the lifetime of each multiplicity tends to increase as the energy gap between the two spin states increases.

Theoretical evidence suggests that, for diphenylcarbene (DPC), the energies of the triplet and singlet depend to differing extents upon the angle (α) between the two aryl carbon–carbenic carbon bonds and the torsional angle (θ) by which the two aryl rings are twisted with respect to a plane defined by the bonds on the carbenic center. It is predicted that the singlet state should be stabilized with respect to the triplet as a function of decreasing α and increasing θ.[38] This prediction has been tested by studying the effect of ring size on the reactivity of cyclophane diarylcarbenes (CPC), where the phenyl rings are linked by alkyl chains of 9 to 12 methylene units (Table 5).[39,40] As n decreases, the bond angle at the carbene center

Table 5. Effect of Ring Size on Reactivities of Cyclophane Diarylcarbenes (CPC)[39]

| Carbene | $k_q(M^{-1}s^{-1})$ | | | ΔG_{ST} (kcal/mol) |
	Isoprene[a]	1,4-Cyclohexadiene[a]	Methanol[b]	
DPC ($n = 0$)	1.4×10^6	3.4×10^6	6.3×10^6	
CPC ($n = 12$)	8.6×10^6	7.3×10^6	2.0×10^8	2.0
CPC ($n = 11$)	9.7×10^6	7.0×10^6	2.4×10^8	1.8
CPC ($n = 10$)	1.5×10^7	1.2×10^7	6.8×10^8	1.2
CPC ($n = 9$)	3.8×10^7	2.3×10^7	2.0×10^9	0.5
spread[c]	27	7	320	—

Notes: [a]In hexane solvent.
[b]In tetrahydrofuran solvent.
[c]Rate constant for carbene CPC($n = 9$) divided by rate constant for DPC.

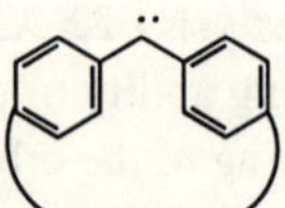

becomes smaller and the phenyl rings are oriented toward each other. Thus, the singlet states should become stabilized with respect to their triplet states as a function of decreasing ring size (i.e., ΔG_{ST} should decrease). Electron spin resonance (ESR) spectra are observed for the case where $n = 12$, 11, and 9, indicating that their ground states are triplet. Interestingly, D values reveal a systematic increase with the reduction of angle α and with increasing fixation of the aryl rings in a perpendicular configuration to the $C\alpha$-plane. No ESR spectra can be detected at 93 K with the $n = 10$ and 8 isomers. This suggests that either the singlet is the ground state or the triplet is slightly below the singlet. Since the triplet is not significantly populated at this temperature, irreversible chemical consumption of the carbene occurs. In any event, it suggests that ΔG_{ST} decreases with decreasing ring size.[40]

The absolute reactivities of the carbenes toward alcohols and dienes have been investigated using the technique of laser-flash photolysis.[39] The results summarized in Table 5 show that decreasing the chain length increases the bimolecular quenching rate constants. Characteristic singlet reactions, i.e., reaction with alcohols, are more sensitive to structural perturbations than triplet reactions, i.e., reactions with dienes. Thus, with methanol, the rate constants jump by factors of 320 upon lowering n. If one assumes that the reactions of the carbenes (CPC) with alcohols proceed by the same mechanism as that for DPC (Eq. 1–3), then the rate of quenching, k_q, is related to k_S and k_{TS}/k_{ST}. Consequently, the increase in reactivity must be due to either an increase in k_S or an increase in k_{TS}/k_{ST}. A higher value of k_S cannot account for the 320-fold increase in reactivity. Thus, it is likely that k_{TS}/k_{ST} increases due to a smaller free energy difference between the ground triplet states of these carbenes and their singlet states as a function of decreasing size of the methylene chain. Assuming that k_S for each carbene is the same as that for singlet DPC, values of ΔG_{ST} are calculated (Table 5). As predicted, the singlet–triplet energy gap decreases as the chain length and the bond angle at the carbene center decrease. The rate of quenching of the carbenes with 1,4-cyclohexadiene, presumably by the hydrogen abstraction reaction, increases also with decreasing ring size, but by only a factor of seven from DPC to CPC ($n = 9$). This suggests that the overall reactivity of the triplet state is not significantly raised, at least by the rather small change in ΔG_{ST}. The increased rate of quenching of the carbenes by isoprene may reflect some participation of the singlet state, in addition to the triplet state, for those carbenes with a sufficiently small energy gap.

Effects of *ortho* substituents on the reactivity of diphenylcarbene are also expected to provide some information on the role of the carbene bond angle on ΔG_{ST}. A series of diphenylcarbenes having methyl groups at the *ortho* positions have been studied (Table 6).[41,42] Steric interaction of the *o*-methyl substituents are expected to force the carbene center to adopt a greater bond angle, while the methyl groups are electron-donating in comparison with hydrogen.

The values of $|D/E|$ obtained by matrix EPR spectroscopy indicate that all three carbenes are considerably less bent than diphenylcarbene. Increased steric bulk in

Table 6. Effects of Ortho Methyl Groups on Reactivities of
Diphenylcarbenes[41a]

Carbene	k_q^a $(M^{-1}s^{-1})$	τ^b (μs)	ΔG_{ST} $(kcal/mol)$	D (MHz)	E (MHz)	E/D
DPC	6.4×10^6	1.7	5.9	12200	570	0.0467
MPC	8.2×10^4			11530	470	0.0408
MTC	2.1×10^4			11200	400	0.0357
DMC	$<2 \times 10^3$	>200	>8.0	10440	240	0.0230

Notes: [a]Bimolecular rate constant for the reaction of triplet diarylcarbenes with methanol in isooctane at
298 K.
[b]In cyclopentane at 298 K.

MPC   MTC   DMC(=**11a**)

the carbene substituents results in structures that approach a linear C–C–C bond
angle at the carbene center and orthogonality between the aryl rings. This in turn
should result in predominant triplet chemistry and wide singlet–triplet gaps.

Laser-flash photolysis (LFP) of the appropriate diazo precursors in isooctane
allow observation of the transient spectra of all three carbenes. The absorption
spectra of the triplet carbenes are quenched by addition of methanol and the rate
constants (k_q) with methanol are obtained. The data clearly show that the rate
constants decrease as values of the spectroscopic E and D parameters decrease. This
correlation suggests that the methanol reaction becomes more facile as the triplet–
singlet energy gap becomes smaller. If one assumes that the triplet carbene reacts
with methanol according to the pre-equilibrium mechanism, $k_q = k_S/K_{eq}$, where
$K_{eq} = k_{ST}/k_{TS}$, from the values of k_S determined for DPC and DMC, K_{eq} values are
estimated to be 2.0×10^4 and $> 7.0 \times 10^5$ for DPC and DMC, respectively. These
values give a ΔG_{ST} to be 5.9 and > 8.0 kcal/mol, respectively.[41] This result is clearly
in accord with the prediction that the singlet–triplet energy gap becomes greater as
the C–C–C angle of the carbene is expanded and suggests that the effect of the bond
angle change is apparently greater than that caused by electron-donation, which
should preferentially lower the energy of the singlet state.

Similar dramatic effects of a central angle on the stability of triplet states are also
noted for dialkylcarbenes. The triplet states of alkyl- and dialkylcarbenes are
rendered inaccessible since the reactive singlet states are effectively trapped by
available intramolecular reactions. Several alkylcarbenes without α-H atoms were
generated and spectroscopically characterized in matrices. For instance, di-*tert*-
butylcarbene (DBC) is sufficiently stable to be observed in 2-methyltetrahydro-

furan (MTHF) glass at 20 K by ESR (electron-spin resonance) spectroscopy. At temperatures above 70 K the carbene disappeared irreversibly. Work-up produces mainly products typical of a singlet carbene.[43]

Diadamantylcarbene (DAC) is even more stable; triplet diadamantylcarbene is long-lived in ethanol matrices between 4 and 75 K as shown by ESR. In this respect, it is even less reactive than diphenylcarbene. Unlike di-*tert*-butylcarbene, it shows substantial triplet reactivity even in solution at 25 °C. The wide angle enforced by steric interaction stabilizes the triplet state of DAC compared to the singlet. The hyperfine components lead to a bond angle of 152° at the carbene center, substantially wider than that in methylene and essentially the same as that in diphenylcarbene.[17]

DBC   DAC

IV. PERSISTENT TRIPLET DIPHENYLCARBENES

Examination of the relationship between structure and reactivity thus far reported has shown that electronic effects usually play an important role in stabilizing the singlet state and that the singlet state undergoing thermodynamic stabilization becomes less reactive due to the contribution of ylidic character to such an extent that the species can be isolated under ambient conditions. While the triplet states are also stabilized by electronic effects with respect to the singlet state and their reactions through upper-lying singlet state are suppressed, their intrinsic reactivities are not affected. In this light, kinetic stabilization using steric protectors should be more effective to generate persistent triplet carbenes. Introduction of sterically bulky groups around the carbenic center increases the carbene bond angle and, therefore, must make ΔG_{ST} larger. Diphenylcarbene is an ideal prototypic carbene for such a study for the following reasons. First, it is known to undergo significant expansion of the carbene bond angle by the substituents at the *ortho* positions, which should result in wide singlet–triplet gaps. Second, spin delocalization is less extended compared to the other polynuclear arylcarbenes,[14] which must decrease possible diversity of the reactivities and complexity of the electronic structure. Finally, it is less sensitive to the electronic effects due to the flexibility of the two aromatic rings, which might reduce an undesirable decrease in ΔG_{ST} by introducing a bulky but electron-donating group. Thus, we have prepared diphenyldiazomethanes bearing a series of substituents primarily at the *ortho* positions and investigated the reactivity of the generated carbenes not only by product analysis but also by direct spectroscopic observation.

A. Polychlorinated Diphenylcarbenes

2,2′,4,4′,6,6′-Hexachlorodiphenylcarbene (**2a**) was first generated by Zimmerman and Paskovich in 1964 in their attempts to envisage a complete lack of reactivity of the hindered divalent species with external species.[9] Although this carbene was not stable enough to be isolated, it exhibited unusual chemical properties. Thus, in solution at room temperature, **2a** did not react with the parent diazo compound (**1a**) to give azine but dimerized instead to give tetrakis(2, 4, 6-trichlorophenyl)ethylene (**3a**) in 70 to 80% yield (Scheme 1). The formation of olefinic dimerization products as main products is rare in the decomposition of diazo compounds while the formation of ketazines is virtually omnipresent. This carbene also did not undergo carbon–hydrogen insertion or hydrogen abstraction from the cyclopentane solvent. This also contrasts with the behavior of diphenylcarbene in hydrocarbon solvents. The authors explained these results by assuming that the hindered diarylmethylenes, having bent geometries, do not have accessible singlet counterparts since the singlet would require a smaller carbene angle and incur severe aryl–aryl repulsion, and that as a result of severe steric hindrance and, consequently, resistance to external attack by solvent, the hindered triplet diarylcarbene concentration builds up to the point where dimerization occurs.

While the product analysis studies clearly indicate that the carbene is fairly persistent in its triplet state, even in solution at room temperature, if not isolated, no further efforts have been made to get more quantitative insights into the persistence of the triplet states and/or to extend the lifetime by modifications. For instance, one would naturally expect that the carbenic center must become more crowded as one introduces four chlorine groups at the *meta* positions, which can buttress the four *ortho* chlorine atoms. Thus, we prepared diphenyldiazomethanes **1** having 6 and 10 chlorine atoms and investigated the reactivities of the polychlorinated diphenylcarbenes generated.[44,45]

Irradiation of **1** in benzene afforded the chlorinated tetraphenylethylenes **3** almost exclusively, regardless of the number of chlorine atoms on the phenyl rings. However, when the carbenes were generated in benzene in the presence of methanol, the OH insertion product was produced at the expense of the dimer, which sharply decreased as the chlorine atoms on the rings decreased (Scheme 2, Table 7). The difference in reactivities between **2a** and **2b** can be interpreted as reflecting

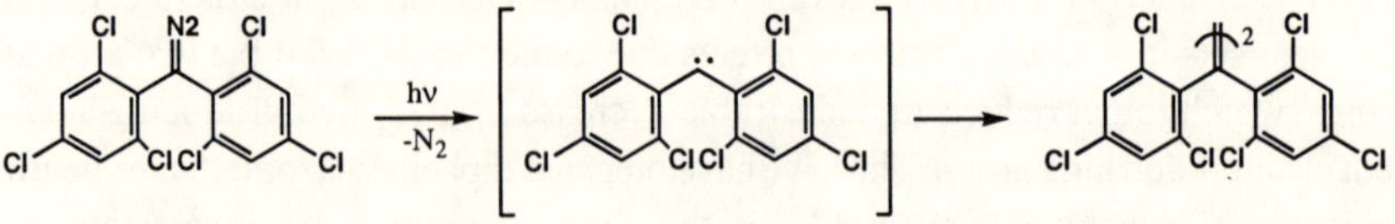

Scheme 1.

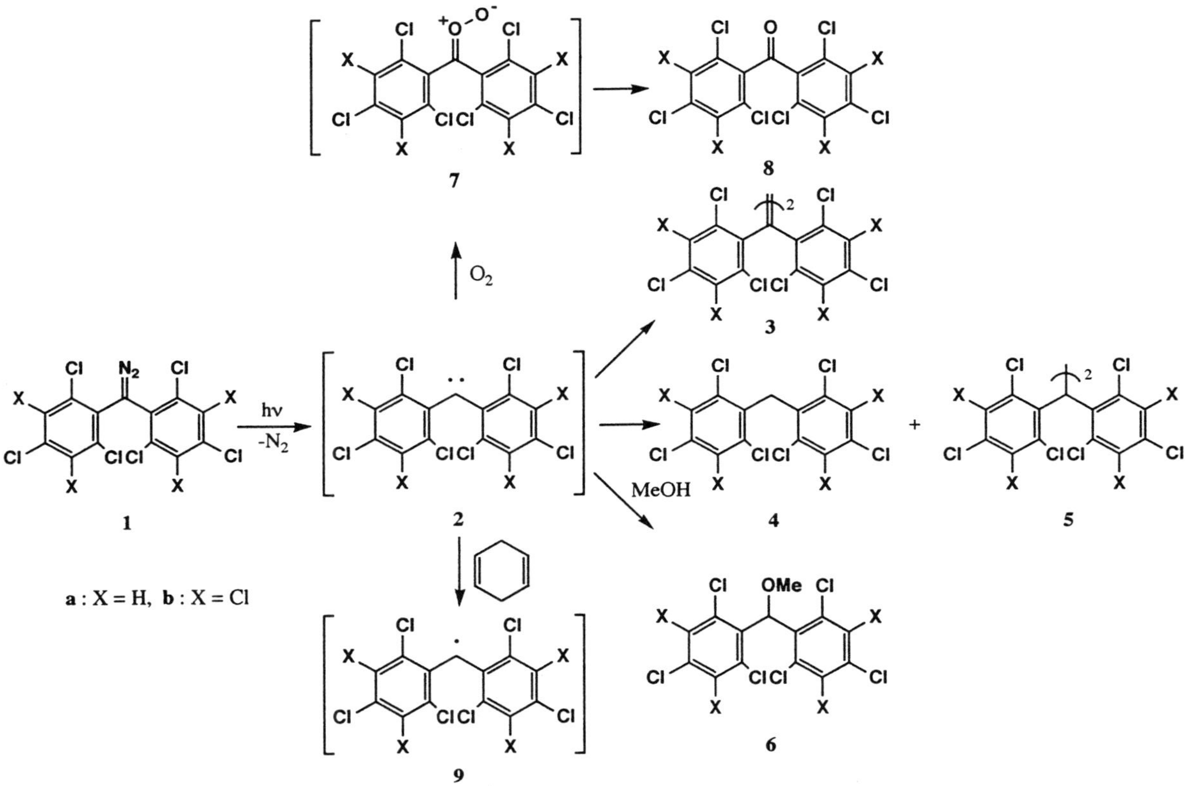

5
8
3
4
6
7
2
9
1
MeOH
O_2
hv
-N_2
a : X = H, b : X = Cl
Scheme 2.

Table 7. Product Distributions in Reactions of Polychlorinated
Diphenylcarbenes (**2**) Generated by Photolysis of Diazomethanes (**1**)[45]

| | | Yield(%) | | | |
Diazo	Solvent	3	4	5	6
1a	PhH	>99	<1	<1	—
1b		>99	<1	<1	—
1a	c-C_6H_{12}	26	29	17	—
1b		80	17	3	—
1a	MeOH-PhH	14	24	1	61
1b		50	20	9	21

the difference in the extent of steric crowdedness around the carbenic centers
between these two carbenes.

It is well known that in 1,2-disubstituted benzene derivatives introduction of
substituents at the 3-position exerts a very large effect on the rate of appropriate
reactions and the results are considered in light of the importance of bond bending;
the 3-substituents buttress the 2-substituents.[46,47] Thus, in **2a**, each of four *ortho*
chlorine atoms around the carbene center is buttressed by one of four *meta* chloro
atoms and, therefore, the carbene center in **2a** is surrounded by the chloro atoms
more rigidly than that in **2b**.

Flash photolysis of a degassed benzene solution of 2,2′,4,4′,6,6′-hexachloro-
diphenyldiazomethane (**1a**) at room temperature with the output of a xenon flash
lamp (pulse width of 10 μs) produced a transient species absorbing at 344 nm, which
appeared coincidently with the xenon pulse and disappeared within approximately
50 ms. The final stable products formed from irradiation of **1a** under these
conditions consist mainly of the carbenic dimer **3a**. The absorption spectrum
obtained 50 μs after the excitation is similar to that measured during the photolysis
of **1a** in a 2-methyltetrahydrofuran (MTHF) glass at 77 K. The spectrum consists
of two identifiable features, a sharp, relatively intense UV band with a maximum
at 338 nm and a broad, weak absorption extending from 450 to 500 nm. These
absorption bands are stable for hours at 77 K, but warming the sample to room
temperature leads to their disappearance. The product analysis of the spent solution
again showed the presence of the dimer as the major product. The optical absorption
spectra of several triplet aromatic carbenes in frozen media have been assigned.
Typically, they consist of an intense UV absorption and a weak visible transition.[48]
These features are present in the spectrum obtained from the photolysis of **1a**.
Moreover, photolysis of **1a** in MTHF at 77 K exhibited ESR signals characteristic
of randomly oriented triplet molecules attributable to triplet hexachlorodiphenylcarbene ³**2a**. The ESR signals were stable for at least several hours at this temperature. On the basis of the low-temperature spectrum and the chemical analysis, the

transient product, showing the absorption maximum at 344 nm from the photolysis of **1a** in benzene, was assigned to triplet hexachlorodiphenylcarbene ^{3}**2a**.

Support is lent to this assignment by trapping experiments using oxygen. When flash photolysis measurements were carried out on a non-degassed benzene solution of **1a**, the half-life of ^{3}**2a** was greatly decreased and a broad absorption band with a maximum at 390 nm appeared. The rate of increase in the absorbance at 390 nm was practically the same as that of the decay of the peak at 344 nm, showing that ^{3}**2a** was quenched with oxygen to form a new species. Analysis of the spent solution showed the presence of a large amount of hexachlorobenzophenone **8a** at the expense of the carbenic dimer. It is well-documented that diarylcarbenes with triplet ground states are readily trapped by oxygen to give the corresponding diarylketone oxides, which are observed directly either by matrix isolation techniques or by flash photolysis. These carbonyl oxides usually show a rather broad absorption band centered around 390 to 450 nm.[49] Thus, the observation is interpreted to mean that the ^{3}**2a** is trapped with oxygen to generate carbonyl oxide **7a**, thus confirming that the transient absorption quenched by oxygen is due to ^{3}**2a**.

The decay of the transient absorption due to ^{3}**2a** was found to be second order, in accordance with the product analysis data, showing that dimerization to form the dimer is the main pathway for ^{3}**2a** under these conditions. Second-order fitting of transient absorption data yields 2 $k_d/\varepsilon l$ values to be 8.9×10^2 s^{-1}, where k_d is the dimerization reaction rate constant, ε the extinction coefficient, and l the optical path. The lifetime in the form of half-life was estimated from the decay curve to be roughly 18 ms. Thus ^{3}**2a** is shown to be four orders of magnitude longer-lived than the parent diphenylcarbene. This quantitatively confirms expectations for the role of steric effects in extending the triplet carbene established in 1964.[9]

In order to learn about the reactivities of the persistent triplet carbene toward typical quenchers, the absolute rate constants of the reaction of ^{3}**2a** with several selected substrates were determined. The kinetics of carbonyl oxide **7a** formation were studied by monitoring their formation at the corresponding absorption maximum following laser excitation at 308 nm. The build-up of the signal followed pseudo-first-order kinetics (k_{obs}) and was monitored for several concentrations of oxygen. The bimolecular rate constant, k_{O_2}, was determined from plots of k_{obs} vs. [O$_2$], according to Eq. (4),

$$k_{obs} = k_o + k_{O_2}[O_2] \tag{4}$$

where k_o represents the rate of decay of ^{3}**2a** in the absence of oxygen in the solvent employed for the experiment, and k_{O_2} is the rate constant for the reaction of the carbene with oxygen. The rate constant for generation of the carbonyl oxygen is 7.4×10^7 M^{-1}s^{-1} in benzene. This is about one order of magnitude smaller than for the parent DPC.[49b]

When a solution of **1a** in a degassed benzene solution containing 1,4-cyclohexadiene (CHD) was excited, a new species was formed showing a strong absorption with $\lambda_{max} = 376$ nm, as the 344 nm signals of ^{3}**2a** decayed. The decay of ^{3}**2a** is shown to be kinetically correlated with the growth of the new species. The new signal was attributable to the hexachlorodiphenylmethyl radical **9a** formed as a result of H abstraction of ^{3}**2a** from the diene, since it is well-documented that triplet arylcarbenes generated in good hydrogen donor solvents undergo H abstraction leading to the corresponding radical.[18] Indications based on the reported data for diphenylcarbene suggest that the radicals are usually red-shifted by about 30 nm with respect to the corresponding triplet carbenes.[50] If the same rule of thumb is applied to ^{3}**2a**, the corresponding radical would be expected in the 370 to 380 nm region. A plot of the observed pseudo-first-order rate constant of the formation of the radical against [cyclohexadiene] is linear and the slope of this plot yields the absolute rate constant for H atom abstraction reaction of ^{3}**2a** from CHD, $k_{CHD} = 3.5 \times 10^3$ M^{-1}s^{-1}.

Essentially similar measurements were done with perchlorodiphenyldiazomethane **1b**. Here the absorption due to the carbonyl oxide **7b** appeared at 390 nm in the presence of oxygen. The absorption which can be assigned to the perchlorodiphenylmethyl radical **9b** appeared at 376 nm in the presence of the diene, as triplet decachlorodiphenylcarbene ^{3}**2b** signals at 357 nm decayed. Kinetic data obtained for ^{3}**2a** as well as ^{3}**2b** are summarized in Table 8, which also includes the data for the parent DPC for comparison purposes. Inspection of the data immediately indicates that the dimerization rate is sharply decreased as one simply introduces four chlorine atoms at the *ortho* positions, but is further decreased as four additional chlorine atoms are introduced at the *meta* positions. This confirms the explanation based on the product analysis and supports the idea that the buttressing effect should play an important "buttressing" role in protecting the reactive center. On the other hand, k_{CHD} is also decreased dramatically in going from DPC to ^{3}**2a**, but it increases, albeit slightly, as one introduces the four more chlorine atoms. This is somewhat surprising in the light of a decrease in the dimerization rate by the four *meta*-buttressing chloro substituents as noted above and indicates that the electrophilicity of the carbenic center is increased as one introduces more chlorine atoms on the phenyl rings. Thus, as more chloro substituents are introduced, the carbenic center becomes more rigidly protected and, on the other hand, becomes more reactive. The attack of a bulky substrate, such as a

Table 8. Kinetic Data for Polychlorinated Diphenylcarbenes Ar$_2$C: (2)[45]

| | | k_q(M^{-1}s^{-1}) | | | |
| | Ar in | | | | |
Carbene	Ar$_2$C:	k_d	k_{O_2}	k_{CHD}	τ (ms)
DPC	C$_6$H$_5$	5.4×10^9		1.0×10^7	0.002
2a	2,4,6-Cl$_3$C$_6$H$_2$	2.5×10^6	7.4×10^7	3.5×10^3	18
2b	Cl$_5$C$_6$	1.2×10^6		6.2×10^3	28

chlorinated diphenylcarbene, on the carbenic center, therefore, must be severely restricted, while the rate of abstraction of a very small atom, such as hydrogen, from a very efficient hydrogen donor is still controlled by the electrophilicity of the carbenic center, even in these highly sterically hindered carbenes.

Finally, it should be noted that the decay rate of ^{3}2, in contrast to other diarylcarbenes, was not changed appreciably by addition of methanol. This is true at least in the range of concentration where ^{3}DPC was quenched efficiently. However, methanol reduced the quantum yield for triplet formation, suggesting that its precursor reacted readily with methanol. If one assumes that the precursor is the singlet carbene, this is explained by proposing that the triplet ground state cannot convert into the singlet state so reaction with methanol cannot take place. In other words, singlet-to-triplet intersystem crossing is extremely efficient, while the reverse process is not. The sharp distinctions between the chemistries of the singlet and triplet states imply that there is a substantial free energy separation between them.

B. Polymethylated Diphenylcarbenes

Dimesitylcarbene (DMC) **11a** (Scheme 3) was also first generated by Zimmerman and Paskovich in 1964.[9] Although this carbene was also not stable enough to be isolated, the congestion at the carbene center in DMC **11a** is reflected in its unique behavior. Thus, in solution at room temperature, **11a** did not react with the parent diazo compound but dimerized instead to give tetrakis(dimesityl)ethylene (**12a**) as the main product. **11a** was also unreactive toward secondary carbon-hydrogen bonds in solvent molecules and, at high temperatures, decayed by attack at an *ortho*-methyl group to form benzocyclobutene **13a**, a reaction that is not observed for 2-methyldiphenylcarbenes under similar conditions. The chemistry found for **11a** is again in sharp contrast with that found for other diarylcarbenes and is interpreted in terms of steric effects, as described with the hexachloro analogue.

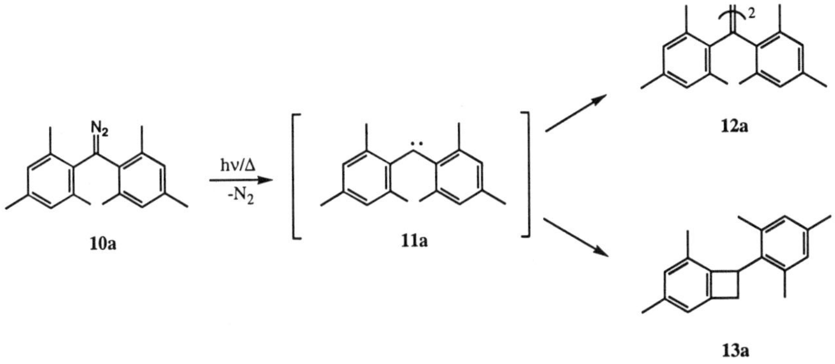

Scheme 3.

More recent ESR studies showed that the carbene had a structure that was significantly less bent than that of diphenylcarbene and in which the aryl groups were orthogonal.[41] This geometry effectively maximizes the triplet–singlet energy gap and protects the reactive site so that intermolecular reactions are not favored.

In spite of those very promising features of DMC as a potentially isolable triplet carbene, almost no further elaborate work has been made to characterize and/or modify it. Thus, we have generated a series of polymethylated diphenylcarbenes and investigated their reactivities not only by product analysis studies but also by flash photolysis techniques, in hope of extending the lifetime by buttressing effects.[51–53]

The polymethylated diphenylcarbenes **11a–c** (Scheme 4) comprising a series of methyl groups were generated either by thermolysis at 140 °C or by photolysis of a benzene solution of the precursor diazomethanes **10**. The reactions observed with didurylcarbene **11b** and decamethyldiphenylcarbene **11c** were analogous to those observed with dimesitylcarbene **11a**.[9] A key difference is found in the formation of benzocyclobutene **13**. In the case of **11a**, for instance, while **13a** was produced as main product in the thermolytic run, **11a** underwent mainly dimerization when generated at 15 °C. On the other hand, in the case of **11b**, **13b** was produced not only by thermolysis but also by photolysis along with the dimer **12b**. Carbene **11c** produced **13c** as the main product at the expense of **12c**, even in a photolytic run at 15 °C (Scheme 4, Table 9).

It is generally accepted that triplet carbenes, like radicals, dimerize in a very fast and exothermic reaction.[54] On the other hand, the benzocyclobutenes **13** could be produced either from the singlet or the triplet. Therefore, the significant difference in the product distribution can be interpreted in terms of electronic effects and/or steric factors either on the reactivities of the triplet or on the singlet–triplet energy gap. However, if one assumes that *meta*- and *para*-methyl groups on one phenyl ring would induce the observed difference in the product distribution simply by electronic effects, the sum of Hammett σ constants decreases in the order of **11b** > **11a** > **11c**. This, however, is not reconciled with the observed trends in the product distribution changes.

The difference in the product distribution can be better interpreted as reflecting the difference in the extent of steric hindrance at the carbene centers between these carbenes. Thus, in **11b**, each of the four *ortho*-methyl groups around the carbene is buttressed by that of four *meta*-methyl groups. The *ortho*-methyl groups in **11c** must be buttressed more effectively since four *meta*-methyl groups are buttressed by two *para*-methyl groups in this case. Therefore, the carbene center in **11c** is most effectively blocked by the *ortho*-methyl groups from external reagents. Thus, carbene **11c** is forced to undergo CH insertion to afford **13c**. Alternatively, it is possible to assume that, as the *ortho* methyl groups are buttressed more effectively, they are brought much closer to the carbene, and hence the carbene is more easily trapped intramolecularly by the methyl groups before it undergoes dimerization.

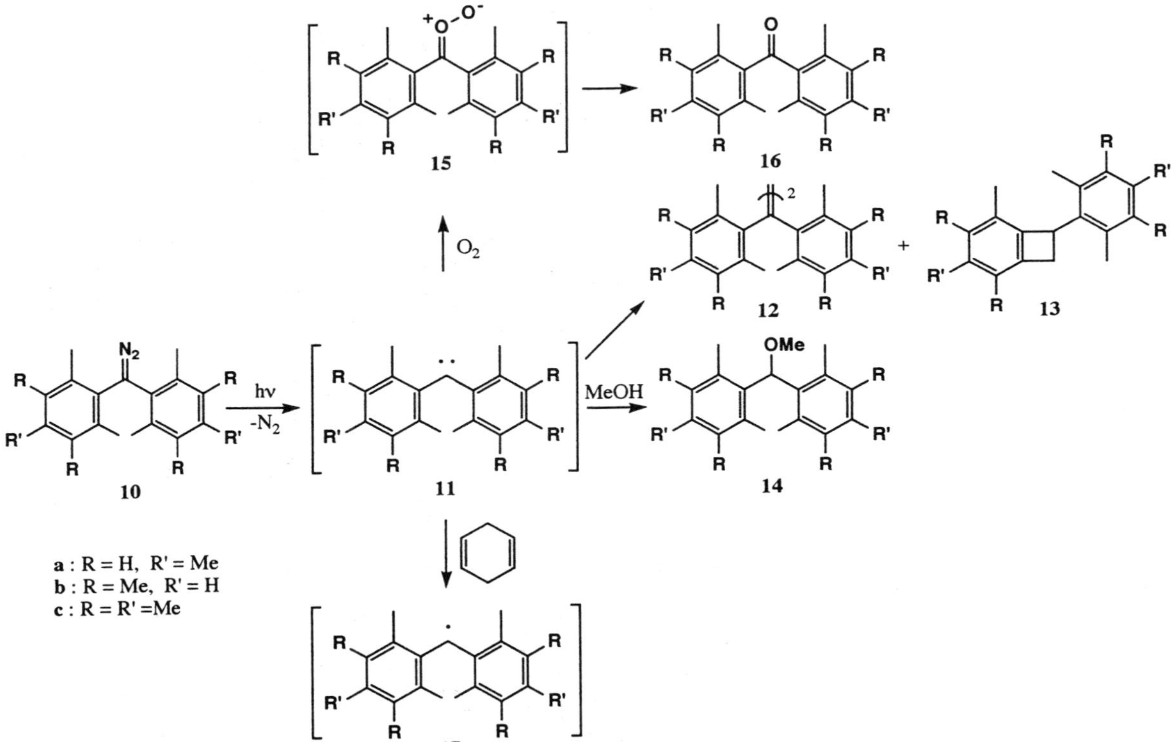

R
R'
R
R
R'
R
13
+
R
R'
R
O
16
R
R'
R'
R
R
R'
R
2
12
R
R'
R'
OMe
R
R'
R
14
R
R'
R'
O
O⁻
15
O₂
R
R'
R
:
11
R
R'
R'
R
R'
R
17
R
R'
R'
MeOH
hv
-N₂
N₂
R
R'
R
10
R
R'
R'
a : R = H, R' = Me
b : R = Me, R' = H
c : R = R' = Me

Scheme 4.

Table 9. Product Distributions in Reactions of Polymethylated
Diphenylcarbenes (**11**) Generated from Diazomethanes (**10**)[52]

Diazo	Conditions	Yield (%) 12	13	14
10a	PhH/140 °C	<1	70.0	—
10b		<1	95.8	—
10c		<1	90.5	—
10a	PhH/hv	94	6	—
10b		29.1	40.8	—
10c		5.2	78.8	—
10c-d_{12}		78.0	22.0	—
10a	MeOH/hv	<1	<1	90.0
10b		<1	4.2	84.7
10c		<1	5.9	76.2

These two possibilities, however, cannot be discriminated simply by product analysis (*vide infra*).

Irradiation of the diazo precursors **10** in MTHF glass at 77 K gave paramagnetic species that could readily be characterized from their EPR spectra as triplet diphenylcarbenes **11**. The EPR signals were stable for several hours at this temperature and were analyzed in terms of E and D values, which are reported in Table 10.

Optical spectroscopy in the frozen medium gave analogous but more intriguing results. For instance, irradiation of **10b** in MTHF glass at 77 K gave a spectrum consisting of an intense UV band at 321 and 335 nm and a weak visible transition at 466 and 495 nm, characteristic of triplet diarylcarbenes. The glassy solution did not exhibit any spectral change for several hours if kept at this temperature, but as the matrix temperature was slowly warmed, a new broad absorption at 375 nm appeared and increased as carbene absorption bands decreased. The species responsible for this new absorption was assigned to *o*-xylylene **18b** (Scheme 5) formed as a result of an intramolecular H abstraction of triplet carbene **11b**. Since the

Table 10. ESR Data of Polymethylated Diphenylcarbenes Ar_2C: (**11**)[a 53,55]

Carbene	Ar in Ar_2C:	$D(cm^{-1})$	$E(cm^{-1})$	E/D
DPC	C_6H_5	0.4053	0.0190	0.0469
11a	$2,4,6\text{-}Me_3C_6H_2$	0.3551	0.0116	0.0326
11b	$2,3,5,6\text{-}Me_4CH$	0.3805	0.0106	0.0279
11c	$2,3,4,5,6\text{-}Me_5C$	0.3636	0.0095	0.0260

Note: [a] Measured in 2–methyltetrahydrofuran at 77 K.

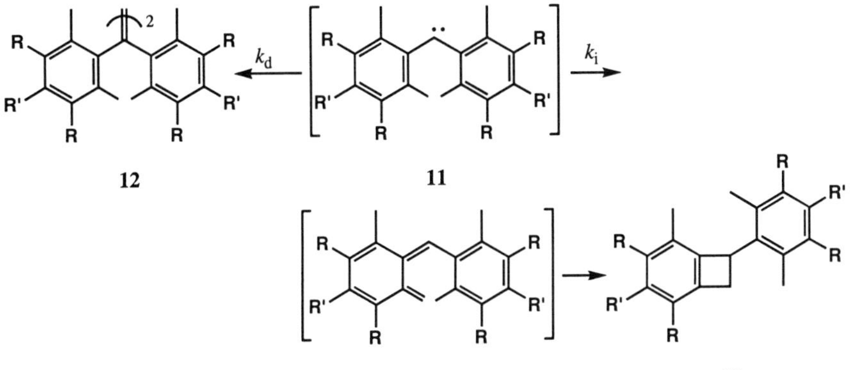

a: R=H, R'=Me, **b**: R=Me, R'=H, **c**: R=R'=Me

Scheme 5.

product analysis of the spent solution showed the presence of dimer **12b** and benzocyclobutene **13b**, the latter compound must be formed from *o*-xylylene **18b** and a subsequent cyclization.

A similar spectrum was obtained in the photolysis of **10a** and **c** in MTHF glass at 77 K. Here **11a** and **11c** showed two sharp, intense UV bands at 320 and 330 nm, and 325 and 338 nm, respectively, and generated the *o*-xylylenes **18a** and **18c**, exhibiting a broad absorption at 390 and 380 nm, respectively.

Laser-flash photolysis of **10b** in a degassed benzene solution at room temperature with a XeCl laser pulse produced a transient species at 320 to 330 nm, coincidental with the laser pulse. The decay kinetics of the transients indicate that the absorption at 330 nm decays within 1 s to generate a new species with an absorption maximum around 370 nm, which is too-long lived to be monitored by our system. Product analysis of the spent solution again showed the presence of **12b** and **13b**. On the basis of the low-temperature spectrum together with chemical analysis, the initially formed transient with a maximum at 340 nm was assigned to carbene **11b**, while the second was assigned to *o*-xylylene **18b**. Support is lent to the assignment of **11b** by trapping experiments using oxygen, where the lifetime of the carbene is dramatically decreased. Moreover a new broad absorption band with a maximum at 390 nm ascribable to the corresponding ketone oxide **15b** appeared at the expense of the carbene absorption.

The decay kinetics also show that the decay of **11b** is kinetically correlated with the growth of **18b**; the decay rate was determined to be 2.1 ± 0.1 s^{-1}, while the growth (k_i) was 2.2 s^{-1}. From the decay curve, a half-life ($t_{1/2}$) of **11b** was estimated to be approximately 410 ms, while the lifetime based on k_i was determined to be

Table 11. Kinetic Data for Polymethylated Diphenylcarbenes Ar_2C: (11)[52]

Carbene	Ar in Ar_2C:	$k_i^H(s^{-1})$	$t_{1/2}(ms)$	$k_{O_2}(M^{-1}s^{-1})$	$k_{CHD}(M^{-1}s^{-1})$	$k_{MeOH}(M^{-2}s^{-1})$
11a	$2,4,6\text{-}Me_3C_6H_2$	1.5	160	2.0×10^8	4.6×10^2	2.0×10^4
11b	$2,3,5,6\text{-}Me_4C_6H$	2.2	410	7.1×10^7	3.4×10	8.9×10^2
11c	$2,3,4,5,6\text{-}Me_5C_6$	4.1	180	1.0×10^8	9.8×10	4.7×10^3

455 ms. Similar measurements were done for carbenes **11a** and **c** and values of k_i and $t_{1/2}$ are summarized in Table 11.

These spectroscopic studies corroborate the trend observed in the product analysis studies (*vide supra*) that led to the same, but more quantitative, conclusions. Thus, triplet dimesitylcarbene **11a** is shown to be some five orders of magnitude longer-lived than the parent DPC, thus confirming again expectations for the role of steric effect in persistence of triplet carbenes 30 years ago.[9] As more methyl groups are introduced to the aromatic rings, the carbenic center is more tightly blocked by the four *ortho*-methyl groups toward external reagents. Thus, diduryl-carbene **11b** is shown to be longer-lived than **11a**. On the other hand, the steady increase in k_i in going from **11a** to **11c** clearly suggests that, as the *o*-methyl groups are brought much closer to the carbenic center by the buttressing methyl groups, the carbenic center comes to interact more easily with the *o*-methyl groups. Thus, decamethyldiphenylcarbene (**11c**) comes to be trapped by the *o*-methyl groups to generate *o*-quinodimethane **18** more efficiently and hence becomes shorter-lived again. Optimized geometries calculated for a series of polymethylated mono-phenylcarbenes with PM3-ROHF/CI (4 × 4) reveal that the distances between the carbenic carbon and the *o*-methyl carbon atom for mesityl, duryl-, and pentamethyl-phenylcarbenes decrease from 285 to 278 pm.

The above study reveals that polymethylated diphenylcarbenes **11** are exceptionally long-lived for arylcarbenes, although they are still not persistent enough to be isolated. The buttressing effects are shown to be effective to strengthen the *ortho* effect that protects the reactive center. However, in case of the methyl group that is potentially reactive toward the carbene, the *o*-methyl groups become a more efficient intramolecular quencher of the carbene as a result of the buttressing effect, thus making the carbene less persistent again. One of the best ways to quench this process is to replace the hydrogens at the *o*-methyl groups with deuterium.

Product analysis studies indicate that the 1,4 H-atom migration in the polymethylated diphenylcarbenes undergo substantial kinetic deuterium isotope effects (KDIEs). Thus, **11c**-d_{12} (Scheme 6), where all hydrogens of the four *o*-methyl groups are replaced with deuterium, afforded carbene dimer **12c**-d_{12} as the major product at the expense of benzocyclobutenes. This is in contrast to the product ratios observed with the protiated analogue **11c** (Scheme 6, Table 9).

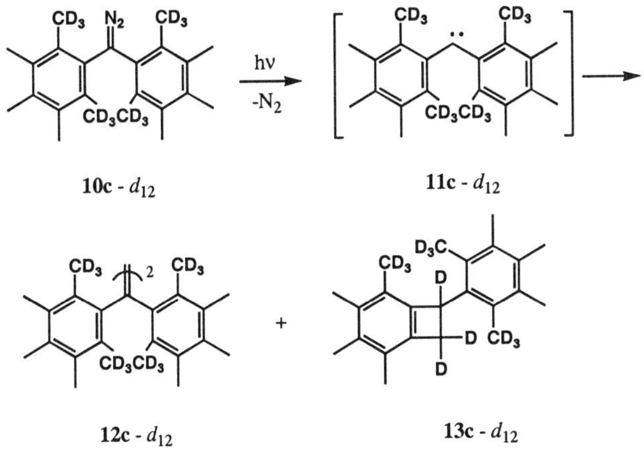

Scheme 6.

LFP of **10c**-d_{12} in a degassed benzene solution at 20 °C resulted in essentially the same transient absorption bands that were observed for the protio analogue (**10c**-d_0); a transient showing a strong absorption at 340 nm due to deuterated triplet carbene **11c**-d_{12} appeared coincidentally with the XeCl pulse, and a second transient at 377 nm due to the deuterated o-quinodimethane **18c**-d_{12} appeared as the initial absorption decayed. However, the kinetic behaviors of these two carbenes are distinctly different. Thus, the growth rate and optical yield of the deuterated o-quinodimethane were very small compared to those of the protio analogue, and the decay of **11c**-d_{12} was found not to fit with a single exponential, as opposed to the almost complete unimolecular decay of **11c**. This is in accordance with the product analysis data showing that **11c**-d_{12} gave the carbenic dimer as main product, while **11c**-d_0 produced the cyclobutene almost exclusively. The lifetime, in the form of half-life, was estimated from the decay curve to be 1.1 s in benzene at 20 °C. This is, therefore, the first triplet carbene showing a half-life of over 1 s in solution phase at room temperature.

The growth rate of o-xylylene from **11c**-d_{12}, which is equal to the intramolecular D abstraction rate constant (k_i^D) was then determined to be 0.85 s^{-1} at 20 °C. KDIE based on the absolute rate constants, i.e., k_i^H/k_i^D, is calculated to be 4.8. In order to get more insight into the nature of the 1,4 H migration, the rate of the migration was measured as a function of temperature over a temperature range of 15 to 75 °C. The Arrhenius plot of the data is almost completely linear ($r = 0.99$) for both **11c**-d_0 and **11c**-d_{12} in the temperature range studied. No tendency of deviation from the linearity at lower temperature is seen. The Arrhenius parameters for H and D migration are summarized in Table 12. The differential activation energy ($E_a^D - E_a^H = 1.0$ kcal/mol) and the ratio of preexponential factors ($A_H/A_D = 0.9$) are

Table 12. Arrherius Parameters for H Migration in Carbenes **11c** and
11c-d_{12}

Carbene	Temp (K)	E_a (kcal/mol)	A (s^{-1})	$[\Delta E_a]_D^H$	A_H/A_D
11c	338–288	9.9	9.8×10^7	1.0	0.9
11c-d_{12}	338–290	10.9	1.1×10^8		

consistent with a classical atom-transfer reaction.[56,57] Thus, moderate k_i^H/k_i^D, linear Arrhenius plots and normal differential Arrhenius parameters all suggest that there is no contribution of quantum mechanical tunneling to the 1,4 H migration in **11c,** at least within the temperature range studied.

In order to learn about the intrinsic reactivities of these highly congested carbenes, the absolute rate constants with typical carbene quenchers, i.e., oxygen, 1,4-cyclohexadiene (CHD) and methanol were determined and are included in Table 11. In spite of Griller's observation that the triplet lifetime of dimesitylcarbene **11a** is essentially unaffected by addition of methanol (1.2 M),[42c] our time-resolved experiments have shown that the ground-state triplets of **11** are quenched by methanol (< 0.5 M). Plots of k_{obs} vs methanol concentration were curved, but linear plots were obtained when k_{obs} were plotted vs the concentration of methanol dimers. The reactivities are found to increase in the order of **11b** < **11c** < **11a**. The increase in reactivity must be due to either an increase in k_S or an increase in k_{TS}/k_{ST}. The values of E/D obtained by matrix EPR spectroscopy decrease in going from **11a** to **11b** to **11c**, suggesting that ΔG_{ST} must be increasing in this order. The distances between the carbenic carbon and the *ortho*-methyl groups, judged from k_i values and semiempirical calculations, also increase in this order. From these data one may suppose that the reactivity must be decreased in this order. On the other hand, the electron-donating ability ($\Sigma\sigma$) of the ring methyl groups is increased in the order of **11b** < **11a** < **11c**. If one supposes that ΔG_{ST} is decreased in this order by stabilizing the singlet state or that k_S is increased by stabilizing the putative cationic intermediate in the OH insertion reaction,[32] one would expect that the reactivity will increase in this order. The observed changes in reactivities toward methanol is then interpreted to mean that the reactions in this case are controlled by combinational effects of the substituents on ΔG_{ST} and k_S. It is to be noted that the H absorption rate (k_{CHD}) also increases in the order of **11b** < **11c** < **11a**.

C. [2,4,6-Tri(*tert*-butyl)phenyl]phenylcarbenes

The *tert*-butyl group has been recognized as one of the most effective protecting groups and has been successfully employed to protect many reactive centers in organic molecules. For instance, the divalent center of silylenes and germylenes, the heavy atom analogues of carbenes, has been shown to be blocked by *tert*-butyl groups.[58] One would wonder, however, whether the *o-tert*-butyl group should be acting as an equally effective protector of the carbenic center in light of the voracious appetite of carbenes for electrons by which they react, even with very weak sources of electrons, e.g., σ electrons of CH bonds. On the other hand, the rather large KDIEs observed for **11c** suggest that the *ortho tert*-butyl group might be able to exert a large depressing effect on the reactivities of the carbene, if it is perdeuterated.

Irradiation of a degassed solution of tri(*tert*-butyl)diphenyldiazomethane (**19**) in benzene at room temperature afforded almost exclusively 4,6-di-*tert*-butyl-1,1-di-methyl-3-phenylindan (**21**), which must be produced from the photolytically gen-erated [2,4,6-tri-(*tert*-butyl)diphenyl]carbene (**20**) as a result of insertion into the CH bonds of the *tert*-butyl group at the *ortho* position (Scheme 7). Similar irradiation of **19** in more reactive solvents such as cyclohexane and methanol gave also indan **21** as main product, and no solvent insertion products were detected, even in methanol. It is rather surprising to note that carbene **20** is not quenched at all, even by methanol, which is known as an excellent scavenger for most carbenes. This indicates that the carbenic center in **20** is almost completely blocked toward external reagents. The perdeuterated carbene **20**-d_{27} also gave essentially the same result. Thus, the indan **21**-d_{27} was formed as main product, irrespective of the solvents used (Scheme 8).

Laser-flash photolysis of a degassed solution of **19** in cyclohexane or acetonitrile at room temperature with the fourth harmonic wavelength (266 nm) of a Nd/YAG laser produced no transient species, to the limit of time resolution of the LFP system (5 ns). Similar LFP of **19** in the presence of 1,4-cyclohexadiene also produced no transient absorption. However, when LFP measurements were carried out in an

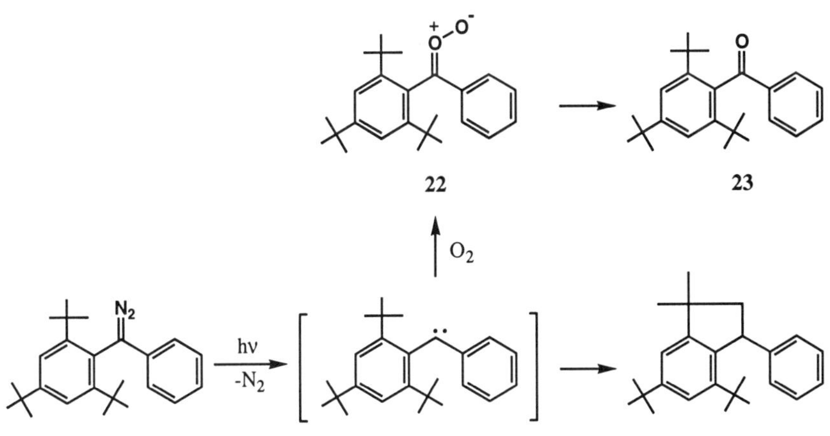

22   **23**

O_2

19   **20**   **21**

Scheme 7.

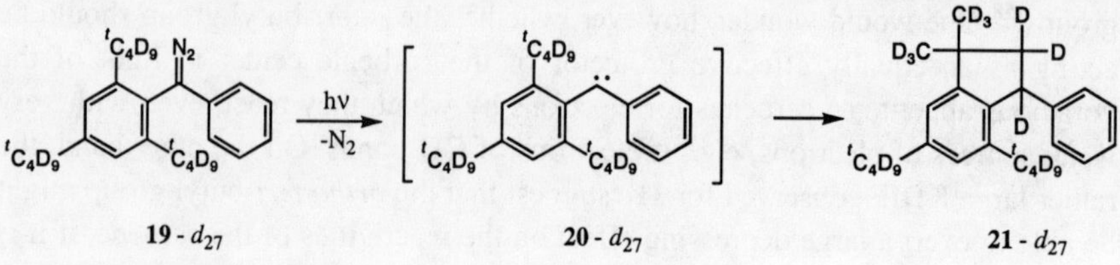

Scheme 8.

oxygen-saturated acetonitrile solution of **19**, a weak and broad absorption band with a maximum at 400 nm appeared, and the growth rate of this transient absorption increased as the concentration of oxygen was increased. This observation can be again interpreted as indicating that the triplet carbene is trapped with oxygen to generate carbonyl oxide **22**. The bimolecular rate constant for the reaction of **20** with oxygen, k_{O_2}, is determined to be $(1.2 \pm 0.2) \times 10^8 \ M^{-1}s^{-1}$. The observations indicate that the nascent singlet carbene **20** is trapped almost instantaneously by the *tert*-butyl group before it undergoes intersystem crossing to an extent that the triplet can be observed directly.

LFP of a degassed benzene solution of benzophenone (BP) as a triplet sensitizer at room temperature with the third harmonic wavelength (355 nm) of a Nd/YAG laser produced a transient species showing maxima at 340 and 530 nm, apparently due to the excited triplet state of BP.[59] When LFP of BP was carried out in the presence of **19**, a similar absorption was observed. However, the absorption band at 530 nm disappeared much faster than that at 340 nm and the rate of the 530-nm decay increased with increasing concentration of **19**, suggesting that triplet BP is quenched by **19**. The bimolecular rate constant for triplet BP quenching by **19** was estimated to be $k_{BP} = (4.7 \pm 0.8) \times 10^9 \ M^{-1}s^{-1}$, indicating that energy transfer from triplet BP to **19** occurs. The residual spectrum at 340 nm is similar to that obtained during the photolysis of **19** in MTHF glass at 77 K, and assigned to the triplet carbene generated from the excited triplet state of **19**. The decay was found to be first order $(k_i = 8.2 \times 10^3 \ s^{-1})$, in accordance with the product analysis, showing that intramolecular CH insertion to form **21** is the main decay pathway for triplet carbene **20** under these conditions. Thus, the lifetime was determined to be approximately 120 µs, which is some 60 times longer than that observed for the parent DPC. It is probable that indane **21** must be produced from the triplet carbene by a stepwise mechanism by way of a 1,5-biradical. However, as the signal due to **20** decayed, no new transient absorption was detected to the limit of our system. Since the 1,5-biradical formed as a result of δ-H abstraction of the *o-tert*-butyl benzophenone triplet state is shown to have a lifetime of only 4 ns in non-polar solvents,[60] it is likely that the 1,5-biradical is too short-lived to be detected by our system.[61]

Sensitized LFP of **19**-d_{27} also provided a transient absorption at 340 nm assignable to **20**-d_{27}. The decay of the absorption was also found to be first-order, again in accordance with the product analysis data, and k_i^D was determined to be 6.3 × 10^3 s^{-1} ($\tau = 160$ μs). The KDIE estimated for intramolecular H abstraction by triplet carbene **20** based on k_i is, therefore, only 1.3. This is an extremely small value when compared to that observed for the similar H abstraction by triplet carbene **11c**.

In order to gain more insight into the nature of this H migration, the rate measurements were carried out as a function of temperature over a temperature range of 7 to 25 °C. The Arrhenius plot of the data again is almost completely linear ($r = 0.99$) for both the **20** and **20**-d_{27} decays in the temperature range studied. The Arrhenius parameters for H and D abstraction are summarized in Table 13. Thus, the differential activation energy ($E_a^D - E_a^H = 0.3$ kcal/mol) and the ratio of pre-exponential factors ($A_H/A_D = 0.7$) again are all consistent with a classical atom-transfer reaction.[52]

It is very interesting to compare the kinetic parameters for the H-abstraction reaction between the two carbene systems, **11c** and **20**. First, while the nature of hydrogens being abstracted is very different between the two carbenes, i.e., benzylic hydrogens for **11c** and primary aliphatic ones for **20**, there is only a small difference in their E_a values. This can be rationalized by the suggestion that the transferring hydrogen in **11c** is in the plane of the aromatic ring and, hence, little benzylic stabilization is expected in the transition state (TS). On the other hand, the very large A value for **20** over that for **11c** can easily be understood in terms of a large difference in the number of the transferring hydrogens and the distance between the hydrogens and the carbenic center, and is also responsible for the large difference in k_H between the two systems with nearly equal E_a values.

Second, the significant difference in k_H/k_D and $\Delta[E_a]_D^H$ should also be noted. These values are usually considered to provide information about the transition state geometry and the possibility of a tunneling mechanism. It is generally accepted, for instance, that values of $\Delta[E_a]_D^H$ larger than the difference in C–H and C–D bond zero point energies (≥ 1.2 kcal/mol) and an unusually low pre-exponential ratio of A_H/A_D (< 0.7) provide strong evidence for the occurrence of tunneling.[56,57] By these criteria, the data obtained with **11c** and **20** are consistent with a classical atom transfer. Concerning transition state (TS) geometries, it is suggested that a "linear" TS experiences a higher KIE than a "bent" TS.[62] The closest possible

Table 13. Arrhenius Parameters for H Migration in Carbenes **20** and **20**-d_{27}[52]

Carbene	Temp (K)	E_a (kcal/mol)	A (s^{-1})	$[\Delta E_a]_D^H$ (kcal/mol)	A_H/A_D
20	298–280	11.0	1.1×10^{12}	0.3	0.7
20-d_{27}	298–280	11.3	1.6×10^{12}		

distance between the carbenic carbon and the transferring H atom for triplet 2-methyl- and 2-*tert*-butylphenylcarbenes is calculated to be 244 and 167 pm, respectively (PM3-ROHF/CI, 4 × 4). It is then conceivable that the transferring H in **20** should have more freedom in the TS than in **11c**. Thus, the large difference in k_H/k_D for the H transfer between the two carbenes can be understood in terms of difference in the TS structure related to the ground state structure.

The *tert*-butyl group, successfully employed as one of the most effective protecting groups for many other reactive centers in organic chemistry, was found to be ineffective, much less even than methyl and chlorine groups in protecting triplet DPC.

D. Polybrominated Diphenylcarbenes

Among the protecting groups thus far examined, the methyl group is most effective in protecting the carbenic center. Thus, didurylcarbene **11b** is shown to be some five orders of magnitude longer-lived than DPC. However, when two more methyl groups are introduced at the *para* positions, the carbenic center is more easily trapped by the *ortho* methyl groups which are brought closer to the carbene center. The carbene, therefore, becomes shorter-lived again. In order to realize a much longer-lived triplet DPC, we need to explore a new protecting group that must be bigger and less reactive than a methyl group.

Bromine atoms appear to be promising as protecting groups of a triplet carbene center because the van der Waals radius is similar to that of methyl (Br: 195 pm, Me: 200 pm), and the C–Br bond length (185 pm) is longer than C–C(Me) (150 pm).[63] This suggests that the *ortho*-bromine atoms must protect the reactive site more effectively. Moreover, C-halogen bonds are usually reactive toward singlet carbenes to form halonium ylides, but not reactive with triplet states.[64] In addition, a value of ξ_l, which gives a measure of the strength of the spin–orbit interaction, is increased dramatically in going from chlorine ($370\ \mathrm{cm^{-1}}$) to bromine ($2460\ \mathrm{cm^{-1}}$),[65] suggesting that intersystem crossing from the nascent singlet carbene to the triplet should be accelerated by introducing a bromine atom. Thus, bromine atoms are expected to stabilize triplet DPC in several ways, especially when introduced at the *ortho* positions.

Product analysis studies show that the reaction patterns observed with polybrominated diphenylcarbenes are completely different from those observed with other sterically congested diarylcarbenes. Thus, irradiation of a degassed solution of 2,2′,4,4′,6,6′-hexabromodiphenyldiazomethane (**24a**) in benzene at room temperature afforded rather complex product mixtures containing only small amounts of carbenic dimers and radical dimers (Scheme 9). Photolysis in methanol solution did not result in an appreciable formation of the expected O–H insertion product. This is in marked contrast especially with the results observed for the hexachlorodiphenylcarbene **2a**, which gave the carbene dimer in benzene and the methyl ether in methanol, respectively, as almost exclusive products. These findings

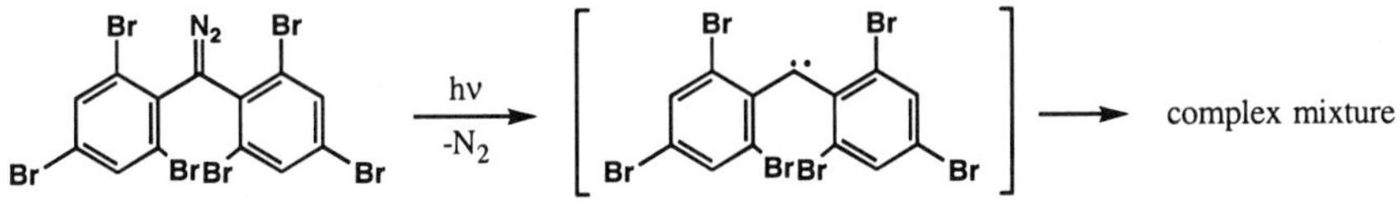

Scheme 9.

suggest that the carbenic center in hexabromodiphenylcarbene **25a** is highly sterically congested.

Irradiation of **24a** in MTHF glass at 4 K gave a fine-structured ESR line shape characteristic of randomly oriented triplet molecules with a large D value attributable to triplet carbene **25a** ($D = 0.3598$ cm^{-1}, $E = 0.0295$ cm^{-1}). The ESR signals not only were stable at this temperature, but also survived even at 130 K. However, as the samples were warmed, the x and y lines of the spectrum moved closer together, resulting in an essentially zero E value ($D = 0.3656$ cm^{-1}, $E = 0$ cm^{-1}). Cooling the sample did not reverse this change. This indicates that, on warming, the carbene relaxes to a structure with an expanded C–C–C angle, presumably to gain relief from steric compression. Similar geometrical changes upon annealing of the matrix have been observed especially for sterically congested carbenes.[41,42] Thus, when a carbene is formed at low temperature, the rigidity of the matrix prevents it from assuming its minimum energy geometry. However, when the matrix is softened on annealing, the carbene relaxes to a stabler structure. Photolysis of single crystals of **24a** at liquid helium temperature also generated ESR spectra with fine-structure, due to a typical triplet diarylcarbene with a large D value. Considering the effect of the Boltzmann factor, the signals survived at room temperature without loss of intensity.

Irradiation of **24a** in MTHF glass at 77 K resulted in the appearance of new absorption bands in the UV/Vis spectrum at the expense of the original absorption due to **24a**. The new spectrum consists of two identifiable features that are usually present in the spectra of triplet diarylcarbenes. The glassy solution did not exhibit any spectral changes for several hours when kept at 77 K, but the characteristic bands disappeared on warming the matrix to room temperature. On the basis of these observations, coupled with the ESR data, the absorption spectrum can be attributed to triplet hexabromodiphenylcarbene **25a**.

Monitoring of the spectral changes as a function of temperature shows two interesting features. First, the absorption maxima shifted slightly but distinctly when the matrix was warmed from 90 to 95 K. When the samples were cooled again, this change was not reversed. This is interpreted in terms of the geometrical changes carbenes can undergo. It should be noted here that such a distinct shift in a UV/Vis spectrum has not previously been observed for other relatively congested diarylcarbenes undergoing geometrical changes. This suggests that the magnitude of the change must be fairly large, as noted in the ESR spectral changes. Second, while absorption bands of most diarylcarbenes disappear in the temperature range of 100 to 105 K, where viscosity of the matrix changes dramatically from 10^7 to 10^3 P, no appreciable changes were observed for the absorption bands ascribable to **25a** in MTHF at 120 K, where the samples were completely fluid. Significant decomposition began only at 130 K (viscosity is 1.6 P) where the "first-order" half-life ($t_{1/2}$) was approximately 15 min. In more viscous solvents, **25a** was able to survive at a much higher temperature; 190 K in dibutyl phthalate ($t_{1/2} = 37$ min), and 210 K in triacetine ($t_{1/2} = 33$ min).

LFP of **24a** in a degassed benzene solution at room temperature with a 308 nm pulse from a XeCl excimer laser produced a transient species showing a strong absorption at 353 nm and a weak absorption extending from 420 to 500 nm, which appeared coincidentally with the pulse. The transient signals disappeared rather gradually, persisting for at least 30 s under these conditions. On the basis of the low-temperature spectrum, the transient product was assigned to triplet **25a**. Trapping experiments using oxygen supports this assignment. Thus, the lifetime of **25a** was dramatically decreased and a new broad band with a maximum at 410 nm, ascribable to the corresponding carbonyl oxide, appeared at the expense of the carbene absorption when LFP was carried out on a non-degassed solution of **24a**. Decay of triplet **25a** in the absence of oxygen is found to be second-order ($2\,k_d/\varepsilon l = 8.9\ \text{s}^{-1}$). From the decay curve, the rough lifetime of **25a** is estimated in the form of its half-life, $t_{1/2}$, to be 1 ± 0.1 s.[66]

Intrinsic reactivities of hexabromodiphenylcarbene **25a** toward typical carbene quenchers were determined by LFP and are summarized in Table 14, which reveals rather surprising results. While the quenching rate constant by oxygen, for instance, is somewhat decreased compared to that observed for the hexachloro **2a** and hexamethyldiphenylcarbene **11a**, the H-abstraction rate constant from CHD is found to be a little bit greater than those for **2a** and **11a**. The signals were not quenched by methanol, indicating that this carbene does not react with methanol, as shown by the product analysis study.[67]

More interestingly, the lifetimes of triplet *ortho*-tetrabrominated diphenylcarbenes in benzene are found to be affected by the substituents at the *para* positions

Table 14. Kinetic Parameters for Polybrominated Diphenylcarbenes Ar$_2$C: (25)[67]

Carbene	Ar in Ar$_2$C:	$2k/\varepsilon l$ (s^{-1})	$k_q (M^{-1}s^{-1})$		$t_{1/2}$ (s)
			k_{O_2}	k_{CHD}	
25a	2,4,6-Br$_3$C$_6$H$_2$	8.9	1.1×10^7	7.4×10^2	1.0
25b	2,6-Br$_2$-4-MeC$_6$H$_2$	—	2.3×10^7	3.7×10^2	0.22
25c	2,6-Br$_2$-4-^{t}BuC$_6$H$_2$	0.35	2.1×10^7	5.3×10^2	16

Table 15. ESR Data of Polybrominated Diphenylcarbenes Ar$_2$C: (**25**)[a67]

Carbene	Ar in Ar$_2$C:	$D(cm^{-1})$	$E(cm^{-1})$	E/D
25a	2,4,6-Br$_3$C$_6$H$_2$	0.3958	0.0295	0.0745
25b	2,6-Br$_2$-4-MeC$_6$H$_2$	0.3957	0.0275	0.0691
25c	2,6-Br$_2$-4-^{t}BuC$_6$H$_2$	0.3966	0.0311	0.0783

Note: [a]Measured in 2–methyltetrahydrofuran at 77 K.

in a striking manner. Thus, the halflife of the triplet state was decreased when bromine atoms at the *para* positions are replaced by methyl groups, but increased rather dramatically when more bulky groups such as *tert*-butyl groups are introduced. Thus, in the case of 2,2′,6,6′-tetrabromo-4,4′-di-*tert*-butyldiphenylcarbene (**25c**), the transient absorption survived over several minutes in degassed benzene at room temperature. The halflife of **25c** was estimated to be 16 s. Surprisingly, however, the reactivities of these highly long-lived triplet carbenes toward oxygen and CHD were not changed so dramatically (Table 14). In order to obtain insight into the effect of *para* substituents on the structure of these tetrabromodiphenylcarbenes, ESR spectra were measured. Zero field splitting parameters summarized in Table 15 clearly suggest that changes of substituents at the *para* positions result in almost no meaningful change in their structures.

V. SUMMARY

Our studies have shown that the introduction of four substituents at all *ortho* positions of diphenylcarbenes is a very effective way to increase the lifetime of triplet states without resulting in an undesirable decrease in ΔG_{ST}. The extent of stabilization is, of course, dependent not only upon the van der Waals radius and bond length but also the reactivity of the substituents. The bromine group is found to be the most effective kinetic stabilizer for a triplet carbene center. In light of the fact that C–F bonds are generally believed to be the least reactive bonds toward carbenes, trifluoromethyl groups seem to be the most effective and, hence, attractive stabilizers for triplet carbenes. The idea to generate 2,2′,6,6′-tetra-(trifluoromethyl)substituted diphenylcarbenes is, however, encountered by the problem of preparing the corresponding carbene precursor.[68,69]

Effects of *meta* substituents on the reactivity of 1,2-disubstituted benzenes are usually explained in terms of a steric effect exerted on a functional group by a more distant substituent through another immediately adjoining substituent (buttressing effect). Originally, the buttressing effect was demonstrated by a comparison of 2,3- and 2,5-dimethyl derivatives in order to eliminate possible polar effects.[70,71] The work along this line has not been carried out yet. On the other hand, it has been shown by the isodesmic reactions calculated for a series of methylbenzoic acids at the semiempirical level (AMI) that this effect is not connected with the word "buttress." It does not say more than that the interaction energy in trisubstituted systems is generally larger than predicted from disubstituted systems, since this effect is not zero, even in 2,5-dimethyl derivatives.[72] We have not yet analyzed the effect of *meta* substituents in terms of isodesmic reactions. However, not only product analysis and kinetic data, but also theoretical analysis for polymethylated diphenylcarbenes geometries, fits well with the classical picture based on space-filling models, at least in the range of the methyl substitution patterns we studied.

Substituents at *para* positions are usually considered to exert little steric effect to strengthen the *ortho* effect, although the isodesmic reactions suggest that the interaction with the *ortho* substituents is certainly not zero.[72] The examination of the effect of methyl groups at the *para* positions on the reactivity toward methanol reveals that polar effects obviously play more important roles. However, *para* effects are obviously not understood simply by polar effects. Peculiar effects of *p*-alkyl groups on the reactivity of 2,2′,6,6′-tetrabromodiphenylcarbenes (**25**) are especially noteworthy; large effects on the lifetime in benzene contrast with small effects on k_{CHD} and k_{O_2}. The fact that the reactions with highly efficient triplet-trapping reagents are little affected by *para* substituents can be interpreted to mean that *para* substituents exert little effect on the steric congestion around the carbene center. The reactions of these carbenes in benzene appear to be complicated since no significant products are isolated from the reaction mixtures.

However, since benzene is generally recognized as a very unreactive solvent, especially for triplet carbenes and, hence, the hindered triplet diarylcarbene concentration must build up, it is highly likely that the most reactive counterparts under these conditions must be the triplet carbenes themselves. However, the products obtained from the photolysis in benzene consist of a highly complex mixture containing only small amounts of carbene dimers. This is in marked contrast with the results observed in the similar reaction of the hexachloro- and hexamethyldiphenylcarbenes, which both produce the corresponding dimers as main products. It is then probable that simple dimerization of hexabromodiphenylcarbene at the carbenic center must suffer from severe steric repulsion and, therefore, the carbene is forced to react at the aromatic rings where spin can be delocalized.

In this light, it is important to note that trityl radicals are known to undergo either "head-to-tail" or methyl–*para*, or "tail-to-tail" or *para–para* couplings depending on the substitution patterns.[73] Thus, it is likely that the brominated diphenylcarbenes also undergo similar couplings. The coupling reactions of trityl radicals are not suppressed by "reactive" *para* substituents. The anomalous reactivity of halogen at the *para* position is noted; for instance, *p*-bromophenyldiphenylmethyl undergoes methyl–*para* coupling.[74] On the other hand, tri-*p*-tolylmethyl undergoes rapid disproportionation to yield tris-*p*-tolylmethane and a quinoid structure that rapidly polymerizes.[75] However, the coupling at the *para* position is retarded by *tert*-butyl groups not only at the *para*[76] but also at *meta* positions.[77] Thus, the rather large effects of *tert*-butyl groups at the *para* positions on the lifetime of the brominated diphenylcarbenes in benzene compared to that of bromo and methyl groups are compatible with the effect of substituents observed in the coupling reactions of trityl radicals. The complexity of the products observed in the reactions of the brominated diphenylcarbenes must be partly due to the complexity associated with the coupling reactions. The α-*para* coupling of triplet carbenes, for instance, unlike that of trityl radicals, does not lead to a final stable product, but gives rise to an intermediate open-shell molecule, whose subsequent reactions will be complicated by possible ambient reactivity.

Finally, triplet diarylcarbenes with polynuclear aromatic rings should also be briefly mentioned here in terms of persistency. The chemistry of these carbenes has still not been well studied and mainly examined by spectroscopical methods. ESR studies of 1,1′-dinaphthyl- (**26**)[78] and 9,9′-dianthrylcarbenes (**27**)[79] clearly suggest that the triplet states come to take a less bent structure with more extensive spin delocalization as the number of the aromatic rings is increased.

9,9′-Dianthrylcarbene is especially interesting in this connection. This sterically highly hindered species exhibits unusual properties, which mainly derive from its linear geometry with a nearly completely delocalized spin within the molecule. The ZFS parameters given below were measured at 4 K and represent the upper limits; at higher temperatures, E is essentially zero. In an anthracene matrix, the carbene can be detected at temperatures as high as 160 °C.[79] However, the stability is completely due to the rigidity of the environment and not to an intrinsic lack of reactivity. For instance, flash photolysis studies show that the carbene undergoes self-reaction at the diffusion-controlled limit in benzene at room temperature.[80]

26

D=0.2609 cm^{-1}, E=0.0051 cm^{-1}

(15K/MTHF)[78]

27

D=0.113 cm^{-1}, E=0.0011 cm^{-1}

(4K/DAK)[79]

ACKNOWLEDGMENTS

Many thanks go to Mr(s) Hidetsumu Okada, Kazunori Komatsu, Takehito Nakayama, Tetsuya Watanabe, Hidehiko Mizuno, Junichi Nakajima, and Makoto Hattori for their time-consuming and painstaking efforts to prepare the highly congested diazomethanes; Dr. Katsuyuki Hirai for his very careful spectroscopic measurements; and Professors Shigeru Murata, Takeji Takui, and Koichi Itoh for their enlightening discussions.

We are also grateful for the financial support from the Ministry of Education, Science and Culture, Japan (Monbusho) through a Grant-in-Aid that has made this work possible.

REFERENCES

1. Gomberg, M. *J. Am. Chem. Soc.* **1900**, *22*, 757; Gomberg, M. *Ber. Dtsch. Chem. Ges.* **1900**, *33*, 3150.

2. Pryor, W.A., Ed.; *Organic Free Radicals*; American Chemical Society: Washington, DC, 1977.

3. Leffler, J.E. *An Introduction to Free Radicals*; Wiley: New York, 1993.

4. For reviews see: (a) Rajca, A. *Chem. Rev.* **1994**, *94*, 871; (b) Iwamura, H. *Adv. Phys. Org. Chem.* **1990**, *26*, 179.

5. For a brief history of carbene chemistry, see: Kirmse, W. *Carbene Chemistry*; Academic Press: New York, 1964, p 5.

6. (a) Igau, A.; Grützmacher, H.; Baceiredo, A.; Bertrand, M. *J. Am. Chem. Soc.* **1988**, *110*, 6463; (b) Igau, A.; Baceiredo, A.; Trinqueir, G.; Bertrand, G. *Angew. Chem., Int. Ed. Engl.* **1989**, *28*, 621; (c) Gilette, G.R.; Baceiredo, A.; Bertrand, G. *Angew. Chem., Int. Ed. Engl.* **1990**, *29*, 1429; (d) Dixon, D.A.; Dobss, K.D.; Arduengo, A.J., III; Bertrand, G. *J. Am. Chem. Soc.* **1991**, *113*, 8782; (e) Soleilhavoup, M.; Baceiredo, A.; Treutler, O.; Ahlrichs, R.; Nieger, M.; Bertrand, G. *J. Am. Chem. Soc.* **1992**, *114*, 10959.

7. (a) Arduengo, A.J., III; Harlow, R.L.; Kline, M. *J. Am. Chem. Soc.* **1991**, *113*, 361; (b) Arduengo, A.J., III; Kline, M.; Galabrese, J.C.; Davidson, F. *J. Am. Chem. Soc.* **1991**, *113*, 9704; (c) Arduengo, A.J., III; Kline, M.; Galabrese, J.C.; Davidson, F. *J. Am. Chem. Soc.* **1991**, *113*, 9704; (d) Dixon, D.A.; Arduengo, A.J., III. *J. Phys. Chem.* **1991**, *95*, 4180; (e) Arduengo, A.J., III; Rasika Dias, H.V.; Harlow, R.L.; Kline, M. *J. Am. Chem. Soc.* **1992**, *114*, 5530; (f) Arduengo, A.J., III; Bock, H.; Chen, H.; Denk, M.; Dixon, D.A.; Green, J.C.; Herrman, W.A.; Jones, N.L.; Wagner, M.; West, R. *J. Am. Chem. Soc.* **1994**, *116*, 6641; (g) Arduengo, A.J., III; Rasika Dias, H.V.; Dixon, D.A.; Harlow, R.L.; Klooster, W.T.; Koetzle, T.F. *J. Am. Chem. Soc.* **1994**, *116*, 6812.

8. (a) Regitz, M. *Angew. Chem., Int. Ed. Engl.* **1991**, *30*, 674; (b) Dagani, R. *Chem. Eng. News* **1991**, *Jan 28*, 19; **1994**, *May 2*, 20; (c)Heinemann, C.; Müller, T.; Apeloig, Y.; Schwartz, H. *J. Am. Chem. Soc.* **1996**, *118*, 2023; (d) Beohme, C.; Frenking, G. *J. Am. Chem. Soc.* **1996**, *118*, 2039.

9. Zimmerman, H.E.; Paskovich, D.H. *J. Am. Chem. Soc.* **1964**, *86*, 2149.

10. For reviews, see: (a) Regitz, M., Ed.; *Carbene(oide), Carbine;* Houben-Weyl, Thieme: Stuttgart, 1989; Vol. E19b; (b) Wentrup, C. *Reactive Molecules*; Wiley: New York, 1984; Ch. 4; (c) Moss, R.A., Jones, M., Jr., Eds.; *Carbenes*; Wiley: New York, 1973, 1975; Vols. I and II; (d) Kirmse, W. *Carbene Chemistry*, 2nd ed.; Academic Press: New York, 1971.

11. Fleming, I. *Frontier Orbitals and Organic Chemical Reactions*; Wiley: New York, 1976. See also Rauk, A. *Orbital Interaction Theory of Organic Chemistry*; Wiley: New York, 1994.

12. Murray, K.K.; Leopold, D.G.; Miller, T.M.; Lineberger, W.C. *J. Chem. Phys.* **1988**, *89*, 5442.

13. Koda, S. *Chem. Phys. Lett.* **1978**, *55*, 353; *Chem. Phys.* **1982**, *66*, 383.

14. Sander, W.; Bucher, G.; Wierlacher, S. *Chem. Rev.* **1993**, *93*, 1583.

15. Harrison, J.F. *J. Am. Chem. Soc.* **1971**, *93*, 4112.

16. Reisenauer, H.P.; Maier, G.; Riemann, A.; Hoffmann, R.W. *Angew. Chem.* **1984**, *96*, 596.

17. Myers, D.R.; Senthilnathan, V.P.; Platz, M.S.; Jones, M., Jr., *J. Am. Chem. Soc.* **1986**, *108*, 4232.

18. For reviews see: Platz, M.S., Ed.; *Kinetics and Spectroscopy of Carbenes and Biradicals*; Plenum: New York, 1990; Jackson, J.E.; Platz, M.S. *Advances in Carbene Chemistry*, Brinker, U.H., Ed.; JAI Press: Stamford, CT, 1994, Vol. 1, p 87.

19. Moss, R.A. *Acc. Chem. Res.* **1989**, *22*, 15; **1980**, *13*, 58.

20. (a) Cox, D.P.; Gould, I.R.; Hacker, N.P.; Moss, R.A.; Turro, N.J. *Tetrahedron Lett.* **1983**, *24*, 5313; (b) Moss, R.A.; Fan, H.; Hadel, L.M.; Shen, S.; Wloshtowska, J.; Wlostowski, M.; Krogh-Jes-

persen, K. *Tetrahedron Lett.* **1987**, *28*, 4779; (c) Moss, R.A.; Perez, L.A.; Turro, N.T.; Gould, I.R.; Hacker, N.P. *Tetrahedron Lett.* **1983**, *24*, 1983.

21. See also: Moss, R.A.; Turro, N.J. *Kinetics and Spectroscopy of Carbenes and Biradicals*, Platz, M.S., Ed.; Plenum: New York, 1990; p 213.

22. (a) Kim, S.-J.; Hamilton, T.P.; Schaefer, H.F., III *J. Chem. Phys.* **1991**, *94*, 2063; (b) Russo, N.; Sicilia, E.; Toscano, M. *J. Chem. Phys.* **1992**, *95*, 5031.

23. Du, X.-M.; Fan, H.; Goodman, J.L.; Kesselmayer, M.A.; Krogh-Jespersen, K.; Lavilla, J.A.; Moss, R.A.; Shen, S.; Sheridan, R.S. *J. Am. Chem. Soc.* **1990**, *112*, 1920.

24. Rondan, N.; Houk, K.N.; Moss, R.A. *J. Am. Chem. Soc.* **1980**, *102*, 1770.

25. Chateauneuf, J.E.; Johnson, R.P.; Kirchhoff, M.M. *J. Am. Chem. Soc.* **1990**, *112*, 3217.

26. (a) Carr, R.W., Jr.; Peterson, D.G.; Smith, F.K. *J. Phys. Chem.* **1986**, *90*, 607; (b) Chowdhury, P.K.; Rama Rao, K.V.S.; Mittal, J.P. *J. Phys. Chem.* **1988**, *92*, 102.

27. Moss, R.A.; Shen, S.; Hadel, L.M.; Kmiecik-Lawrynowicz, G.; Wlostowska, J.; Krogh-Jespersen, K. *J. Am. Chem. Soc.* **1987**, *109*, 4341.

28. Moss, R.A.; Wlostowski, M.; Shen, S.; Krogh-Jespersen, K.; Matro, A. *J. Am. Chem. Soc.* **1988**, *110*, 4443.

29. Arduengo, A.J., III; Goerlich, J.R.; Marshall, W.J. *J. Am. Chem. Soc.* **1995**, *117*, 11027.

30. Alder, R.W.; Allen, P.R.; Murray, M.; Orpen, A.G. *Angew Chem., Int. Ed. Engl.* **1996**, *35*, 1121; see also, Alder, R.W.; Allen, P.R.; Williams, S.J. *J. Chem. Soc., Chem. Commun.* **1995**, 1267.

31. Hadel, L.M.; Maloney, V.M.; Platz, M.S.; McGimpsey, W.G.; Scaiano, J.C. *J. Phys. Chem.* **1986**, *90*, 2488.

32. For review of O-H insertion reactions of carbenes, see Kirmse, W. *Advances in Carbene Chemistry*, Brinker, U.H., Ed.; JAI Press: Stamford, 1994; Vol. 1, p 1.

33. Bethell, D.; Hayes, J.; Newall, A.R. *J. Chem. Soc., Perkin Trans. 2* **1974**, 1307.

34. Griller, D.; Nazran, A.S.; Scaiano, J.C. *J. Am. Chem. Soc.* **1984**, *106*, 198; (b) Griller, D.; Nazran, A.S.; Scaiano, J.C. *Tetrahedron Lett.* **1985**, *41*, 1527.

35. Miller, R.J.; Shechter, H. *J. Am. Chem. Soc.* **1978**, *100*, 7920.

36. Schuster, G.B. *Adv. Phys. Org. Chem.* **1986**, *22*, 311.

37. (a) Lapin, S.C.; Brauer, B.-E.; Schuster, G.B. *J. Am. Chem. Soc.* **1984**, *106*, 2092; (b) Rak, S.F.; Lapin, S.C.; Falvey, D.E.; Schuster, G.B. *J. Am. Chem. Soc.* **1987**, *109*, 5003; (c) Brauer, B.-E.; Grasse, P.B.; Kaufmann, K.J.; Schuster, G.B. *J. Am. Chem. Soc.* **1982**, *104*, 6814; (d) Grasse, P.B.; Zupancic, J.J.; Lapin, S.C.; Hendrich, M.P.; Schuster, G.B. *J. Org. Chem.* **1985**, *50*, 2352; (e) Chuang, C.; Lapin, S.C.; Schrock, A.K.; Schuster, G.B. *J. Am. Chem. Soc.* **1985**, *107*, 4238; (f) Lapin, S.C.; Schuster, G.B. *J. Am. Chem. Soc.* **1985**, *107*, 4243.

38. Hoffmann, R.; Zeiss, G.D.; Van Dine, G.W. *J. Am. Chem. Soc.* **1968**, *90*, 1485; Metcalfe, J.; Halevi, E.A. *J. Chem. Soc., Perkin Trans. 2* **1977**, 634.

39. Alt, R.; Gould, I.R.; Staab, H.A.; Turro, N.J. *J. Am. Chem. Soc.* **1986**, *108*, 6911.

40. Staab, H.A.; Alt, R. *Chem. Ber.* **1984**, *117*, 850; Alt, R.; Staab, H.A. *Tetrahedron Lett.* **1984**, *25*, 633.

41. Gilbert, B.C.; Griller, D.; Nazran, A.S. *J. Org. Chem.* **1985**, *50*, 4738.

42. (a) Nazran, A.S.; Gabe, E.J.; LePage, Y.; Northcott, D.J.; Park, J.M.; Griller, D. *J. Am. Chem. Soc.* **1983**, *105*, 2912; (b) Nazran, A.S.; Griller, D. *J. Chem. Soc., Chem. Commun.* **1983**, 850; (c) Nazran, A.S.; Griller, D. *J. Am. Chem. Soc.* **1984**, *106*, 543; (d) Nazran, A.S.; Lee, F.L.; Gabe, E.J.; LePage, Y.; Northcott, D.J.; Park, J.M.; Griller, D. *J. Phys. Chem.* **1984**, *88*, 5251.

43. Gano, J.E.; Wettach, R.H.; Platz, M.S.; Senthilnathan, V.P. *J. Am. Chem. Soc.* **1982**, *104*, 2326.

44. Tomioka, H.; Hirai, K.; Fujii, C. *Acta Chem. Scand.* **1992**, *46*, 680.

45. Tomioka, H.; Hirai, K.; Nakayama, T. *J. Am. Chem. Soc.* **1993**, *115*, 1285.

46. Westheimer, F.H. in: *Steric Effects in Organic Chemistry*; Newman, M.S., Ed.; Wiley: New York, 1956; p 523.

47. For buttressing effects in carbene chemistry, see: Tomioka, H.; Kimoto, K.; Murata, H.; Izawa, Y. *J. Chem. Soc., Perkin Trans. 1* **1991**, 471.

48. (a) Trozzolo, A.M. *Acc. Chem. Res.* **1968**, *1*, 329; (b)Trozzolo, A.M.; Wasserman, E. in: *Carbenes*, Vol. II, Moss, R.A.; Jones, M., Jr., Eds; Wiley: New York, 1975; p 185.

49. (a) Sander, W.W. *Angew. Chem., Int. Ed. Engl.* **1990**, *29*, 344; (b) Scaiano, J.C.; McGimpsey, W.G.; Casal, H.L. *J. Org. Chem.* **1989**, *54*, 1612.

50. Closs, G.L.; Rabinow, B.E. *J. Am. Chem. Soc.* **1976**, *98*, 8190.

51. Tomioka, H.; Okada, H.; Watanabe, T.; Hirai, K. *Angew. Chem., Int. Ed. Engl.* **1994**, *33*, 873.

52. Tomioka, H.; Okada, H.; Watanabe, T.; Banno, K.; Komatsu, K.; Hirai, K. *J. Am. Chem. Soc.* **1997**, *119*, 1582.

53. Furukawa, K.; Teki, Y.; Takui, T.; Itoh, K.; Watanabe, T.; Hirai, K.; Tomioka, H. *Mol. Cryst. Liq. Cryst.* **1995**, *271*, 183.

54. Braun, W.; Bass, A.M.; Pilling, M. *J. Chem. Phys.* **1970**, *52*, 5131; Laufer, A.H. *Res. Chem. Intermed.* **1981**, *4*, 225.

55. Tomioka, H.; Hattori, M.; Hirai, K. unpublished observations.

56. Melander, L.; Saunders, W.H. *Reaction Rates of Isotopic Molecules*; Wiley: New York, 1980.

57. Bigeleisen, J. *J. Phys. Chem.* **1952**, *56*, 823.

58. (a) Puranik, D.B.; Fink, M.J. *J. Am. Chem. Soc.* **1989**, *111*, 5951; (b) Lange, L.L.; Meyer, B.; du Mont, W.-W. *J. Organomet. Chem.* **1987**, *329*, C17.

59. Encinas, M.W.; Scaiano, J.C. *J. Am. Chem. Soc.* **1981**, *103*, 6393.

60. Wagner, P.J.; Giri, B.P.; Scaiano, J.C.; Ward, D.L.; Gabe, E.; Lee, F.L. *J. Am. Chem. Soc.* **1985**, *107*, 5483.

61. Hirai, K.; Komatsu, K.; Tomioka, H. *Chem. Lett.* **1994**, 503.

62. Sorokin, A.; Robert, A.; Meunier, B. *J. Am. Chem. Soc.* **1993**, *115*, 7293.

63. Gordon, A.J.; Ford, R.A. *The Chemist's Companion*; Wiley: New York, 1972.

64. Roth, H.D. *Acc. Chem. Res.* **1977**, *10*, 85.

65. Murov, S.L., Carmichael, I.; Hug, G.L. *Handbook of Photochemistry*; Marcel Dekker: New York, 1993.

66. Tomioka, H.; Watanabe, T.; Hirai, K.; Furukawa, K.; Takui, T.; Itoh, K. *J. Am. Chem. Soc.* **1995**, *117*, 6376.

67. Tomioka, H.; Hattori, M.; Hirai, K. *J. Am. Chem. Soc.* **1996**, *118*, 8723.

68. Diazotetrakis(tifluoromethyl)cyclopentadiene was prepared by a rather tedious method. Very interesting reactivities of the carbene generated from the diazo compound were revealed. See: Janulis, E.P., Jr.; Anduengo, III, A.J. *J. Am. Chem. Soc.* **1983**, *105*, 3563, 5929; Janulis, E.P., Jr.; Wilson, S.R.; Arduengo, III, A.J. *Tetrahedron Lett.* **1984**, *25*, 405.

69. Perfluorodiphenyldiazomethane was shown to be prepared only by the base cleavage of the corresponding N-nitroso-N-carbamate. See: Tomioka, H.; Mizutani, K.; Matsumoto, K,; Hirai, K. *J. Org. Chem.* **1993**, *58*, 7128.

70. Dippy, J.F.J.; Hughes, S.R.C.; Laxton, J.W. *J. Chem. Soc.* **1954**, 1470.

71. Peltier, D. *C.R. Acad. Sci., Ser. C* **1955**, *241*, 57.

72. Decouzon, M.; Ertl, P.; Exner, O.; Gal, J.-F.; Maria, P.-C. *J. Am. Chem. Soc.* **1993**, *115*, 12071.

73. Reviews of triarylmethyls: (a) Sholle, V.D.; Rozantsev. E.G. *Russ. Chem. Rev.* **1973**, *42*, 1011; (b) McBride, J.M. *Tetrahedron* **1974**, *30*, 2009.

74. Bowden, S.T.; Watkins, T.F. *J. Chem. Soc.* **1940**, 1249.

75. Marvel, C.S.; Rieger, W.H.; Mueller, M.B. *J. Am. Chem. Soc.* **1939**, *61*, 2769; Marvel, C.S.; Mueller, M.B.; Himel, C.M.; Kaplan, J.F. *J. Am. Chem. Soc.* **1939**, *61*, 2771.

76. Rajca, A.; Utamupanya, S. *J. Org. Chem.* **1992**, *113*, 2552.

77. Neumann, W.P.; Uzick, W.; Zarkadis, A.K. *J. Am. Chem. Soc.* **1986**, *108*, 3762.

78. Tukada, H.; Sugawara, T.; Murata, S.; Iwamura, H. *Tetrahedron Lett.* **1986**, *27*, 235.

79. Wasserman, E.; Kuck, V.J.; Yager, W.A.; Hutton, R.S.; Greene, F.D.; Abegg, V.P.; Weinshenker, N.M. *J. Am. Chem. Soc.* **1971**, *93*, 6355.

80. Astles, D.J.; Girard, M.; Griller, D.; Kolt, R.J.; Wayner, D.D.M. *J. Org. Chem.* **1988**, *53*, 6053.

LASER FLASH PHOTOLYSIS STUDIES OF CARBONYL CARBENES

John P. Toscano

Advances in Carbene Chemistry
Volume 2, pages 215–244
Copyright © 1998 by JAI Press Inc.
All rights of reproduction in any form reserved.
ISBN: 1-55938-837-4

I. INTRODUCTION

The chemistry of carbonyl carbenes is closely associated with the Wolff rearrangement[1] (Scheme 1), a versatile reaction that has found wide application in diverse areas of research and technology. For example, the Arndt–Eistert synthesis utilizes the Wolff rearrangement of α-diazoketones to ketenes in the homologation of carboxylic acids.[2] Wolff rearrangement of cyclic diazoketones leading to ring contraction has been particularly useful in the synthesis of strained small-ring compounds.[3] In addition, the majority of photoresist materials involved in the lithographic production of integrated circuits are α-diazoketone derivatives.[4] Upon photolysis these compounds form ketenes that react with water to produce carboxylic acids. The difference in carboxylic acid and diazoketone solubility is then utilized to pattern electronic devices.

A persistent question concerning the Wolff rearrangement is whether or not an excited state of the diazo precursor[5] is involved, as well as relaxed carbene and oxirene intermediates. Carbonyl carbenes have been detected in low-temperature matrices by electron spin resonance (ESR)[6] and FTIR[7] spectroscopies. Evidence for the participation of oxirenes[8] comes from experiments where diazoketones labeled at the carbonyl carbon rearranged to form ketenes that possessed a label at both carbon atoms (Scheme 2).[9] Similar results were obtained in studies of unsymmetrically substituted diazoketones.[10] The direct observation of oxirene intermediates, however, has been more elusive. Initial reports of oxirene detection[11,12] were later called into question.[13,14] Recently, by low-temperature FTIR methods, Bodot and co-workers have advanced evidence for the existence of dimethyloxirene and its deuterated analogue.[15]

The intermediates involved in the Wolff rearrangement have also been probed theoretically. Very recently, Schaefer, Radom, and co-workers have calculated the C_2H_2O potential energy surface, which includes formylmethylene, oxirene, and

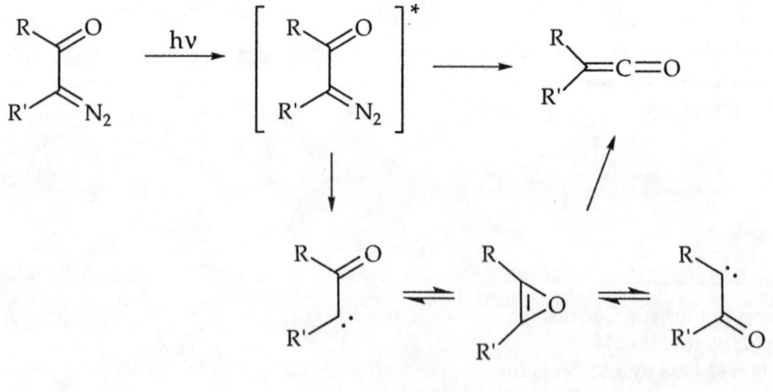

Scheme 1.

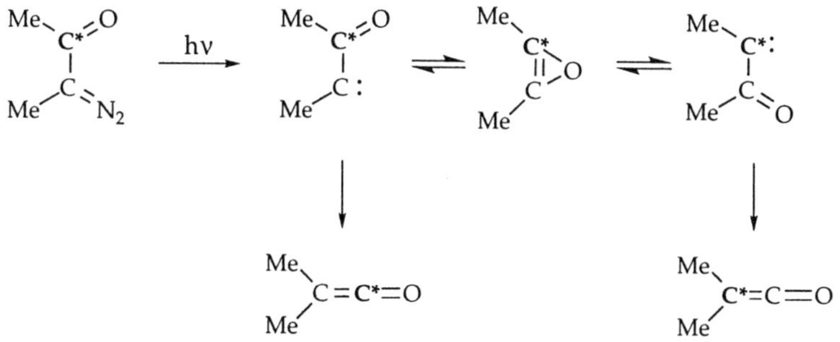

Scheme 2.

ketene, at several levels of *ab initio* theory.[16] They conclude that there is little or no barrier separating singlet formylmethylene and oxirene, but that both of these intermediates must pass over a 5 kcal/mol energy barrier to form ketene (Scheme 3). In accord with previous results,[17] they find that the singlet carbene is nonplanar with the methylene hydrogen perpendicular to the H–C=O plane. In addition, it has been predicted that the triplet carbene is the ground state and that it possesses a planar geometry.[17] Again, this is in agreement with experimental ESR studies[6] that have detected carbonyl carbenes in two rotameric forms (*syn* and *anti*).

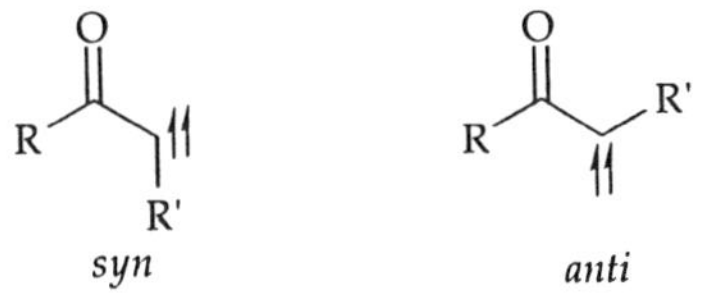

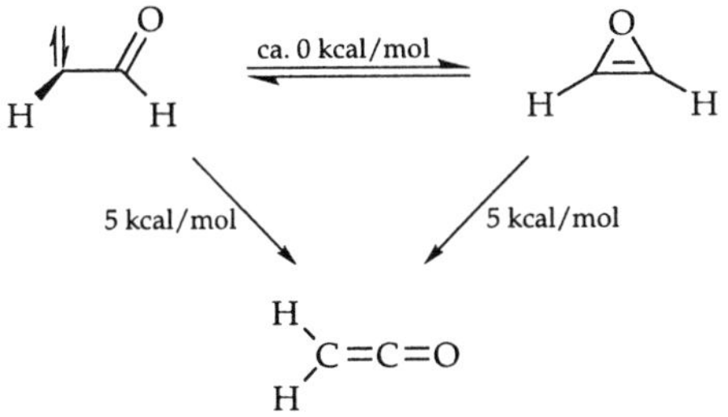

Scheme 3.

II. SINGLET *VS* TRIPLET REACTIVITY

Previous classical studies have clearly indicated that the Wolff rearrangement takes place in the singlet rather than triplet manifold.[18] For instance, Jones and Ando[18b] found that photolysis of diazoacetone **1** in olefinic solutions leads only to Wolff rearrangement products (Scheme 4). Carbene adducts, however, were isolated upon

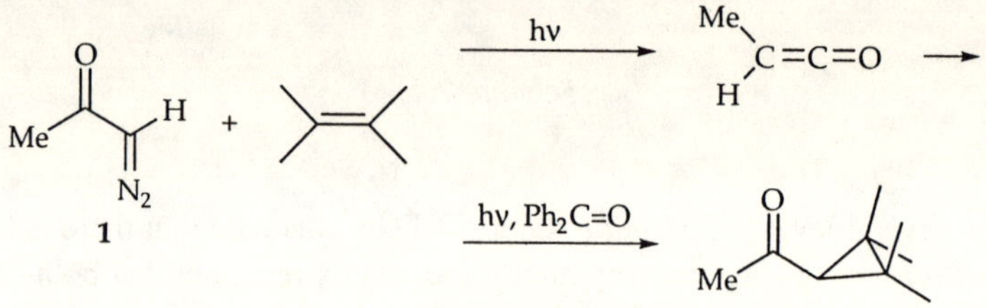

Scheme 4.

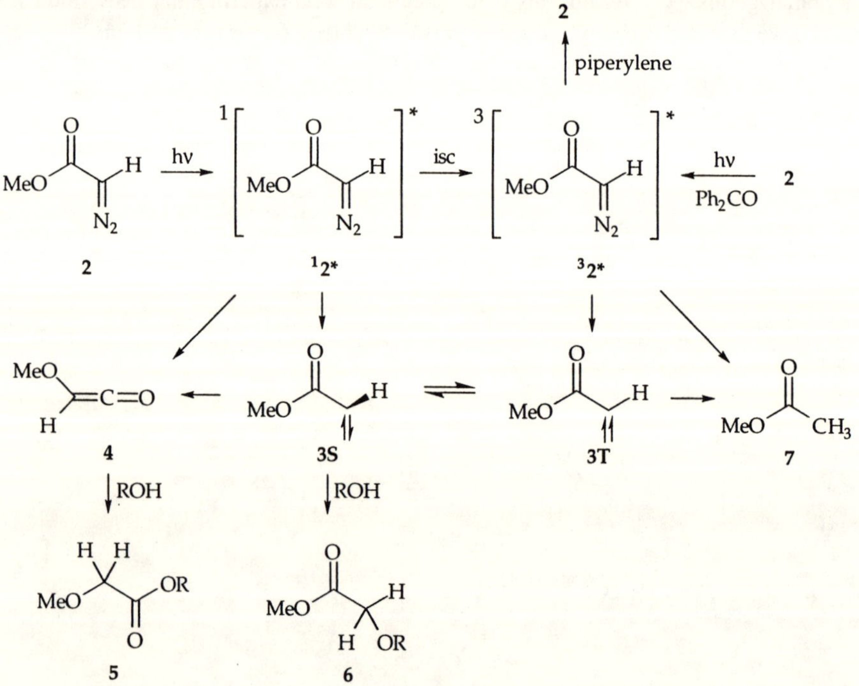

Scheme 5.

triplet-sensitized photolysis, demonstrating that the singlet state of either the diazo precursor or the carbene, is involved in the Wolff rearrangement.

Further insight into the multiplicity of the species involved has been obtained by Tomioka and co-workers in their thorough study of the photochemistry of methyl diazoacetate **2** (Scheme 5).[19] They demonstrated that direct photolysis of **2** produces an excited singlet state (^{1}**2***) that may suffer Wolff rearrangement concurrent with nitrogen extrusion to yield ketene **4**, lose nitrogen to form the singlet carbene **3S**, or intersystem cross to give the triplet diazo ester ^{3}**2***. The ketene and singlet carbene intermediates are both trapped by alcohol, but give different products, **5** and **6**, respectively. Reduction product **7** is clearly derived from a triplet species, ^{3}**2***, **3T**, or both, since its yield is enhanced upon triplet-sensitized photolysis, but is suppressed by the triplet quencher piperylene, following direct photolysis.

The alcohol insertion product **6** is still formed upon sensitized photolysis of **2**, demonstrating that the triplet carbene **3T** can interconvert to the singlet carbene **3S**. Hutton and Roth previously demonstrated by low-temperature ESR spectroscopy that **3T** is the ground state of the carbene.[6b] Thus, Tomioka's results show that the more reactive singlet carbene is close in energy to the triplet ground state. In addition, since formation of ketene **4** is suppressed upon triplet sensitization of **2** in neat alcohol, any singlet carbene that is formed must react with neat alcohol faster than it can undergo Wolff rearrangement.

III. THE ROLE OF CONFORMATION IN THE WOLFF REARRANGEMENT

Several studies have indicated that the conformation of the diazo precursor plays a critical role in the pathway of the Wolff rearrangement. Kaplan and co-workers were the first to demonstrate that diazocarbonyl compounds exist as an equilibrium mixture of *syn* and *anti* forms and that concerted rearrangement is facile in the *syn* conformation, but less likely in the *anti* form (Scheme 6).[20] They found that photolysis of 3,3,6,6-tetramethyl-2-diazocyclo-1-hexanone (**8**), which is locked in the *syn* conformation, leads almost exclusively to the Wolff rearrangement product. On the other hand, photolysis of 2,2,5,5-tetramethyl-4-diazo-3-hexanone (**9**), which exists predominantly in the *anti* conformation, produces no ketene.[21] Instead, carbene-derived products, mainly **10** and **11**, are obtained (Scheme 7).

In general, Kaplan found that the *syn* form is highly preferred in alkyl diazoke-tones, but that the *syn* and *anti* forms are present in approximately equal amounts

8   9

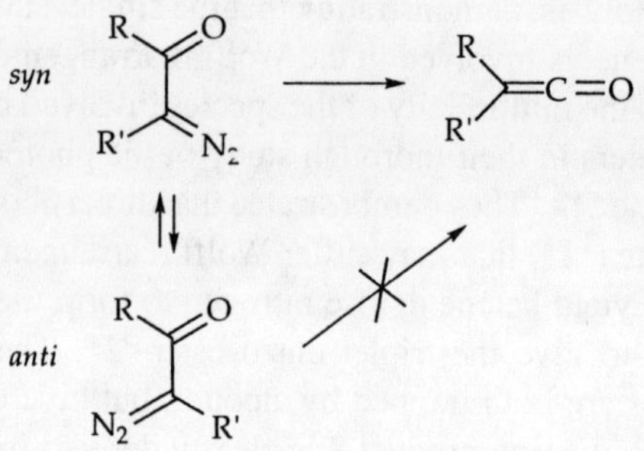

Scheme 6.

in the case of diazoacetic esters.[20] These conformational properties may be the reason that intermolecular carbene chemistry is much more prevalent in carbene esters than in simple ketocarbenes.[1]

Tomioka and co-workers expanded on Kaplan's work in their investigation of a series of compounds, including the diazoketones **12** and **13**, which exist predominantly in the *syn* conformation.[19] Direct photolysis of these compounds in methanol results almost exclusively in products derived from the Wolff rearrangement. However, triplet-sensitized photolysis results in a significant decrease in Wolff rearrangement, along with the appearance of carbene-derived products. These results support Kaplan's hypothesis that diazocarbonyl conformation is preserved in the singlet excited state and that rearrangement occurs in concert with nitrogen loss from the *syn* form of this species.

An alternative to Kaplan's conformational arguments has been advanced by Torres et al., who assert that the conformational control does not originate in the diazo excited state, but rather in the ketocarbene.[22] This, however, seems unlikely since the rearrangement is known to occur from the singlet state and calculations have indicated that the singlet carbene is nonplanar. Thus, once formed, the singlet carbene no longer possesses *syn/anti* conformational identity.

Bodot and co-workers have also argued against concerted rearrangement from a diazo excited state, since in the gas phase photolysis of *anti* diazoketone **9** a significant amount of the ketene rearrangement product was observed.[23] This is in

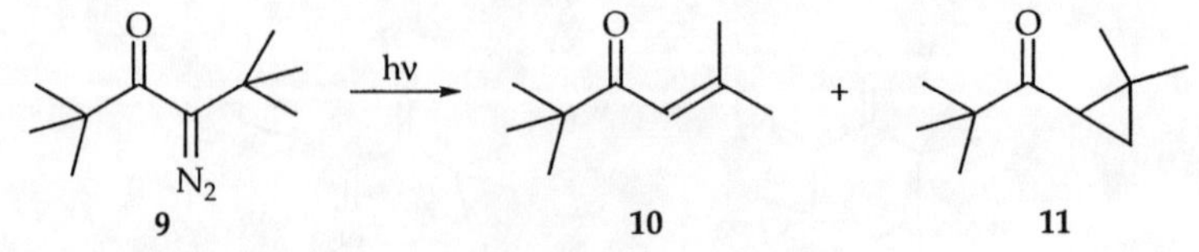

Scheme 7.

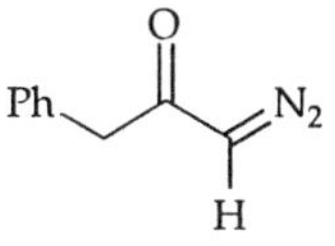

12 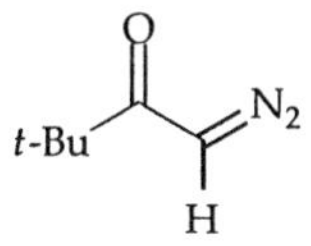   13

contrast with Kaplan and Mitchell's observation that no ketene is observed following photolysis in solution.[20b] Bodot proposes that, in the gas phase, rearrangement may take place from an excited vibrational state of either **[19]*** or the corresponding carbene.

IV. LASER-FLASH PHOTOLYSIS METHODOLOGIES

Recently, laser-flash photolysis (LFP) studies have made significant contributions to understanding the chemistry of carbonyl carbenes and to clarifying the mechanism of the Wolff rearrangement. LFP with UV–VIS detection (transient absorption spectroscopy) is an established tool that is now routinely applied to the study of carbene chemistry.[24] Obviously, in order to study an intermediate of interest by absorption spectroscopy, it must possess a useful chromophore in the UV–VIS region. Unfortunately, many non-aryl carbenes do not possess such chromophores and, therefore, cannot be detected directly by this technique. As a remedy to this problem, in 1988 Platz and co-workers introduced the pyridine-ylide probe methodology to the study of spectroscopically "invisible" carbenes.[25] Here, carbenes without useful absorptions react with probes such as pyridine to form strongly absorbing ylides from which kinetic data can be derived.[24f,25]

The carbene reacts with pyridine in competition with all other possible routes of decay such as rearrangement or reaction with solvent, trace impurities, or carbene precursor. Usually conditions are used such that reactions with solvent, impurities, or precursor are minimized so that the pathway of interest can be studied. The pyridine ylide will be formed in an exponential process following the laser pulse as shown in Eq. 1.

$$[\text{Ylide}]_t = [\text{Ylide}]_\infty \, [1 - \exp(-k_{obs}t)] \tag{1}$$

Here, $[\text{Ylide}]_\infty$ is the concentration of ylide after its formation is complete, but its decay is still insignificant. The time constant k_{obs} is the observed pseudo-first-order rate constant of formation of the ylide. The magnitude of k_{obs} is given in Eq. 2, where k_o is the sum of all rate constants of all first-order and pseudo-first-order processes that consume the carbene in the absence of pyridine.[26]

$$k_{obs} = k_o + k_{pyr}[\text{pyridine}] \tag{2}$$

Thus, a plot of k_{obs} versus [pyridine] reveals both the carbene lifetime ($1/k_o$) and the rate constant for carbene reaction with pyridine (k_{pyr}).

This method can also be used to measure the absolute rate constants for intermolecular carbene reactions with a variety of quenchers. In the presence of an additional quencher, the observed rate constant for the formation of the pyridine ylide is given by Eq. 3. Thus, the observed pseudo-first-order rate constant for ylide formation will increase with increasing quencher concentration and a plot of k_{obs} versus [quencher] at constant [pyridine] will be linear with slope k_q.

$$k_{obs} = k_o + k_{pyr}[\text{pyridine}] + k_q[\text{quencher}] \tag{3}$$

Estimates of carbene lifetimes may still be made even in cases where the rate constant of ylide growth is too fast to be measured. The quantum yield of ylide formation ϕ_y is given in Eq. 4 where ϕ_c is the quantum yield of carbene formation.

$$\phi = \frac{\phi_c k_{pyr}[\text{pyridine}]}{k_o + k_{pyr}[\text{pyridine}]} \tag{4}$$

The optical yield of ylide A_y is related to ϕ_y by Eq. 5 where A_y^∞ is the maximum yield of ylide that can be obtained when the concentration of pyridine is sufficiently large to capture every carbene that is produced in the laser pulse.

$$A_y = \phi_y A_y^\infty \tag{5}$$

Combining and rearranging Eq. 4 and Eq. 5 leads to Eq. 6.

$$\frac{1}{A_y} = \frac{1}{\phi_c A_y^\infty} + \frac{k_o}{\phi_c A_y^\infty k_{pyr}[\text{pyridine}]} \tag{6}$$

Thus, plots of $1/A_y$ versus $1/[\text{pyridine}]$ are predicted to be linear. Division of the intercept by the slope of such plots yields the ratio k_{pyr}/k_o. Carbene lifetimes may then be deduced by assuming a reasonable value for k_{pyr} based on previous work. Studies of a number of carbenes[24f,25] have indicated that k_{pyr} is on the order of 10^9 $\text{M}^{-1}\text{s}^{-1}$.

For carbenes where ylide growth is too fast to be measured intermolecularly, quenching rate constants can still be obtained. If the concentration of pyridine is sufficiently large such that $A_y = A_y^\infty$, then in the presence of a second carbene trap k_q can be obtained from the Stern–Volmer equation below.

$$\frac{A_y^\infty}{A_y} = 1 + \frac{k_q[\text{quencher}]}{k_{pyr}[\text{pyridine}]} \tag{7}$$

By monitoring the decrease in the yield of ylide as a function of [quencher] at constant [pyridine], the ratio $k_q/k_{pyr}[\text{pyridine}]$ can be derived. Quenching rate constants may then be deduced by again assuming a value for k_{pyr} based on previous work.

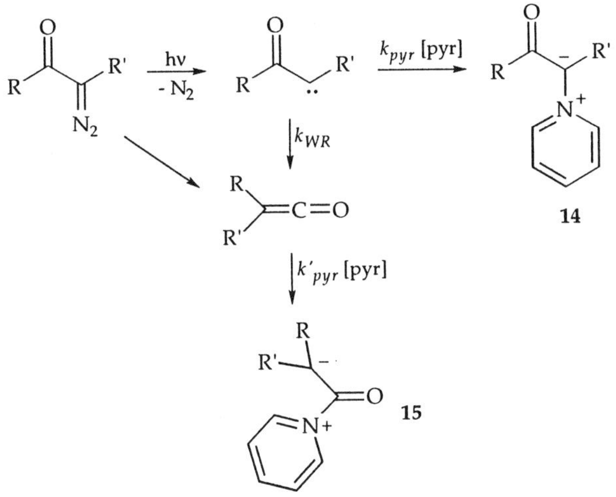

Scheme 8.

As recently pointed out by Scaiano and coworkers,[14,27] a complication that may be faced in LFP studies of carbonyl carbenes is that pyridine may also react with ketenes to form strongly absorbing ketene-pyridine ylides (Scheme 8). Thus, one needs to be concerned about this possibility, and a method of distinguishing carbene-pyridine ylides **14** from ketene-pyridine ylides **15** is required. Such methods will be discussed in the following sections. We will first discuss acyclic carbonyl carbenes, and then cyclic carbonyl carbenes that have been studied by LFP methods.

V. ACYCLIC CARBONYL CARBENES

A. Phenylcarbomethoxycarbene

Itoh, Tomioka, and co-workers have examined methyl diazophenylacetate (**16**) by LFP (Scheme 9).[28] An absorption band attributed to triplet phenylcarbomethoxycarbene (**17T**) with a maximum at approximately 270 nm and a shoulder at 350 nm was observed directly following photolysis in deaerated solutions. At long delay times (> 20 μs), a band with $\lambda_{max} \approx 290$ nm was detected and assigned to radical **18**. These assignments were supported by ESR spectroscopy in low-temperature matrices. In addition, as shown in Table 1, the carbene lifetime was dramatically reduced in aerated solvents consistent with a triplet multiplicity. The triplet carbene was not detected following LFP in toluene, cyclohexane, or 2-methyltetrahydrofuran, suggesting that hydrogen abstraction reactions take place very efficiently in these solvents.

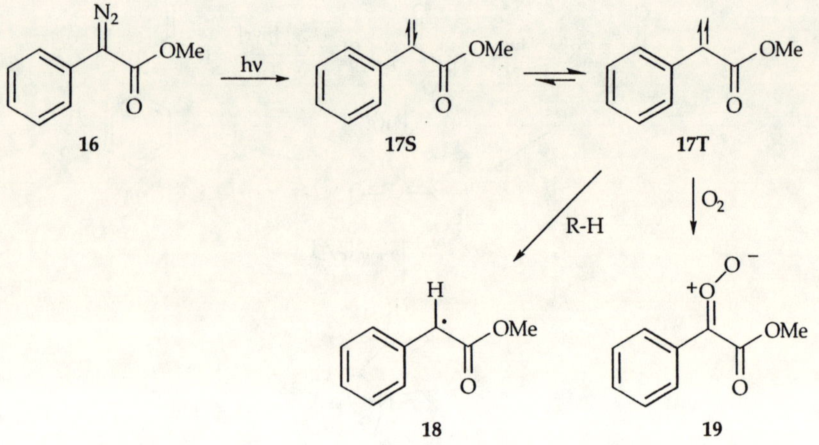

Scheme 9.

Additional evidence for the production of the triplet carbene is provided by the observation of a new transient band (λ_{max} = 410 nm) following LFP in aerated solvents. This band is generated with a risetime of 145 ns in Freon-113 (1,1,2-trichlorotrifluoroethane), which agrees fairly well with the rate of carbene decay (161 ns) in this solvent. Thus, the new band is assigned to carbonyl oxide **19** by analogy to previous studies.[29] By examining the rate of growth of **19** as a function of oxygen concentration, the absolute rate constant for reaction of **17T** with oxygen was determined to be 8.6×10^8 M^{-1}s^{-1}.

The reactivity of carbene **17** with representative quenchers such as 2-methyl-2-butene, methanol, and 1,4-cyclohexadiene was also examined by monitoring the

Table 1. Lifetimes of Phenylcarbomethoxycarbene at 293K in a Variety of Solvents[28]

Solvent	τ (deaerated, ns)	τ (aerated, ns)
Freon-113	461	161
hexafluorobenzene	433	232
benzene	137	80.7
3-methylpentane	126	a
carbon tetrachloride	139	79.2
chloroform	35.1	27.7
acetonitrile	85.6	57.9

Note: [a]No signal for **17T** was observed.

Table 2. The Reactivity of Carbene **17**
with Representative Quenchers[28]

Quencher	k_q $(M^{-1}s^{-1})$
2-methyl-2-butene	7.0×10^7
methanol	2.7×10^8
1,4-cyclohexadiene	2.9×10^8

effect of quencher concentration on the decay of carbene or the growth of carbonyl oxide. Derived rate constants are given in Table 2. Although hydrogen abstraction from 1,4-cyclohexadiene likely occurs through the ground state carbene **17T**, large rate constants for reaction with alkene and methanol are characteristic of a singlet decay channel. Thus, the singlet carbene **17S** must be low-lying and easily accessible at room temperature. The singlet carbene reaction with methanol has recently been confirmed for this system.[30]

B. Carboethoxycarbene

Platz and co-workers have recently examined the photochemistry of ethyl diazoacetate **20** by LFP (Scheme 10).[31] Unlike the aryl diazocarbonyl **16**, when **20** is flashed in organic solvents, a UV–VIS (300 to 750 nm) active transient intermediate is not observed. However, LFP of **20** in the presence of either pyridine or acetone produces intense transients with λ_{max} = 430 nm and 415 nm, respectively. These transients are formed by trapping of the singlet carbene **21S** and are attributed to pyridine ylide **22** and acetone ylide **23**.

As discussed previously in Section IV, Scaiano and co-workers have raised the concern that in addition to effectively trapping carbenes, pyridine may also react with ketenes to form UV–VIS active ylides.[14,27] Thus, how can we be sure that carbene **21** is trapped, rather than ketene **24**? Based on Scaiano's work[14,27] ketene **24** is expected to have a lifetime on the order of tens to hundreds of microseconds in the presence of dilute ($< 10^{-2}$ M) pyridine. The fact that in the presence of as little as 0.015 M pyridine, ylide **22** was observed to be formed faster than 20 ns ($k_{growth} > 5 \times 10^7$ s^{-1}) argues against ketene trapping. In addition, the lack of production of a pyridine ylide when [pyridine] < 0.01 M provided further evidence that pyridine is not trapping a long-lived transient. Furthermore, while acetone may well react with ketene **24** via a cycloaddition process (Scheme 11), the product of this reaction **25** is not expected to be a UV–VIS active species. Thus, the experimental evidence is most consistent with carbene **21** being the precursor to ylides **22** and **23**.

In fact, no evidence for the formation of a ketene-pyridine ylide was found during the study of ethyl diazoacetate, although based on Scaiano's work,[14,27] it seems reasonable that the pyridine ylide of ketene **24** should be long-lived and UV–VIS

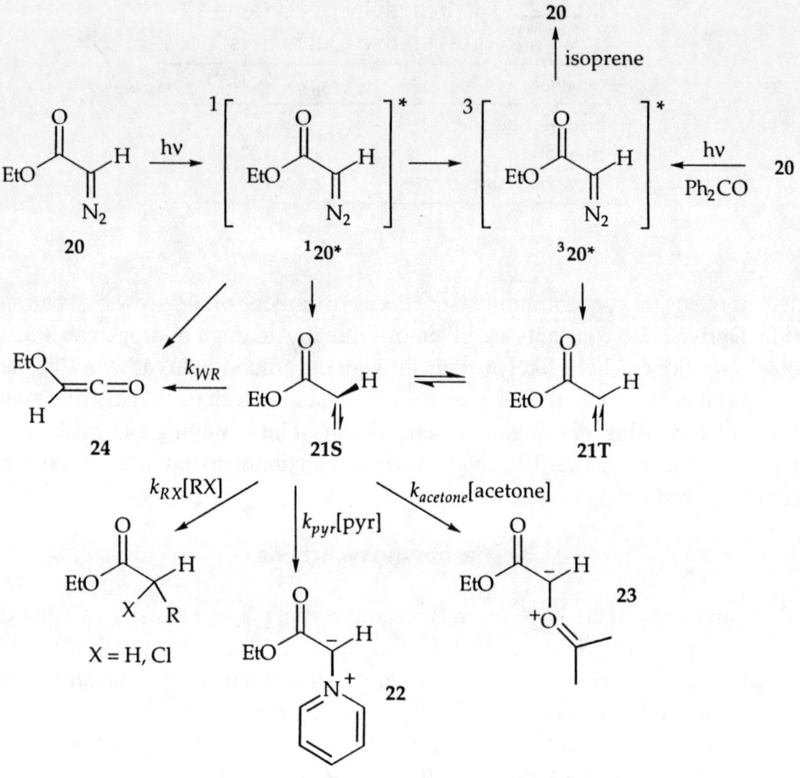

Scheme 10.

active. However, our studies have indicated that only cyclic ketenes, such as those studied by Scaiano, produce detectable pyridine ylides.[32] We have been unable to detect a ketene ylide derived from an acyclic diazocarbonyl compound.

A plot of the yield of ylide **22** as a function of [pyridine] in Freon-113 is shown in Figure 1. The yield of ylide (A_y) grows as the concentration of pyridine increases until [pyridine] $\approx$ 4 M. Above 4 M pyridine in Freon-113, the ylide yield is saturated (A_y^∞), implying that every carbene **21** generated in the laser pulse is captured by

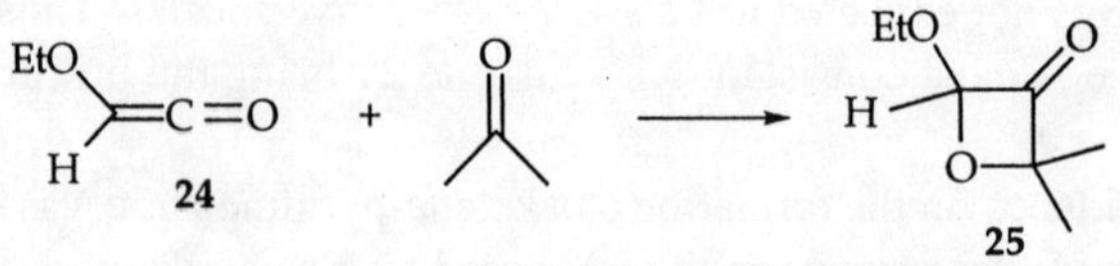

Scheme 11.

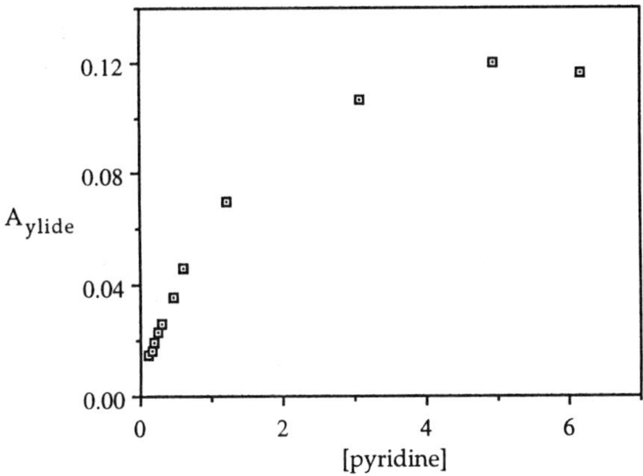

Figure 1. A plot of the yield of ylide **22** as a function of [pyridine] in Freon-113.

pyridine and that $4k_{pyr} \gg k_{WR} + k_{RX}[\text{RX}]$, where k_{pyr}, k_{WR} and k_{RX} are the rate constant for carbene reaction with pyridine, the rate constant for carbene Wolff rearrangement, and the rate constant for carbene reaction with solvent, respectively.

As discussed in Section IV (Eq. 6) an estimate of the carbene lifetime may be deduced from a plot of $1/A_y$ vs. $1/[\text{pyridine}]$ (Figure 2). Division of the intercept by the slope of this plot gives the ratio $k_{pyr}/(k_{WR} + k_{RX}[\text{RX}])$, assuming that all of the decay processes of the carbene proceed through its singlet state. This ratio is given in Table 3 for a variety of solvents. Since k_{pyr} is expected to range from 1 to $5 \times 10^9\ \text{M}^{-1}\text{s}^{-1}$ for singlet carbene **21S**,[24f,25] very large values ($> 10^9\ \text{s}^{-1}$) were

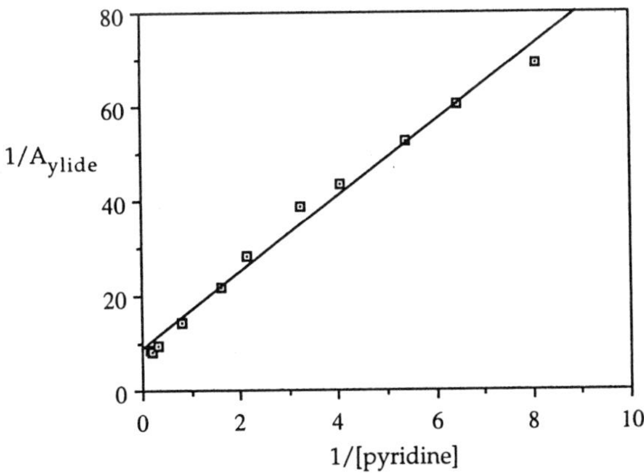

Figure 2. Double reciprocal treatment of the data of Figure 1.

calculated for $(k_{WR} + k_{RX}[RX])$. If the rate of intersystem crossing for **21** is comparable to that of arylcarbenes (approximately 3×10^9 s^{-1}),[33] then these results indicate that spin relaxation to the ground state triplet carbene **21T** is likely incomplete. Slower intersystem crossing, however, is probably due to the difference in geometry of the singlet and triplet carbene.[16,17] The **21S** lifetimes of Table 3 were deduced by assuming $(k_{WR} + k_{RX}[RX]) >> k_{isc}$.

Derived carbene lifetimes in cyclohexane and cyclohexane-d_{12} were very similar. Since singlet carbene C-H insertion reactions with cyclohexane typically show isotope effects of $k_H/k_D = 2$,[34] the small isotope effect observed in this solvent indicated that reaction of carbene **21S** by C-H insertion or hydrogen-atom abstraction does not limit the lifetime of the carbene. Thus, the carbene lifetime is substantially controlled by Wolff rearrangement and $k_{WR} \approx 1/\tau$ for the solvents listed in Table 3.

An analysis of the yield of acetone ylide analogous to that described above for pyridine was also conducted. This revealed that in Freon-113 the quantity $k_{acetone}\tau = 0.48$ (vs. 1.10 for the pyridine ylide). Thus, $k_{acetone} \approx 0.5k_{pyr}$ or $(0.5-2.5) \times 10^9$ M^{-1}s^{-1}.

Since the lifetimes derived for carboethoxycarbene **21** are hundreds of times shorter than those measured for phenylcarbomethoxycarbene **17**, aromatic substituents as expected must play a major role in stabilizing the carbene. Because of the very short derived lifetimes for **21**, however, one must be concerned that pyridine and acetone are trapping an excited state of the carbene precursor rather than the carbene itself. There is now considerable evidence in the literature that diazo excited states can undergo intra- and intermolecular reactions that mimic carbene processes.[35]

In order to eliminate the possibility of a diazo excited state reaction with pyridine, a triplet-sensitized LFP experiment was conducted. Sensitized flash photolysis (Scheme 10) of **20** in Freon-113 and in the presence of pyridine produced an ylide

Table 3. The Lifetime of Carboethoxycarbene **21** in a Variety of Solvents[31]

Solvent	$k_{pyr}/(k_{WR} + k_{RX}[RX])$	Lifetime[a] (ns)
Freon-113	1.10	0.2–1.1
pentane	0.71	0.1–0.7
cyclohexane	0.50	0.1–0.5
cyclohexane-d_{12}	0.59	0.1–0.6
cyclopropyl cyanide	0.29	0.06–0.3
benzene	0.26	0.05–0.3
acetonitrile	0.18	0.04–0.2
acetonitrile-d_3	0.23	0.05–0.2

Note: [a]Assuming $k_{pyr} = (1 - 5) \times 10^9$ M^{-1}s^{-1}.

with $\lambda_{max} = 430$ nm. An analysis of $1/A_y$ *vs* 1/[pyridine] yielded an estimated carbene lifetime that was, within experimental error, identical to that derived following direct photolysis of **20**. Since ^{3}**20*** is not expected to reform ^{1}**20***, the excited singlet diazo compound was excluded as the immediate precursor of pyridine ylide **22**. As will be discussed below, ^{3}**20*** was also eliminated as the sole precursor to ylide **22**, indicating that it is truly carbene **21** that is trapped by pyridine.

The effect of triplet quenchers such as isoprene ($E_{triplet}$ = 59 kcal/mol)[36] and hexatriene ($E_{triplet}$ = 49 kcal/mol)[36] on the yield of pyridine and acetone ylides was examined. With both quenchers, the yield of ylide smoothly decreased as the concentration of polyene was increased from 0 to 1 M. Increases in polyene concentration above 1 M led to only small decreases in ylide yield. A typical Stern–Volmer treatment for the data obtained upon quenching ylide **23** by isoprene in acetone is given in Figure 3.

These results are inconsistent with chemical quenching of the carbene precursor of the ylide. In such a case, the yield of ylide would continue to drop with increasing polyene concentration. The data are also inconsistent with consumption of the diazo precursor by its chemical reaction with polyene. The results are most consistent with a trappable carbene generation proceeding from both the singlet and triplet excited states of **20** and selective quenching of the ^{3}**20*** route by isoprene or hexatriene (Scheme 10). ^{3}**20*** and some of the carbene ylide ultimately derived from it are quenched between 0 and 1 M polyene. Since all of the ^{3}**20*** produced upon direct photolysis is quenched with 1 M polyene, further increase in quencher concentration does not lead to a further reduction in ylide yield. (Carbene **21** may slowly undergo chemical reaction with polyene because the slopes of the Stern–Volmer plots above 1 M polyene are non-zero, but small.) The above conclusions are consistent with Tomioka's study[19] of methyl diazoacetate where product studies

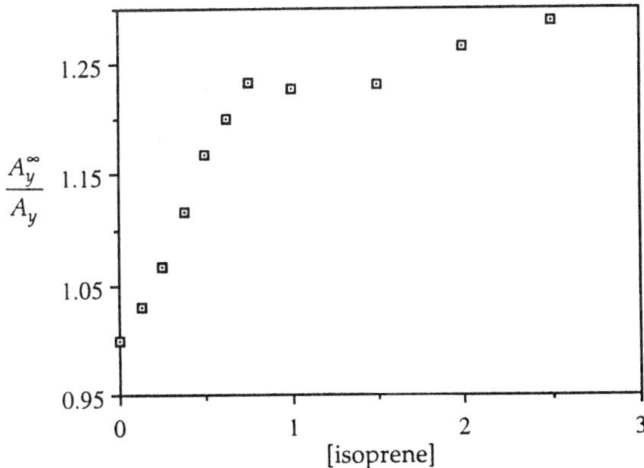

Figure 3. Stern–Volmer treatment of the quenching of ylide **23** yield by isoprene upon LFP of **20** in acetone.

Table 4.   Stern–Volmer Quenching of Carboethoxycarbene-acetone
Ylide **23** in Acetone[31]

Quencher	$k_q/k_{acetone}$ $[acetone]^a$	$k_q{}^b$ $(M^{-1}s^{-1})$	k_H/k_D
dimethyl sulfide	0.23	$(1.6 - 7.8) \times 10^9$	
diethyl ether	0.19	$(1.6 - 6.5) \times 10^9$	
THF	0.17	$(1.2 - 5.8) \times 10^9$	1.1 ± 0.1
THF-d_8	0.16	$(1.1 - 5.4) \times 10^9$	
methanol	0.098	$(0.7 - 3.3) \times 10^9$	1.1 ± 0.1
methanol-d	0.091	$(0.6 - 3.1) \times 10^9$	
2-propanol	0.10	$(0.7 - 3.4) \times 10^9$	
cyclohexene	0.10	$(0.68 - 3.4) \times 10^9$	
cis-1,2-dichloroethylene	0.086	$(0.58 - 2.9) \times 10^9$	

Note: ^{a}neat, [acetone] = 13.6 M.

^{b}Assuming $k_{acetone} = (0.5-2.5) \times 10^9$ $M^{-1}s^{-1}$.

clearly indicated that upon direct photolysis some of the carbene is derived from
the excited triplet diazo compound (Section II).

The yield of ylide is reduced by 20 to 25% by 1 M polyene. Thus, 20 to 25% of
the trappable carbene produced upon direct photolysis of **20** is derived from ^{3}**20***
and 75 to 80% of the trappable carbene originates from ^{1}**20***. The Stern–Volmer
plots obtained from these triplet quenching experiments were typically linear
between 0 and 0.8 M polyene with slopes of approximately 0.4. This slope
corresponds to $k_q\tau$ where k_q is the rate constant of triplet energy transfer and τ is
the lifetime of ^{3}**20***. Since k_q is likely to be 10^{9-10} $M^{-1}s^{-1}$, the lifetime of ^{3}**20*** was
estimated to be in the range of 40 to 400 ps in solution at ambient temperature.

The reactivity of carbene **20** with other quenchers was also examined by Stern–
Volmer methods as discussed in Section IV (Eq. 7). Representative results for the
quenching of acetone ylide **23** are given in Table 4. These data demonstrate that
carbene **21** reacts indiscriminately with alcohols, ethers, alkenes, and dialkyl
sulfides. Unlike many other carbenes that have been examined by LFP methods,[24]
carbene **21** reacts more rapidly with tetrahydrofuran (THF) than with methanol. In
addition, the carbene reacts with THF and THF-d_8 with the same rate constant and

Scheme 12.

there is essentially no isotope effect (OH *vs* OD) on its reaction with methanol. These results demonstrate that carbene **21** reacts with both ethers and alcohols by an ylide mechanism as shown in Scheme 12 for the case of alcohol reaction. Carboalkoxycarbene reaction with alcohols by way of an ylide is in agreement with recent work by Kirmse.[30]

C.   Dicarbomethoxycarbene

Wang et al. have recently studied dimethyl diazomalonate **26** by LFP.[37] As was the case with ethyl diazoacetate, photolysis of **26** in Freon-113, cyclohexane, benzene, or acetonitrile failed to produce a UV–VIS active transient intermediate. However, LFP of **26** in Freon-113 in the presence of pyridine produced an intense transient band with λ_{max} = 450 nm. Weaker transient bands were observed in cyclohexane and benzene, but no signals were observed in acetonitrile. By analogy to the LFP studies[31] of carboethoxycarbene **21**, the transient has been assigned to pyridine ylide **28** obtained by trapping of singlet dicarbomethoxycarbene **27** (Scheme 13).

Since it was not possible to resolve the growth of pyridine ylide **28**, carbene lifetimes were again derived by analyzing the yield of ylide (A_y) as a function of [pyridine]. It was found that this yield is saturated (A_y^∞) when [pyridine] exceeds 0.4 M in Freon-113. Thus, above this concentration of pyridine, every carbene generated in the laser pulse is captured prior to Wolff rearrangement or reaction with solvent. Plots of $1/A_y$ vs $1/$[pyridine] provided the ratio $k_{pyr}/(k_{WR} + k_{RX}[RX])$, again assuming that all of the decay processes of the carbene proceed through its singlet state and that relaxation to the triplet ground state is slow.[16,17]

These assumptions are in accord with previous product studies conducted by Jones and co-workers[38] who found that singlet carbene **27S** likely adds stereospecifically to alkenes, but that the triplet carbene **27T**, produced by triplet-sensitized photolysis of **26**, reacts with alkenes to form cyclopropanes without maintainance of the stereochemistry present in the original alkene. These results demonstrate that carbene intersystem crossing is slow in neat alkene relative to chemical reactions of the carbene.

The ratios $k_{pyr}/(k_{WR} + k_{RX}[RX])$ measured in several different solvents are given in Table 5, where lifetimes of carbene **27** are also deduced, based on the assumption that $k_{pyr} = (1{-}5) \times 10^9$ M^{-1}s^{-1}. There is essentially no solvent isotope effect on the

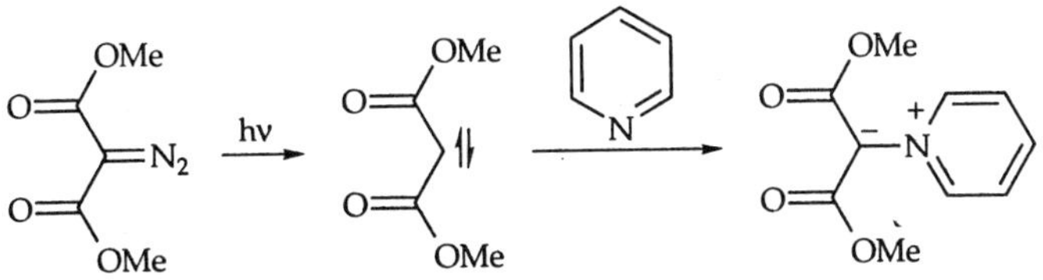

Scheme 13.

Table 5. The Lifetime of Dicarbomethoxycarbene **27** in a Variety of Solvents[37]

Solvent	$k_{pyr}/k_{WR} + k_{RX}[RX])$	Lifetime[a] (ns)	A_y^∞ (rel)
Freon-113	13.2	3 – 13	1.00[b]
benzene	2.5	0.5 – 2.5	c
cyclohexane	3.4	0.7 – 3.5	0.625
cyclohexane-d_{12}	3.5	0.8 – 4.0	0.649

Notes: [a]Assuming $k_{pyr} = (1 - 5) \times 10^9$ M^{-1}s^{-1}.
   [b]By definition.
   [c]It was difficult to saturate the yield of ylide in benzene due to the short carbene lifetime.

carbene lifetime in cyclohexane, indicating that its lifetime in this solvent is largely limited by Wolff rearrangement such that $k_{WR} \approx 1/\tau$, as was the case for carbene **21**.[31]

The rather short lifetime of **27** in benzene is likely due to the addition of the carbene to the aromatic ring, a reaction discovered by Jones and co-workers.[38] These workers also discovered that photolysis of **26** in alkanes leads to C-H insertion adducts. In principle, these products may arise from capture of ^{1}**26*** or **27S** or both species. However, since the carbene lifetime is essentially the same in cyclohexane and cyclohexane-d_{12}, **27S** is not likely the source of the C-H insertion products. Additionally, a comparison of A_y^∞ in Freon-113 with that in cyclohexane (Table 5) indicates that approximately twice as much carbene is produced in Freon-113 per laser flash. Thus, ^{1}**26*** must partition between reaction with C–H bonds of the solvent and decomposition to form carbene **27** as shown in Scheme 14.

Further support for ^{1}**26*** as the source of C-H insertion products was provided by the observation that slightly more trappable carbene is produced in cyclohexane-d_{12} than in cyclohexane per laser pulse. In addition, Jones and co-workers[38]

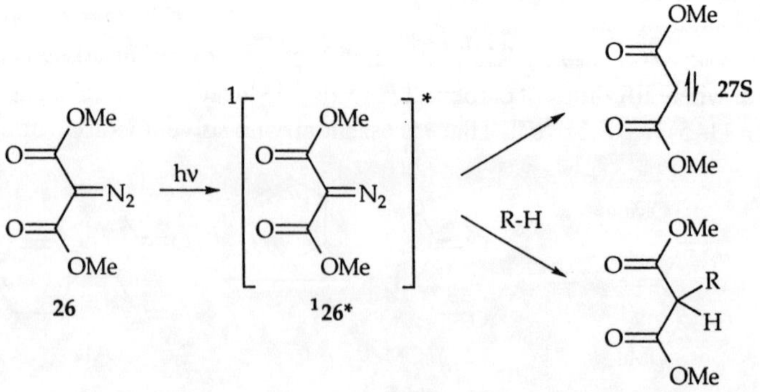

Scheme 14.

have reported that the yield of these adducts is drastically reduced upon triplet sensitization, and LFP of **26** in xylene failed to produce benzylic radicals as would be expected for hydrogen-atom abstraction reactions of triplet- excited diazomalonate. Thus, ^{3}**26*** was eliminated as the origin of the observed C-H insertion adducts.

If ^{1}**26*** can react with alkanes to form products, then there is some concern that pyridine may also be reacting with ^{1}**26*** rather than carbene **27**. In order to address this issue, the reactivity of carbene **27S** with a variety of alkenes was examined by LFP[37] and compared to that observed by Jones and co-workers for the cyclopropanation reaction of the carbene with the same alkenes.[38] This analysis is shown in Table 6.

Perfect stereoselectivity was not observed in the cyclopropanation reaction; about 8% of the wrong isomer was formed. This was attributed to intermolecular reaction of an excited state of the diazo compound (Scheme 15).[38] This view was supported by the observation that dilution of the alkene trap led to a slight increase in stereoselectivity (92% to 95%), presumably due to decreased efficiency in trapping ^{1}**26***. Further dilution of the alkene trap eventually led to decreased

Table 6. Relative Rates of Addition of Singlet Dicarbomethoxycarbene **27S** to Alkenes[37]

Alkene	$k_q{}^a$ $(M^{-1}s^{-1})$	Relative Rates—LFP[37]	Relative Rates— Chemical Analysis[38]
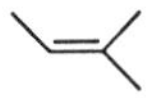	3.5×10^8	1.00^b	1.00^b
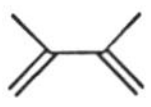	5.2×10^8	1.47	1.30
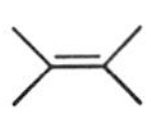	2.6×10^8	0.74	0.88
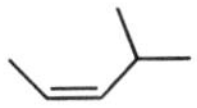	3.1×10^8	0.88	0.55
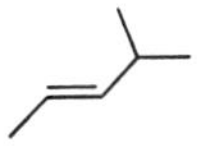	1.43×10^8	0.40	0.23
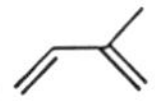	4.3×10^8	1.23	—
	3.8×10^8	1.07	—

Notes: ^{a}From Stern-Volmer analysis of the yield of pyridine ylide **28**, assuming $k_{pyr} = 1 \times 10^9 \, M^{-1}s^{-1}$.
^{b}By definition, relative rates should be compared down vertical columns.

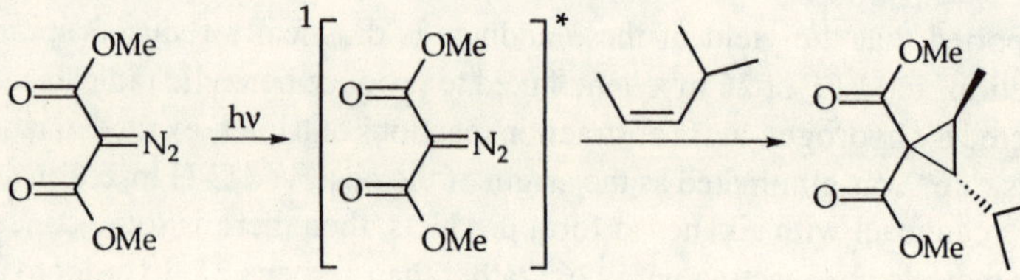

Scheme 15.

stereoselectivity (92% to 87%). This reduction in stereoselectivity arises from a small presence of triplet carbene **27T**, but since the selectivity is not nearly reduced to the extent realized upon triplet sensitization, carbene intersystem crossing (**27S** → **27T**) must be slow even in dilute alkene.

The agreement between the relative reactivities determined chemically in concentrated mixtures of alkenes and by LFP in the presence of relatively dilute alkene demonstrated that the same intermediate is trapped to form cyclopropanes and pyridine ylides. Since the cyclopropane-forming intermediate is stereospecific, it has been postulated to be the singlet carbene **27S**. In addition, since the LFP work was performed in relatively dilute alkene, it is unlikely that either pyridine or alkene has trapped ^{1}**26***. Although these experiments do not unambiguously eliminate trapping of ^{1}**26***, the correspondence of the LFP data with that of a cyclopropane-forming intermediate does rule out pyridine trapping of a ketene.

The reactivity of carbene **27** with a variety of alcohols and ethers was also determined by Stern–Volmer methods (Table 7). Alcohols appear to react slightly more rapidly than ethers, in contrast to the reactivity observed for carboethoxycarbene **21**. This difference in reactivity is not presently understood. The lack of a THF/THF-d_8 isotope effect indicates that carbene **27S** reacts with ethers to form ylides rather than by C-H/C-D insertion. Alcohols, however, apparently do not form ylides with **27S** and react by a different mechanism.

Table 7. Deduced Rate Constants for Reaction of Singlet Dicarbomethoxycarbene **27S** with Alcohols and Ethers[37]

Quencher	$k_q{}^a$ $(M^{-1}s^{-1})$	Relative Reactivity
methanol	1.5×10^9	1.00^b
methanol-d	1.4×10^9	0.93
trifluroethanol	2.2×10^9	1.47
ethanol	1.9×10^9	1.27
THF	4.3×10^8	0.35
THF-d_8	4.5×10^8	0.30

Notes: ^{a}Assuming $k_{pyr} = 1 \times 10^9$ $M^{-1}s^{-1}$.
^{b}By definition.

D. Simple Ketocarbenes

In order to gain additional insight into the role that conformation plays in the Wolff rearrangement, Platz and co-workers have recently examined a series of simple diazoketones $RCON_2H$ **29** (Scheme 16), where R = H (**29a**), Me (**29b**), *i*-Pr (**29c**), and *t*-Bu (**29d**), by LFP methods.[39] Recall that Kaplan and co-workers have found that diazocarbonyl compounds exist as an equilibrium mixture of *syn* and *anti* forms and that for the above series as the steric bulk of R increases, the extent of the *anti* conformer at equilibrium decreases.[20] As discussed in Section III, photolysis of the *syn* conformer is believed to lead directly to ketene **31**, whereas photolysis of the *anti* form releases carbene **30** that can subsequently isomerize to ketene **31** or react with external trapping agents. In addition, carbene **30** may isomerize to oxirene **32**. The LFP experiments described below neither implicated nor disqualified oxirene **32** as an intermediate in the photochemistry of **29a-d**.

As expected, LFP of diazoketones **29a-d** in Freon-113, alkanes, or acetonitrile produced no UV–VIS active transient intermediate. Transient spectra, however, were observed when these compounds were flashed in the presence of pyridine or acetone. By analogy to the previously discussed studies of carboethoxycarbene **21** and dicarbomethoxycarbene **27**, these transients have been assigned to pyridine and acetone ylides, respectively. These ylides can, in principle, be derived from capture of either carbene **30** or oxirene **32** (Scheme 16).

When [pyridine] exceeded approximately 2 M, the yield of ylide was saturated $(A_y = A_y^\infty)$, and thus every trappable intermediate generated in the laser pulse was

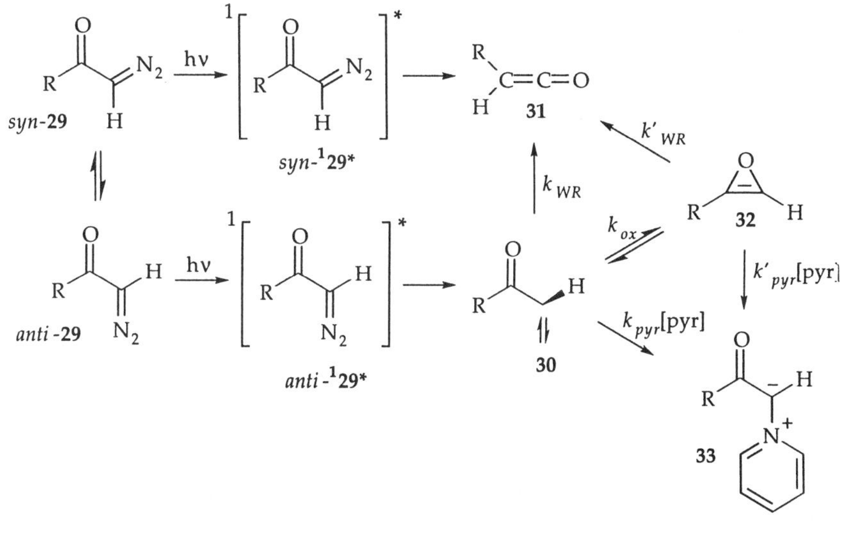

Scheme 16.

captured. A plot of A_y^∞ for either the pyridine or acetone ylides *vs* percent of *anti* conformer of precursor at equilibrium (Figure 4), shows a clear relationship consistent with the predictions of Kaplan and co-workers.[20] This analysis assumes that the ylide extinction coefficient does not vary significantly with R. This is likely a reasonable assumption, since the ylide absorption maximum was constant within each series.

Double reciprocal analysis (Section IV, Eq. 6) of A_y as a function of [pyridine] yielded estimates of the lifetimes of the trappable intermediate (Table 8). The lifetime of this intermediate produced upon LFP of **29a-d** is the same in cyclohexane and cyclohexane-d_{12} at ambient temperature, demonstrating that its lifetime is controlled by the rate of an intramolecular rearrangement rather than by its reaction with solvent. It has been assumed, but not proven, that this intermediate is ketocarbene **30** rather than oxirene **32**.

Activation parameters for the carbene rearrangement were determined by analyzing the value $k_{pyr}\tau$, determined by LFP methods, as a function of temperature (from –26 to 20 °C) in pentane and in Freon-113. The results of this analysis indicated that $E_a = 2.3 \pm 1$ kcal/mol and $A = 10^{10.2}$ $M^{-1}s^{-1}$ in pentane and $E_a = 2 \pm 1$ kcal/mol and $A = 10^{10}$ $M^{-1}s^{-1}$ in Freon-113. Recall (Section I) that theory predicts there is little or no barrier separating carbene **30a** from its corresponding oxirene, but that both intermediates must pass over a 5 kcal/mol barrier to form ketene.[17] Thus, if theory is correct about the rapid isomerization of **30a** and **32a**, then the barrier to Wolff rearrangement of one or both of these species is apparently slightly less than 5 kcal/mol.

Because of the short lifetimes of the carbenes involved, all of the LFP studies described in this section, as well as those of carboethoxycarbene and dicar-

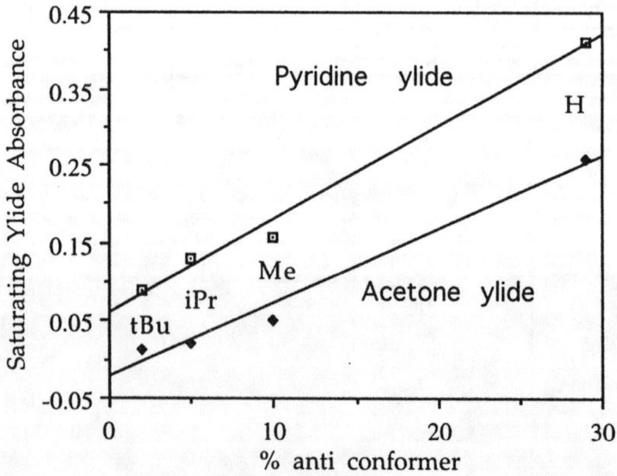

Figure 4. Plot of A_y^∞ (at saturating [pyridine] or in neat acetone) produced by LFP *vs* the equilibrium concentration of the *anti* conformation of the diazoketone precursor determined by NMR.

Table 8. Lifetimes of Simple Ketocarbenes at Ambient Temperature[39]

Ketocarbene	Solvent	Lifetime[a] (ns)
30a (R=H)	Freon-113	2.3–0.5
	cyclohexane	1.0–0.2
	cyclohexane-d_{12}	1.2–0.2
	acetonitrile	0.2–0.04
30b (R=Me)	Freon-113	1.2–0.2
	cyclohexane	1.4–0.3
	cyclohexane-d_{12}	1.4–0.3
	acetonitrile	0.2–0.04
30c (R=*i*-Pr)	Freon-113	0.9–0.2
	cyclohexane	0.8–0.2
	cyclohexane-d_{12}	0.8–0.2
30d (R=*t*-Bu)	Freon-113	0.7–0.1
	cyclohexane	2.1–0.4
	cyclohexane-d_{12}	2.2–0.4

Note: [a] Assuming $k_{pyr} = (1-5) \times 10^9$ M^{-1}s^{-1} and that intersystem crossing is slower than trapping of the singlet carbene.

bomethoxycarbene, rely on the ratio k_{pyr}/k_o where k_o is the sum of all first-order and pseudo-first-order processes that consume the carbene in the absence of pyridine. The carbene lifetime τ is equal to $1/k_o$ and can be derived by making the assumption that k_{pyr} is approximately 1×10^9 M^{-1}s^{-1}. Thus, the derived data are only as good as this assumption.

In order to address this deficiency, Platz, Zimmt, and co-workers have very recently examined diazoketone **29a** in methylene chloride by picosecond transient grating spectroscopy.[40] Transient grating based on calorimetric techniques provide sufficient time resolution to determine transient lifetimes of 1 ns or less.[41] The data obtained from these experiments were adequately fit (Figure 5) using a model containing only two heating terms: a kinetically unresolvable (fast) heat release and one resolvable (slow) exponential heat release. The rate constant for the slow heat release was determined to be 1.1 (± 0.1) $\times 10^9$ s^{-1} or $\tau = 0.9$ ns . This slow heat release was attributed to the singlet carbene **30a** and agrees well with the lifetime determined by LFP methods (0.73 ns) in methylene chloride assuming $k_{pyr} = 1 \times 10^9$ M^{-1}s^{-1}. It is interesting to note that the observed ratio of fast to slow heat release in the transient grating experiments (74:26) corresponds quite nicely to the ratio of *syn* to *anti* conformer for **29a** as determined by Kaplan (70:30).[20]

To further correlate the transient absorption and transient grating experiments, the effect of a quencher on the observed kinetics was studied. Although pyridine would have been the quencher of choice, its strong absorption at 279 nm precluded

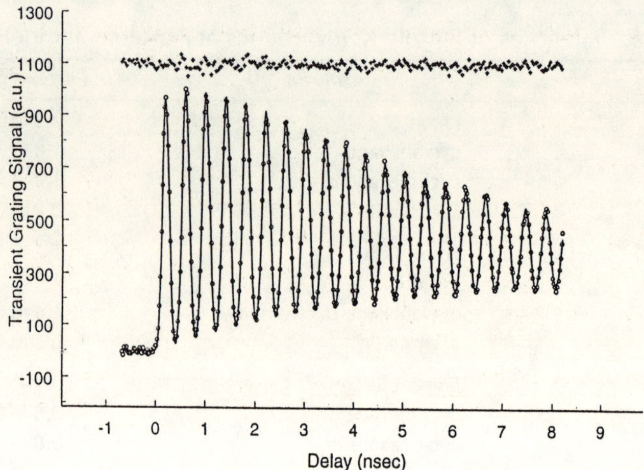

Figure 5. Transient grating waveform generated by irradiation of **29a** in methylene chloride.[40] The circles denote every third data point used in the fitting. The solid line is the calculated fit. Residuals are indicated above. The increase in the peak signal within the first three oscillations is the result of a "slow" heat release.

its use in the grating experiments. Instead, the effect of added methanol on the slow-heat-release rate constant was determined. In the presence of methanol the slow-heat-release rate constant k_{obs} increases according to Eq. 8. Analysis of the data yielded $k_{MeOH} = 4.3 \ (\pm 0.4) \times 10^9 \ \mathrm{M^{-1}s^{-1}}$.

$$k_{obs} = k_o + k_{MeOH}[\mathrm{MeOH}] \tag{8}$$

The value of $k_{MeOH}\tau$ determined from transient grating spectroscopy (3.9) agreed very well with that determined by Stern–Volmer analysis of the methanol quenching of pyridine ylide **33** following LFP of **29a** (3.4). Thus, it was demonstrated that the same species was studied by two independent techniques and previous kinetic assumptions were validated.

VI. CYCLIC CARBONYL CARBENES

A. Derived from 2-Diazo-1-naphthoquinones

Diazonaphthoquinones **34** are commonly used in photoresist materials and, as a result, there has been much interest in their photochemistry (Scheme 17). Recently Sinta, Scaiano, and co-workers have clarified previous work on the production of ketene **35** with a nanosecond LFP study.[14] They found no evidence for carbene-derived products and concluded that Wolff rearrangement from diazocarbonyl **34** is concerted on the nanosecond time scale. They also provided evidence for the intermediacy of carboxylic acid enol **36** in the formation of the final carboxylic acid

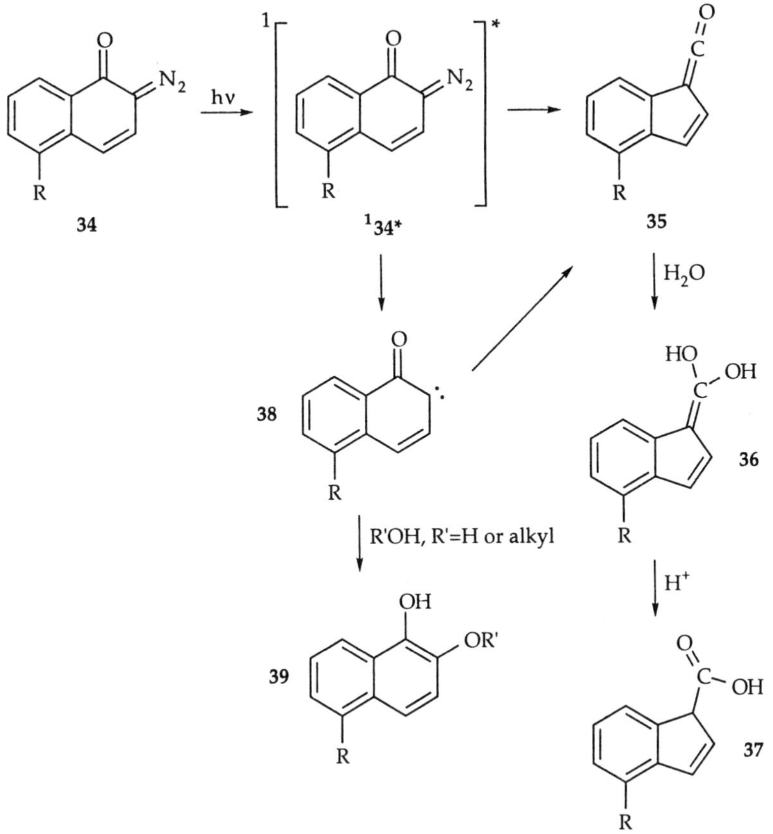

Scheme 17.

product **37**. The presence of this intermediate and the mechanism of its isomerization to **37** has subsequently been confirmed.[42]

A concerted rearrangement and the lack of carbene formation from diazocarbonyl **34**, which is confined to the *syn* conformation, is consistent with Kaplan's predictions[20] (Section III). However, very recently Varma and co-workers have studied this system with picosecond transient absorption spectroscopy and have provided good evidence for the presence of carbene intermediate **38** (for R = H only).[43] In addition, by HPLC analysis of product mixtures following photolysis in water or alcohol, they were able to detect small amounts (up to 15%) of carbene derived product **39** (for R = SO_3 or R = SO_3-C_6H_4-t-Bu). An analysis of the yield of **39** as a function of water or alcohol concentration led to estimates that indicated these rate constants were larger than 3.5×10^8 s^{-1}.

The transient carbene was observed at 323 nm, where the ketene also absorbs. Following laser excitation in methanol, the absorption at this wavelength was found to increase within the time response of the experiment (8 ps). A portion of the absorption (that presumably due to the carbene) decayed with a rate constant of 5×10^{10} s^{-1} ($\tau = 20$ ps). The possibility that this decay was due to the diazo excited state ^{1}**34*** was ruled out since recovery of ground state absorption bands was not observed. Such a recovery was expected if the decay of ^{1}**34*** could be resolved, since the quantum yield for the disappearance of **34** was determined to be 0.54.

Thus, these results seem to be in contradiction to Kaplan's hypothesis for the photochemistry of a diazocarbonyl confined to the *syn* conformation. A possible explanation for this discrepancy is that the carbene is formed by way of the triplet excited state of **34** (as observed in the cases of methyl and ethyl diazoacetate[19,31]) and ^{1}**34*** rearranges directly to ketene **35** in accord with Kaplan's expectations. This possibility could be eliminated by determining that the rate of ketene growth was equal to the rate of carbene decay. However, such a determination was not possible since the ketene absorption band overlapped with that due to the carbene. This scenario would require that intersystem crossing be complete within 5×10^{10} s^{-1}, which is much faster than that observed in arylcarbenes.[33] Clearly more work is necessary to resolve these interesting questions.

B.   Derived from 5-Diazo-2,2-dimethyl-1,3-dioxane-4,6-dione

5-Diazo-2,2-dimethyl-1,3-dioxane-4,6-dione (5-diazo Meldrum's acid **40**) may be thought of as a cyclic analogue of dimethyl diazomalonate **26** and, therefore, a comparison of these two compounds can provide substantial insight into the effect of diazocarbonyl conformation on the observed photochemistry. Cyclic diazocarbonyl **40** is confined to the *syn* conformation, whereas **26** can exist in a conformation in which the diazo functionality is arranged *anti* to one of the carbonyl groups.

Jones and co-workers first examined the products produced upon photolysis of **40** in the presence of alkenes.[38a] They found that direct photolysis afforded very poor yields ($\leq$ 2%) of carbene derived products, but that yields were increased (approximately 20%) upon triplet-sensitized photolysis. This is in contrast to **26**, for which reasonable yields (16 to 40%) of carbene-derived products were detected following direct photolysis.[38]

Platz and co-workers have investigated the photochemistry of **40** by LFP methods (Scheme 18).[37] In the presence of only millimolar concentrations of pyridine, a transient band was observed (λ_{max} = 460 nm in acetonitrile, 530 nm in Freon-113) that grew in over several hundreds of nanoseconds. Since in the case of **26**, carbene-pyridine ylide **28** was formed faster than 20 ns, this transient was attributed to pyridine ylide **43** formed by trapping of ketene **41**. The lifetime of ketene **41** (2.4 μs) and the rate constant of its reaction with pyridine (3.6×10^9 M^{-1}s^{-1}) were determined by examining the growth of ylide **43** in acetonitrile as a function of pyridine concentration (Eq. 2).

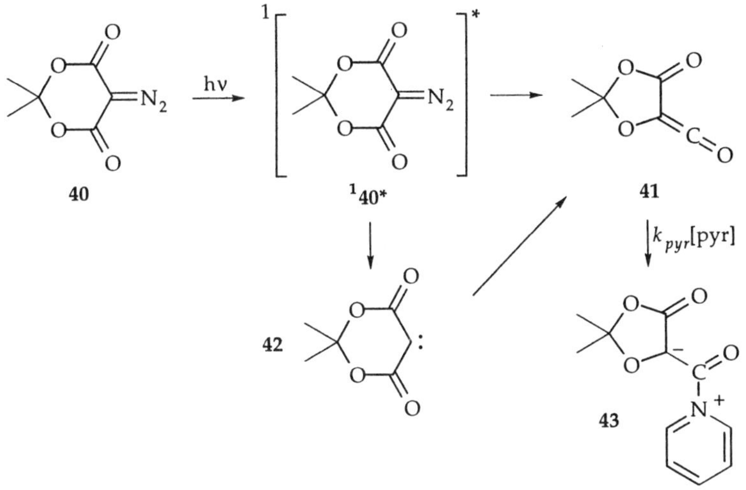

Scheme 18.

No evidence for an ylide derived from carbene **42** was found following direct photolysis in acetonitrile. Since Jones and co-workers found increased yields of carbene adducts upon triplet-sensitized photolysis, triplet-sensitized LFP experiments in neat pyridine were attempted for **40** and seem to indicate formation of a carbene-pyridine ylide.[32]

Cyclic diazocarbonyl **40** has also been examined in poly(methyl methacrylate) (PMMA) matrices by picosecond transient infrared spectroscopy.[44] By monitoring the decay of the diazo band at 2190 cm^{-1} and the growth of the ketene band at 2150 cm^{-1}, these experiments determined that ketene formation was complete within the time response of the apparatus, i.e., 20 ps. Thus, if carbene **42** is formed upon direct photolysis, it must have a lifetime shorter than 20 ps, in marked contrast to its acyclic analogue **27**, which has a lifetime of 1 to 13 ns in organic solution.

VII. CONCLUSIONS AND OUTLOOK

LFP studies of carbonyl carbenes have helped to clarify their chemistry and have confirmed previous hypotheses concerning the mechanism of the Wolff rearrangement. Clearly, it has been shown that the reaction is mediated in the singlet manifold and that the conformation of the diazocarbonyl precursor has a very significant effect on the resulting photochemistry. For acyclic diazocarbonyls, a higher population of the *anti* conformer results in greater carbene production. For cyclic diazocarbonyls, which are confined to the *syn* conformation, either the carbene is not formed or its lifetime is dramatically shortened when compared to analogous acyclic carbonyl carbenes.

However, questions remain. The possibility has been raised that ketocarbene lifetimes are not limited by their rearrangement to ketenes, but rather by ring closure to oxirenes. In addition, the possibility exists that carbene traps such as pyridine react not with carbenes, but with oxirenes. These questions are difficult to address with transient absorption spectroscopy since very little is learned about the structure of the intermediate being probed. Structural questions, in addition to dynamics, can more readily be addressed with time-resolved vibrational studies. Thus, time-resolved infrared and resonance Raman spectroscopy have the potential to provide greater insight into the above mechanistic questions as demonstrated by recent studies.[44]

It is interesting to note that a reaction discovered at the beginning of this century[45] is still receiving attention and stimulating new experiments, even as the century comes to a close.

ACKNOWLEDGMENTS

Much of the laser-flash photolysis work described in this chapter was carried out at Ohio State University in the laboratories of Professor Matthew S. Platz. It is my sincere pleasure to acknowledge Professor Platz for his guidance, expertise, and friendship over the past few years. Financial support from the NIH in the form of a postdoctoral fellowship while at Ohio State is also gratefully acknowledged.

REFERENCES

1. For previous reviews see: (a) Rodina, L.L.; Korobitsyna, I.K. *Russ. Chem. Rev.* **1967**, *36*, 260; (b) Kirmse, W. *Carbene Chemistry*, 2nd ed.; Academic Press: New York, 1971, p. 475; (c) Baron, W.J.; Decamp, M.R.; Hendrick, M.E.; Jones, M., Jr.; Levin, R.H.; Sohn, M.B. In: *Carbenes*; Jones, M., Jr.; Moss, R.A., Eds.; Wiley: New York, 1973; Vol. 1; (d) Marchand, A.P.; Brockway, N.M. *Chem. Rev.* **1974**, *74*, 431; (e) Meier, H.; Zeller, K.-P. *Angew. Chem. Int. Ed. Engl.* **1975**, **14**, 32.

2. March, J. *Advanced Organic Chemistry*; Wiley: New York, 1992, p 1083.

3. Redmore, D.; Gutsche, C.D. *Adv. Alicyclic Chem.* **1971**, *3*, 125.

4. (a) Reiser, A. *Photoreactive Polymers: The Science and Technology of Resists*; Wiley: New York, 1989; (b) Reichmanis, E. In: *Polymers for Electronic and Photonic Applications*; Wong, C.P., Ed.; Academic Press: New York, 1993.

5. For evidence indicating diazo excited state participation in the Wolff rearrangement, see: (a) Rando, R.R. *J. Am. Chem. Soc.* **1970**, *92*, 6706; (b) Rando, R.R. *J. Am. Chem. Soc.* **1972**, *94*, 1629; (c) Wulfman, D.S.; Poling, B.; McDaniel, R.S., Jr. *Tetrahedron Lett.* **1975**, 4519; (d) Tomioka, H.; Kitagawa, H.; Izawa, Y. *J. Org. Chem.* **1979**, *44*, 3072.

6. (a) Trozzolo, A.M. *Acc. Chem. Res.* **1968**, *1*, 329; (b) Hutton, R.S.; Roth, H.D. *J. Am. Chem. Soc.* **1978**, *100*, 4324; (c) Muarai, H.; Torres, M.; Strausz, O.P. *J. Am. Chem. Soc.* **1980**, *102*, 5104; (d) Muarai, H.; Torres, M.; Strausz, O.P. *Chem. Phys. Lett.* **1980**, *70*, 358; (e) Muarai, H.; Ribo, J.; Torres, M.; Strausz, O.P. *J. Am. Chem. Soc.* **1981**, *103*, 6422; (f) Torres, M.; Safarik, I.; Muarai, H.; Strausz, O.P. *Rev. Chem. Intermed.* **1986**, *7*, 243.

7. (a) Hayes, R.A.; Hess, T.C.; McMahon, R.J.; Chapman, O.L. *J. Am. Chem. Soc.* **1983**, *105*, 7786; (b) McMahon, R.J.; Chapman, O.L.; Hayes, R.A.; Hess, T.C.; Krimmer, H.-P. *J. Am. Chem. Soc.* **1985**, *107*, 7597; (c) Murata, S.; Ohtawa, Y.; Tomioka, H. *Chem. Lett.* **1989**, 853; (d) Torres, M.; Clement, A.; Strausz, O.P. *Res. Chem. Intermed.* **1990**, *13*, 1; (e) Bucher, G.; Sander, W. *J. Org. Chem.* **1992**, *57*, 1346.

8. For reviews of oxirene chemistry, see: (a) Torres, M.; Lown, E.M.; Gunning, H.E.; Strausz, O.P. *Pure Appl. Chem.* **1980**, *52*, 1623; (b) Lewars, E.G. *Chem. Rev.* **1983**, *83*, 519.

9. (a) Csizmadia, I.G.; Font, J.; Strausz, O.P. *J. Am. Chem. Soc.* **1968**, *90*, 7360; (b) Thorton, D.G.; Gosavi, R.K.; Strausz, O.P. *J. Am. Chem. Soc.* **1970**, *92*, 1768; (c) Frater, G.; Strausz, O.P. *J. Am. Chem. Soc.* **1970**, *92*, 6654; (d) Zeller, K.-P. *Angew. Chem., Int. Ed. Engl.* **1977**, *11*, 781.

10. (a) Matlin, S.A.; Sammes, P.G. *Chem. Commun.* **1972**, 11; (b) Tomioka, H.; Okuno, H.; Kondo, S.; Izawa, Y. *J. Am. Chem. Soc.* **1980**, *102*, 7123.

11. Torres, M.; Bourdelande, J.L.; Clement, A.; Strausz, O.P. *J. Am. Chem. Soc.* **1983**, *105*, 1698.

12. Tanigaki, K.; Ebbesen, T.W. *J. Phys. Chem.* **1989**, *93*, 4531.

13. Laganis, E.D.; Janik, D.S.; Curphey, T.J.; Lemal, D.M. *J. Am. Chem. Soc.* **1983**, *105*, 7457.

14. Barra, M.; Fisher, T.A.; Cernigliaro, G.J.; Sinta, R.; Scaiano, J. *J. Am. Chem. Soc.* **1992**, *114*, 2630.

15. (a) Debu, F.; Monnier, M.; Verlaquej, P.; Davidovics, G.; Pourcin, J.; Bodot, H.; Aycard, J.-P. *C.R. Acad. Sci. Paris, Ser. 2* **1986**, *303*, 897; (b) Bachmann, C.; N'Guessan, T.Y.; Debu, F.; Monnier, M.; Pourcin, J.; Aycard, J.-P.; Bodot, H. *J. Am. Chem. Soc.* **1990**, *112*, 7488.

16. (a) Scott, A.P.; Nobes, R.H.; Schaefer, H.F., III; Radom, L. *J. Am. Chem. Soc.* **1994**, *116*, 10159; (b) Vacek, G.; Galbraith, J.M.; Yamaguchi, Y.; Schaefer, H.F., III; Nobes, R.H.; Scott, A.P.; Radom, L. *J. Phys. Chem.* **1994**, *98*, 8660. These references also provide an account of the history of calculations related to the Wolff rearrangement.

17. (a) Bouma, W.J.; Nobes, R.H.; Radom, L.; Woodward, C.E. *J. Org. Chem.* **1982**, *47*, 1869; (b) Novoa, J.J.; McDouall, J.J.W.; Robb, M.A. *J. Chem. Soc., Faraday Trans. 2* **1987**, *83*, 1629; (c) Gosavi, R.K.; Torres, M.; Strausz, O.P. *Can. J. Chem.* **1991**, *69*, 1630.

18. (a) Padwa, A.; Layton, R. *Tetrahedron Lett.* **1965**, 2167; (b) Jones, M., Jr.; Ando, W. *J. Am. Chem. Soc.* **1968**, *90*, 2200.

19. Tomioka, H.; Okuno, H.; Izawa, Y. *J. Org. Chem.* **1980**, *45*, 5278.

20. (a) Kaplan, F.; Meloy, G.K. *J. Am. Chem. Soc.* **1966**, *88*, 950; (b) Kaplan, F.; Mitchell, M.L. *Tetrahedron Lett.* **1979**, 759.

21. Diazoketone **9** was first studied by Newman and Arkell: Newman, M.S.; Arkell, A. *J. Org. Chem.* **1959**, *24*, 385.

22. Torres, M.; Ribo, J.; Clement, A.; Strausz, O.P. *Can. J. Chem.* **1982**, *61*, 996.

23. Marfisi, C.; Verlaque, P.; Davidovics, G.; Pourcin, J.; Pizzala, L.; Aycard, J.-P.; Bodot, H. *J. Org. Chem.* **1983**, *48*, 533.

24. (a) Griller, D.; Nazran, A.S.; Scaiano, J.C. *Tetrahedron* **1985**, *41*, 1525; (b) Schuster, G.B. *Adv. Phys. Org. Chem.* **1986**, *22*, 311; (c) Moss, R.A.; Turro, N.J. In: *Kinetics and Spectroscopy of Carbenes and Biradicals*; Platz, M.S., Ed.; Plenum: New York, 1990, p. 213; (d) Platz, M.S.; Maloney, V.M. In: *Kinetics and Spectroscopy of Carbenes and Biradicals*; Platz, M.S., Ed.; Plenum: New York, 1990, p. 239; (e) Moss, R.A. In *Advances in Carbene Chemistry*, Vol. 1; Brinker, U.H., Ed.; JAI Press: Stamford, CT, 1994; p 59. (f) Jackson, J.E.; Platz, M.S. *In Advances in Carbene Chemistry*, Vol. 1; Brinker, U.H., Ed.; JAI Press: Greenwich, 1994; p. 89.

25. (a) Jackson, J.E.; Soundarajaran, N.; Platz, M.S.; Liu, M.T.H. *J. Am. Chem. Soc.* **1988**, *110*, 5595; (b) Platz, M.S.; Modarelli, D.A.; Morgan, S.; White, W.R.; Mullins, M.; Celebi, S.; Toscano, J.P. *Progress in Reaction Kinetics* **1994**, *19*, 93.

26. This kinetic analysis is based on that originally outlined by Scaiano and co-workers: (a) Small, R.D., Jr.; Scaiano, J.C. *J. Phys. Chem.* **1977**, *81*, 828; (b) Small, R.D., Jr.; Scaiano, J.C. *Chem. Phys. Lett.* **1977**, *50*, 431; (c) Small, R.D., Jr.; Scaiano, J.C. *Chem. Phys. Lett.* **1978**, *59*, 246.

27. Boate, D.R.; Johnston, L.J.; Kwong, P.C.; Lee-Ruff, E.; Scaiano, J.C. *J. Am. Chem. Soc.* **1990**, *112*, 8858.

28. Fujiwara, Y.; Tanimoto, Y.; Itoh, M.; Hirai, K.; Tomioka, H. *J. Am. Chem. Soc.* **1987**, *109*, 1942.
29. (a) Werstiuk, N.H.; Casel, H.L.; Scaiano, J.C. *Can. J. Chem.* **1984**, *62*, 2391; (b) Casal, H.L.; Sugamori, S.E.; Scaiano, J.C. *J. Am. Chem. Soc.* **1984**, *106*, 7623; (c) Casal, H.L.; Tanner, M.; Werstiuk, N.H.; Scaiano, J.C. *J. Am. Chem. Soc.* **1985**, *107*, 4616.
30. Kirmse, W. In: *Advances in Carbene Chemistry*; Brinker, U.H., Ed.; JAI: Stamford, CT, 1994; Vol. 1, p 1.
31. Toscano, J.P.; Platz, M.S.; Nikolaev, V.; Popik, V. *J. Am. Chem. Soc.* **1994**, *116*, 8146.
32. Toscano, J.P.; Platz, M.S.; Nikolaev, V.; Popik, V. Unpublished research.
33. (a) Sitzmann, E.; Langan, J.; Eisenthal, K.B. *J. Am. Chem. Soc.* **1984**, *106* 1868; (b) Grasse, P.B.; Brauer, B.-E.; Zupancic, J.J.; Kaufmann, K.J.; Schuster, G.B. *J. Am. Chem. Soc.* **1983**, *105*, 6833.
34. (a) Baer, T.A.; Gutsche, C.D. *J. Am. Chem. Soc.* **1971**, *93*, 5180; (b) Savino, T.G.; Kanakarajan, K.; Platz, M.S. *J. Org. Chem.* **1986**, *51*, 1305; (c) Barcus, R.L.; Hadel, L.M.; Johnston, L.J.; Platz, M.S.; Savino, T.G.; Scaiano, J.C. *J. Am. Chem. Soc.* **1986**, *108*, 3928.
35. (a) Modarelli, D.A.; Morgan, S.; Platz, M.S. *J. Am. Chem. Soc.* **1992**, *114*, 7034; (b) Fox, J.M.; Scacheri, J.E.G.; Jones, K.G.L.; Jones, M., Jr.; Shevlin, P.B.; Armstrong, B.; Sztyrbicka, R. *Tetrahedron Lett.* **1992**, 5021; (c) Moss, R.A.; Ho, G.J. *J. Phys. Org. Chem.* **1993**, *6*, 126; (d) Celebi, S.; Leyva, S.; Modarelli, D.A.; Platz, M.S. *J. Am. Chem. Soc.* **1993**, *115*, 8613.
36. Murov, S.L.; Carmichael, I.; Hug, G.L. *Handbook of Photochemistry*; Marcel Dekker: New York, 1993.
37. Wang, J.-L.; Toscano, J.P.; Platz, M.S.; Nikolaev, V.; Popik, V. *J. Am. Chem. Soc.* **1995**, *117*, 5477.
38. (a) Jones, M., Jr.; Ando, W.; Hendrick, M.E.; Kulczycki, A., Jr.; Howley, P.M.; Hummel, K.F.; Malament, D.S. *J. Am. Chem. Soc.* **1972**, *94*, 7469; (b) Kammula, S.C.; Tracer, H.L.; Shevlin, P.B.; Jones, M., Jr. *J. Org. Chem.* **1977**, *42*, 293.
39. Toscano, J.P.; Platz, M.S.; Nikolaev, V. *J. Am. Chem. Soc.* **1995**, *117*, 4712.
40. Toscano, J.P.; Platz, M.S.; Nikolaev, V.; Cao, Y.; Zimmt, M.B. *J. Am. Chem. Soc.* **1996**, *118*, 3527.
41. (a) Miller, R.J.D. In: *Advances in Spectroscopy*; Clark, R.J.; Hester, R.E., Eds.; Wiley: New York, 1989, Vol. 18, p 1; (b) Morais, J.; Zimmt, M.B. *J. Phys. Chem.* **1995**, *99*, 8863; (c) Ma, J.; Dutt, B.; Waldeck, D.H.; Zimmt, M.B. *J. Am. Chem. Soc.* **1994**, *116*, 10619.
42. (a) Andraos, J.; Chiang, Y.; Huang, C.-G.; Kresge, A.J.; Scaiano, J.C. *J. Am. Chem. Soc.* **1993**, *115*, 10605; (b) Almstead, J.-I.K.; Urwyler, B.; Wirz, J. *J. Am. Chem. Soc.* **1994**, *116*, 954; (c) Andraos, J.; Kresge, A.J.; Popik, V.V. *J. Am. Chem. Soc.* **1994**, *116*, 961; (d) Chiang, Y.; Kresge, A.J.; Popik, V.V. *J. Am. Chem. Soc.* **1995**, *117*, 9165.
43. Vleggaar, J.J.M.; Huizer, A.H.; Kraakman, P.A.; Nijssen, W.P.M.; Visser, R.J.; Varma, C.A.G.O. *J. Am. Chem. Soc.* **1994**, *116*, 11754.
44. Lippert, T.; Koskelo, A.; Stoutland, P.O. *J. Am. Chem. Soc.* **1996**, *118*, 1551.
45. (a) Wolff, L. *Justus Liebigs Ann. Chem.* **1902**, *325*, 129; (b) Wolff, L. *Justus Liebigs Ann. Chem.* **1912**, *394*, 23.

DIAMINO-, AMINO(OXY)-, DIOXY-, AMINO(THIO)-, OXY(THIO)-, AND DITHIOCARBENES

John Warkentin

Advances in Carbene Chemistry
Volume 2, pages 245–295
Copyright © 1998 by JAI Press Inc.
All rights of reproduction in any form reserved.
ISBN: 1-55938-837-4

I. INTRODUCTION

The classical image of carbenes, as voracious and relatively indiscriminate inter-mediates, came out of the properties of electrophilic members of the family, including methylene, alkylidenes, and some alpha halogen-substituted carbenes. That image began to change with the discovery that one or two good π-donor substituents, such as OR or NR_2 at the carbene center, can cause huge reductions of rate constants of both intramolecular and intermolecular reactions. Increasingly high kinetic stability, culminating in the preparation of carbenes that are persistent at room temperature, at least in an inert environment, has been demonstrated in recent years.

This review is focused on a subclass of doubly heteroatom-substituted carbenes, including only amino-, oxy-, and thio-substituents in various combinations.[*] Such carbenes behave as nucleophiles, whereas carbenes without a donor substituent behave as electrophiles (or as radicals) and those with only one donor substituent are generally either ambiphilic or nucleophilic. Metal complexes (carbenoids) of carbenes are not included, despite their importance. Thus, the review is concerned with only a small fraction of carbene chemistry, which has become a very large field. Moreover, the literature on doubly donor-substituted carbenes is substantial, and selectivity had to be exercised to keep this chapter at a reasonable length.

A very substantial earlier review,[1] covering most aspects of carbene chemistry and containing much material on doubly donor-substituted carbenes, should be consulted for a wider appreciation of the subject. Moreover, the references listed by the present author should be viewed, in general, as including references cited therein. The writer apologizes to those scientists who may feel that their work should have received more explicit recognition.

II. GENERAL PROPERTIES OF CARBENES

Carbenes, being dicoordinate at the carbene carbon, have much to gain by increasing the coordination number to three or four. For instance, upon increasing the coordination number from two to four, the number of electrons in the valence shell is increased from six to eight and the number of bonding electrons is increased from four to eight. Intramolecular carbene chemistry most commonly involves a 1,2-migration of an atom or group to the carbene center or reaction at a σ- or π bond more remote from the carbene site (Scheme 1). The coordination number of the carbene carbon increases to three in case of 1,2-migration and to four in ring closures by addition or insertion. The most prevalent intermolecular reactions of carbenes involve an overall increase in coordination number from two to four, as in cycloaddition or insertion processes. Such reactions of singlet carbenes are commonly concerted, but stepwise reactions are also known.

A singlet carbene can act as an ambiphile, as an electrophile, or as a nucleophile, depending not only on the carbene's innate properties but also on those of its reaction partner. That is, an electrophilic carbene can form a single bond to a basic atom, as in formation of a pyridinium ylide from reaction with pyridine.[2] At the other extreme, a highly nucleophilic carbene is likely to form a single bond to an electrophilic species, as in reactions with a protic acid (alcohol, phenol, etc.) by abstraction of the acidic proton to form an ion pair. Between these mechanistic extremes are concerted reactions, not necessarily involving synchronous formation of two bonds, called either cycloadditions or insertions, depending on whether the substrate uses a π bond or a σ bond to react with the carbene.

Nucleophilic behavior of singlet carbenes is possible despite the fact that a carbene has, at least formally, only six electrons in the valence shell of the carbene carbon atom. The electron density required for nucleophilic attack, and for delocalization of the resulting positive charge, comes from π-donor heteroatoms. An important structural feature of such carbenes is the potential for an increased bond order, between the carbene carbon and the heteroatoms bonded to it, that would result from conjugative interaction between the formally-vacant *p* orbital of the carbene singlet and a lone pair orbital of a heteroatom. To the extent that such conjugation contributes to the structure of the ground state, the carbenic carbon

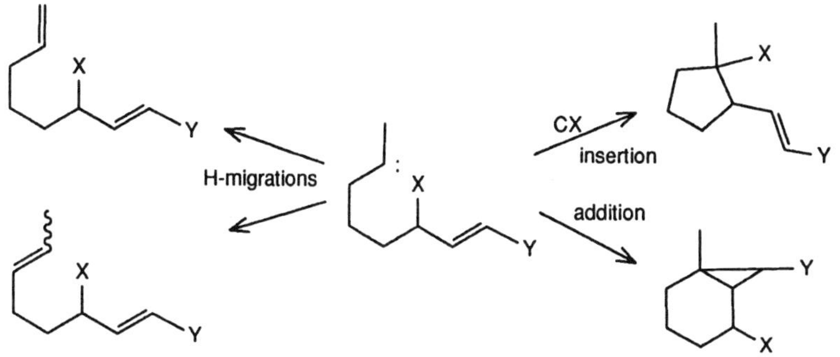

Scheme 1.

atom could become carbanion-like, at the limit approaching the octet electron count illustrated with the dipolar contributors (Scheme 2). The importance of π delocalization, particularly in imidazol-2-ylidenes and in imidazolin-2-ylidenes, is controversial at the present time.[3–5]

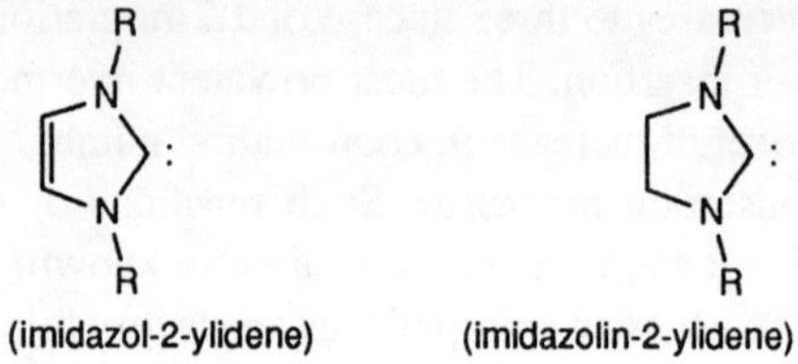

(imidazol-2-ylidene)   (imidazolin-2-ylidene)

Completion of a nucleophilic attack on an electrophilic substrate involves an increase of the donations by donor atoms (D); in other words, the rate of reaction of a nucleophilic carbene involves a composite of effects of the donor substituents on the carbene (ground state) and on the carbene–electrophile complex (transition state). Although this point may appear to be trivial, it is all too easy to focus primarily on the effects of heteroatom substituents on the carbene alone, even though those substituents may have a much larger effect on the energies of transition states for nucleophilic attack. The stabilization by substituents of both the carbene and its reaction partner can be sufficient to permit the dipolar intermediates, similar to those pictured in Scheme 2, to be the actual end products. That does not mean that the carbene forms only one new bond, for the bond order between heteroatoms and carbenic carbon presumably increases on going from the carbene to the dipolar adduct (Scheme 2).

Carbene properties, particularly concerning their reactivities toward alkenes, have also been described in terms of the Frontier Orbital model. Briefly, the dominant interaction in a complex between a nucleophilic carbene and an alkene

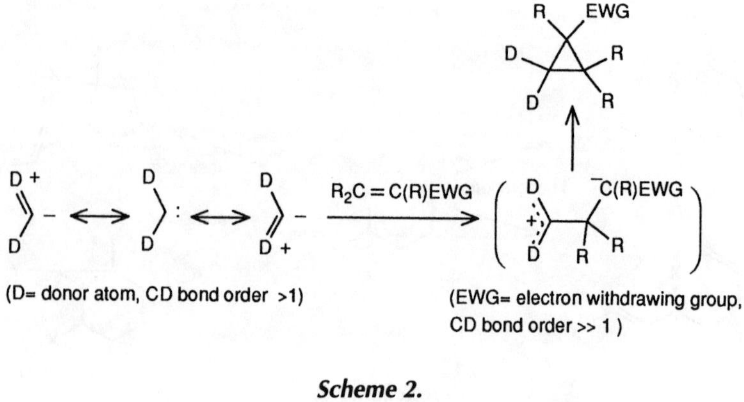

(D= donor atom, CD bond order >1)   (EWG= electron withdrawing group,
CD bond order >> 1)

Scheme 2.

is between the σ orbital of the singlet carbene (its HOMO) and the π^* orbital of the alkene (its LUMO) because those orbitals are most closely matched in energy.[6,7] The resultant model for the transition state is the same as that of the valence-bond model, a dipolar structure resembling the intermediate of Scheme 2.

In addition to the question about dipolarity of ordinary donor-substituted carbenes (Scheme 2) there is a related one concerning electron density distribution in potentially aromatic carbenes of type **I**. Five of them were studied computationally by the *ab initio* method with use of isodesmic comparisons[4] to deduce aromatic stabilization energies (ASE), charges at C2 (Mullikan and NPA), and proton affinities (PA) (Table 1). Additional features[4] are not included here.

A striking conclusion is the fact that the N,N-system, while it has the largest aromatic stabilization energy, has the smallest charge density at C-2 but the largest proton affinity.[4] It appears that 6π-aromatic stabilization, which may have been underestimated initially,[3,5] does not translate into corresponding dipolarity, as the valence bond contributors might imply. Thus, in assessing carbene reactions, it is essential (as is generally true) to consider the effects of donor substituents on both the starting material and the product (equilibria) or the starting material and the transition state (rates). Individual members of the set of carbenes **I** are discussed under the appropriate subheadings.

Another indicator (other than charge density or nucleophilicity) of the importance of conjugative delocalization of heteroatom lone pairs to the carbenic carbon of a singlet carbene is the magnitude of the singlet–triplet energy gap. The extra bonding (Scheme 2) is most important in stabilizing the singlet state and singlet dimethoxy-

Table 1. Computed Properties of Aromatic Carbenes I

(**Ia**: X= Y= NH; **b**: X= NH, Y= O; **c**: X= NH, Y= S; **d**: X= Y= O; **e**: X= Y= S. The 6th combination, X= O, Y= S, was not included)

| | | | C-2 charge | | |
X (kcal/mol)	Y	ASE (kcal/mol)	Mullikan	NPA	PA
NH	NH	21.19	0.098	0.073	262.1
NH	O	13.12	0.174	0.254	247.2
NH	S	17.63	–0.134	–0.252	254.7
O	O	7.65	0.253	0.622	228.2
S	S	13.47	–0.414	–0.707	247.7

carbene is calculated to be more stable than the corresponding triplet[8] by 76.8 kcal mol^{-1}. In the case of dialkylcarbenes, on the other hand, the S-T energy gap is close to zero.[9,10]

III. APPROACHES TO DIAMINO-, AMINO(OXY)-, DIOXY-, AMINO(THIO)-, OXY(THIO)-, AND DITHIOCARBENES

Various methods for generating some of the carbenes listed above have been reviewed in detail.[1] Some of those approaches are included here without detail and without many of the original references. That organization will give the reader some appreciation for the ways in which the newer approaches add to, and complement, the previous array. Reference to the earlier reviews, and to the primary literature cited there, will be necessary for a fuller understanding of the options that are currently available.

A. Diaminocarbenes

Included under "amino" in all sections of this review are those carbenes that have their N-substituent(s) at the carbene carbon with the nitrogen(s) in the lowest oxidation state. Thus, azo, nitroso, and nitro substituents, for example, are excluded but the amide group is included. The first three are not normally in the electron donor category and therefore do not impart nucleophilic properties onto a carbene site. The amido group, bonded to the carbene carbon through N as in β-lactam-4-ylidenes, imparts at least ambiphilic character to the carbene.[11–14]

The parent diaminocarbene, $(H_2N)_2C:$, has been studied computationally.[7c,15–18] It has not been generated in solution but it could be studied by mass spectrometry in the gas phase.[19] Dissociative electron impact ionization of aminoguanidine generated the diaminocarbene radical cation which was subsequently neutralized in a Neutralization-Reionization Mass Spectrometry (NRMS) experiment[19] (Scheme 3). The enthalpy of formation of the carbene was computed with Gaussian-2(G2) theory to be 39.4 kcal/mol,[19] and the barrier to formation of formamidine by 1,2-H migration was estimated to be 50 kcal/mol on the basis of *ab initio* calculations.[15,16]

$$
\begin{array}{ccc}
R_2N \quad\quad NR_2 & & R_2N \\
\diagdown\!\!=\!\!\diagup & \underset{\longrightarrow}{\overset{X}{\rightleftharpoons}} \quad 2 & \diagup \quad : \\
R_2N \quad\quad NR_2 & & R_2N
\end{array}
\tag{1}
$$

By means of crossover experiments, it has been rigorously established that tetraaminoethenes (Eq. 1) do not serve as sources of the corresponding carbenes at temperatures up to 175 °C.[20,21] Transient diaminocarbenes have been generated, as

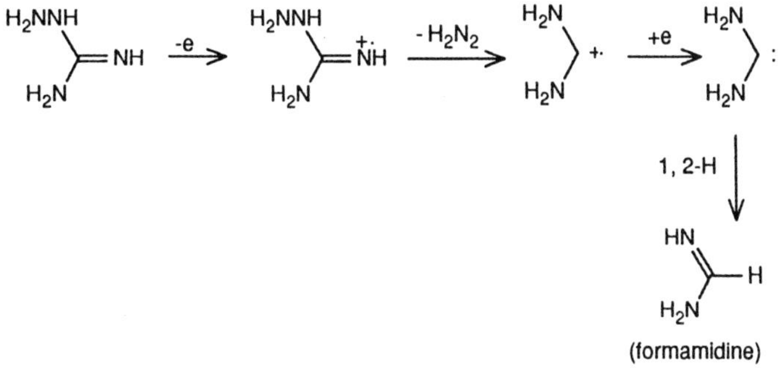

Scheme 3.

inferred from the isolation of corresponding dimers or other products, by a number of methods, five of which are illustrated below with Eqs. 2–6:

1.   From geminal diaminocyclopropanes, by thermal fragmentation.[22]

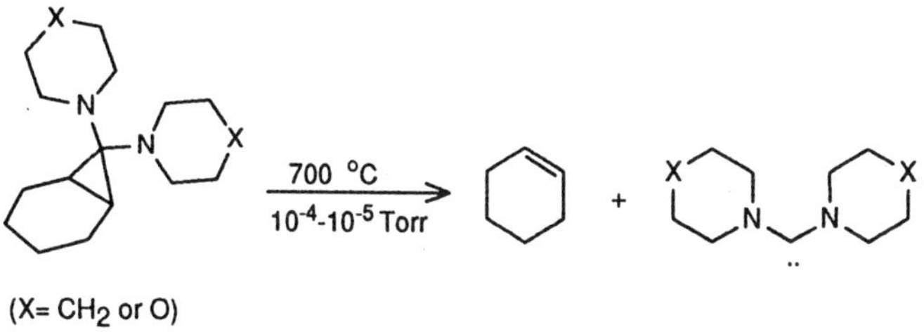

$$\xrightarrow[10^{-4}\text{-}10^{-5}\,\text{Torr}]{700\ ^{\circ}\text{C}} \qquad (2)$$

(X= CH$_2$ or O)

2.   By thermolysis of triaminomethanes.[23,24]

$$(R^2R^1N)_3CH \xrightarrow{160\text{-}250\ ^{\circ}\text{C}} R^2R^1NH + (R^2R^1N)_2C\text{:} \qquad (3)$$

(R^{2}R^1 separate or tethered)

3.   By thermal α-elimination of methanol[25] or chloroform,[26–29] for example.

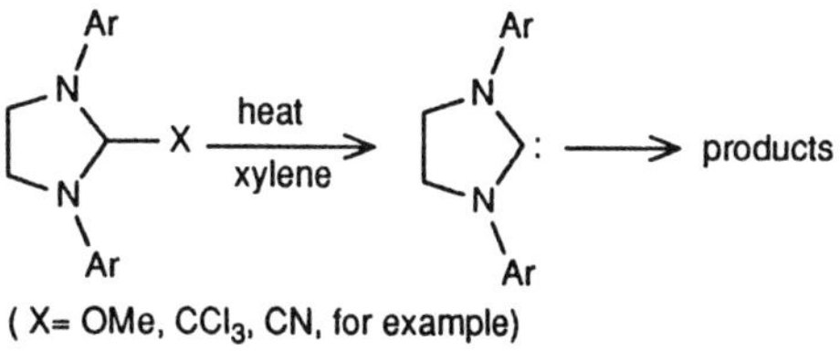

$$\xrightarrow[\text{xylene}]{\text{heat}} \qquad \text{products} \qquad (4)$$

(X= OMe, CCl$_3$, CN, for example)

4. By deprotonation of imidazolium salts.[30,31]

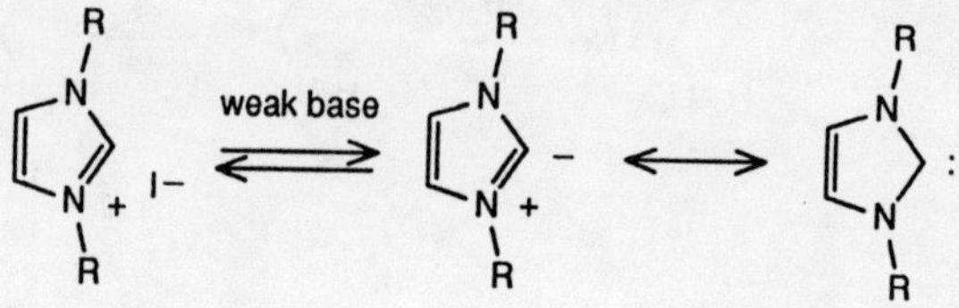
weak base

(5)

5. By reaction of tetraaminoethenes with heterocumulenes such as CS_2.[1,32]

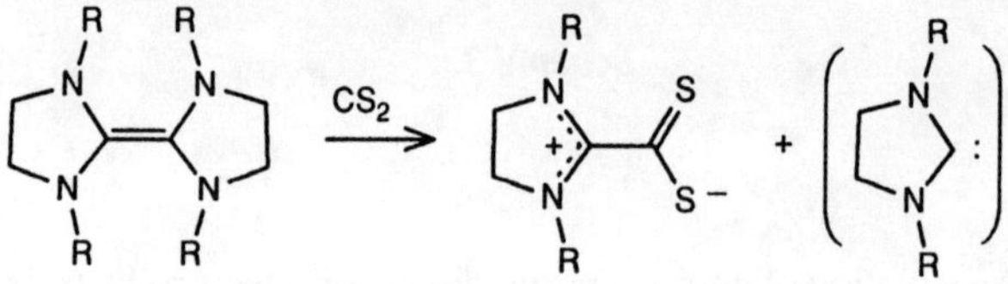
CS_2

(6)

Persistent diaminocarbenes were first announced by Arduengo and co-workers[33] in 1991 and about a dozen have been reported since.[34–39] Some of the approaches are shown below (Eqs. 7–9).

1. By deprotonation of imidazolium salts.[33,35,36]

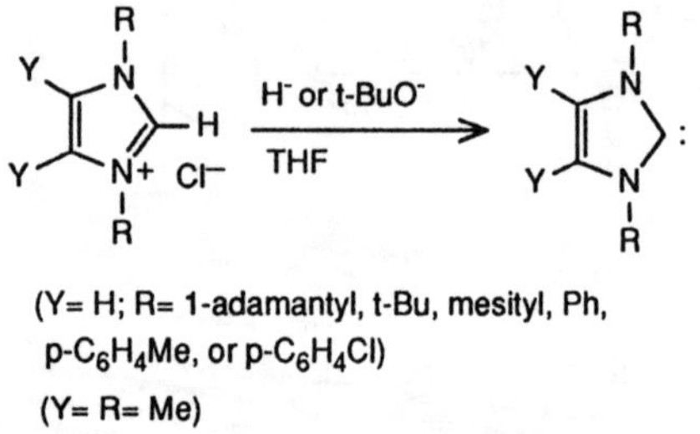
H⁻ or t-BuO⁻
THF
(Y= H; R= 1-adamantyl, t-Bu, mesityl, Ph, p-C_6H_4Me, or p-C_6H_4Cl)
(Y= R= Me)

(7)

2. By desulfurization of imidazole-2(3H)-thiones.[37,39]

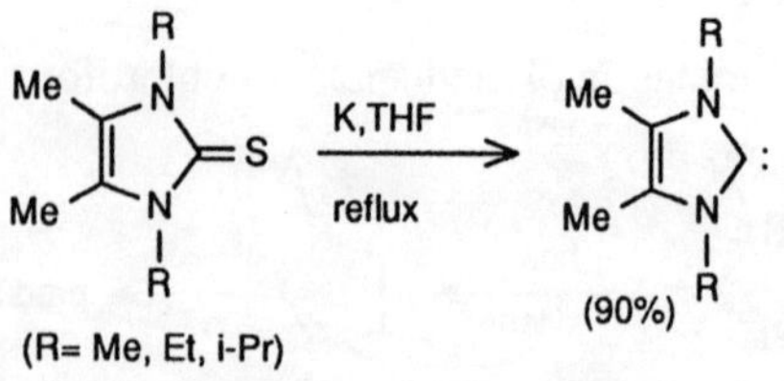
K,THF
reflux
(R= Me, Et, i-Pr)
(90%)

(8)

3. By deprotonation of imidazolinium salts.[34]

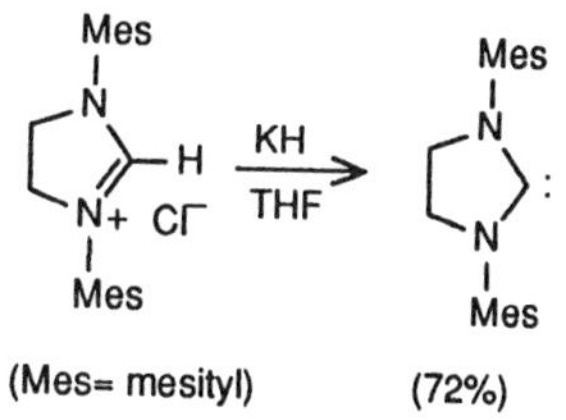

$$\text{(9)}$$

(Mes= mesityl)   (72%)

Several X-ray structures have been determined and these show that the NC bonds to the carbene carbon are essentially single bonds[34] in length and that the electronic charge at the carbene carbon is only 0.08 to 0.09 e.[33,40] This was taken to mean that conjugative delocalization of lone pairs at nitrogen into the formally vacant carbene *p* orbital is a minor effect, in agreement with results of a computation.[16] More recent computations[3,5] indicate that the π bonding is substantial, at least in imidazol-2-ylidenes and imidazolin-2-ylidenes. Nevertheless, the nucleophilic behavior of diaminocarbenes (and of the others included in this chapter) should presumably be ascribed primarily to the effects of donor substituents in stabilizing positive charge at the transition state for nucleophilic attack and, of course, to the ability of the substrate to accommodate negative charge. Photoelectron spectroscopy[36] indicates that the HOMO of 1,3-di-*tert*-butylimidazol-2-ylidene is the σ lone pair at the carbene carbon. The S/T gap for imidazol-2-ylidene was calculated to be $E_{S\text{-}T} = 84.5$[15] and that for 1,3-di-1-adamantylimidazol-2-ylidene about 79.4 kcal/mol.[40] Chemical shielding tensors for a stable carbene have also been determined.[41]

B. Amino(oxy)carbenes

In the alkoxy(amino)carbene series, the parent is aminohydroxycarbene. There are five recent reports of the species[42–45] and hydroxy(methylamino)carbene has been studied as well.[45a] They were generated in the gas phase of a mass spectrometer, from formimidic acid (**1a**) and *N*-methyl formimidic acid (**1b**). First, dissociative ionization formed the corresponding radical cations, which were then neutralized in neutralization–reionization experiments to establish the survival of the neutral carbenes[45a] (Scheme 4). The enthalpy of formation of aminohydroxycarbene (**2a**) was calculated to lie only about 40 kcal/mol above that of formamide.[45a] The existence of hydroxy(methylamino)carbene (**2b**) in the gas phase was also established.[45a]

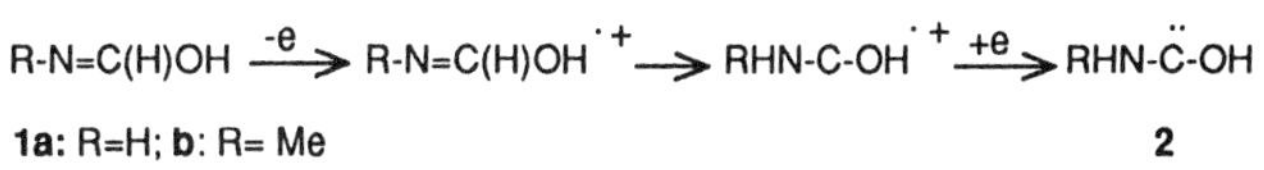

$$\text{R-N=C(H)OH} \xrightarrow{-e} \text{R-N=C(H)OH}^{\cdot\,+} \longrightarrow \text{RHN-C-OH}^{\cdot\,+} \xrightarrow{+e} \text{RHN-}\ddot{\text{C}}\text{-OH}$$

1a: R=H; **b**: R= Me   **2**

Scheme 4.

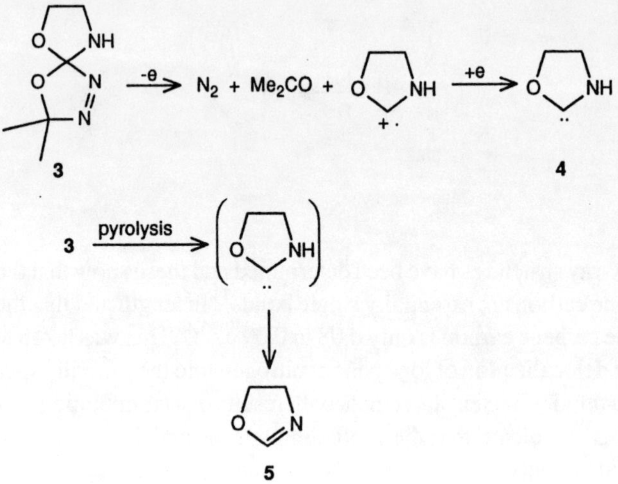

Scheme 5.

Diethylamino(lithiumoxy)carbene has been generated and applied as an acyl anion equivalent.[46] Oxazolidin-2-ylidene (**4**) was generated in the dilute gas phase by dissociative electron impact ionization of oxadiazoline **3** to the radical cation, and neutralization of the latter in a mass spectrometer.[47] Pyrolysis of the same starting material did not afford the carbene but rather **5**, the product of 1,2-hydrogen migration (Scheme 5).

Alkoxy(amino)carbenes can also be generated from aminodialkoxy methanes, apparently via equilibration with an ion pair generated by bond heterolysis and reversible proton transfer from cation to anion.[48–50] The carbene intermediate can be trapped with sulfur (S$_8$), Scheme 6, or with heterocumulenes such as arylisocyanates.[51] An amino(oxy)carbene **6** in which both heteroatoms are in benzylic

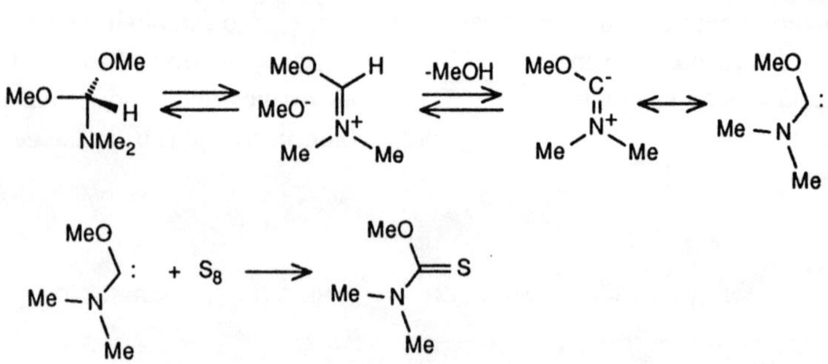

Scheme 6.

positions, was possibly generated as shown in Scheme 7. Intermediacy of the carbene was inferred from isolation of a product originating from reaction of the carbene dimer with the starting salt.[52]

A promising recent approach[53] to some other cyclic amino(oxy)carbenes involves spiro-fused oxadiazoline precursors analogous to **3**. Oxidative cyclization of substituted semicarbazones **7** with $Pb(OAc)_4$ or with $PhI(OAc)_2$ afforded those precursors **8** (Scheme 8). The parent (**8**, R = H) could then be converted to corresponding amides **9** by a standard procedure. It seems to be necessary to tether the amino and oxy groups, acyclic analogs of **8** being difficult to handle. Presumably the advantage of the tethered system is entropic. Hydrolysis is probably slowed because the likely intermediates **10** and **11**, can re-close in competition with further hydrolysis steps.

In the case of **8** the carbene intermediates generated by thermolysis resemble those that fragment in the Corey–Winter reaction,[54,55] and carbene dimers were not obtained. Instead, the evidence for carbene intermediates **12** came from the formation of ethene and isocyanates.[53a,b] The latter were inferred from the products of attack of carbene intermediates on the isocyanates (Section IV. B).

Analogs with a six-membered ring **13** have been prepared by the same route.[53] The parent compound (**13**, R = H) was not isolated, however. Its benzoylation gave the benzamide in low yield, suggesting that the yield of precursor (**13**, R = H) was poor (Scheme 8).

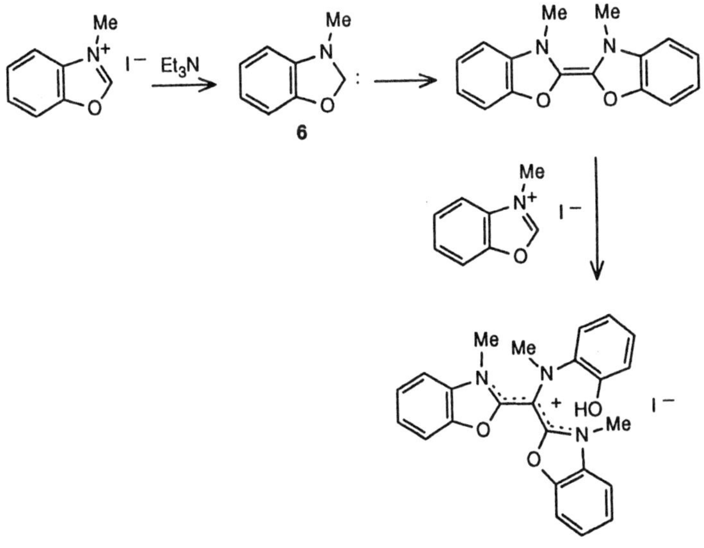

Scheme 7.

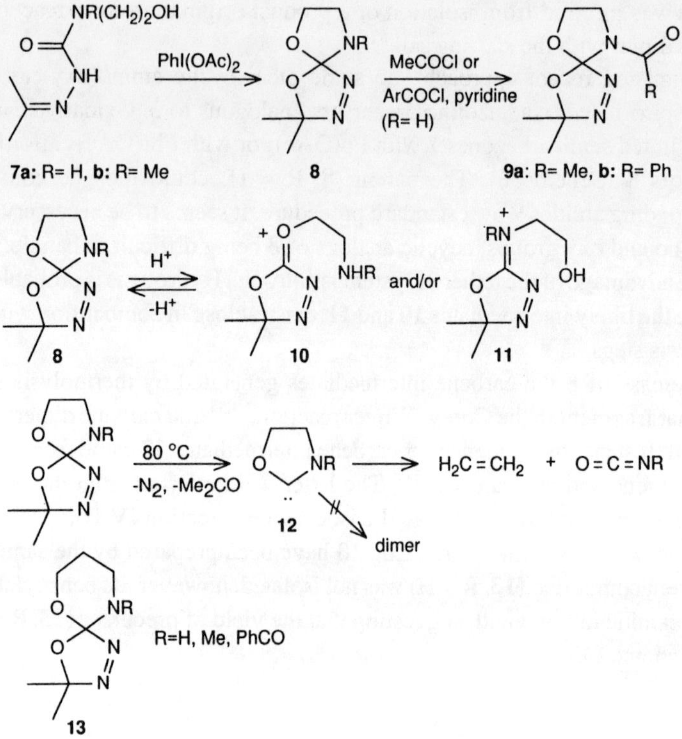

Scheme 8.

C. Dialkoxycarbenes and Other Dioxycarbenes

The parent dioxycarbene, $(HO)_2C:$, has been studied extensively by computation;[7,17,18,56–60] for additional references see Section IV. C. It has also been studied experimentally[58] in the gas phase, by means of mass spectrometry. The precursor was the corresponding radical cation made by dissociative ionization of dihydroxyfumaric acid **14**[58a] (Scheme 9), or oxalic acid.[58b] Dihydroxycarbene is a stable species in the gas phase, but the experimental evidence did not afford a clear-cut answer to the question of multiplicity.[58] The singlet ($\Delta H_f = -50$ kcal/mol) is considerably more likely than the triplet ($\Delta H_f = +6$ kcal/mol).[58a] Dimethoxycarbene has also been examined by computation[7c,8] and spectroscopy.[8]

Some of the numerous earlier approaches to dioxycarbenes are illustrated below:

1. From orthoformates **15** by heating with heterocumulenes[51] (Scheme 10).
2. From dialkoxymethyl acetates by heating[61] (Eq. 10).

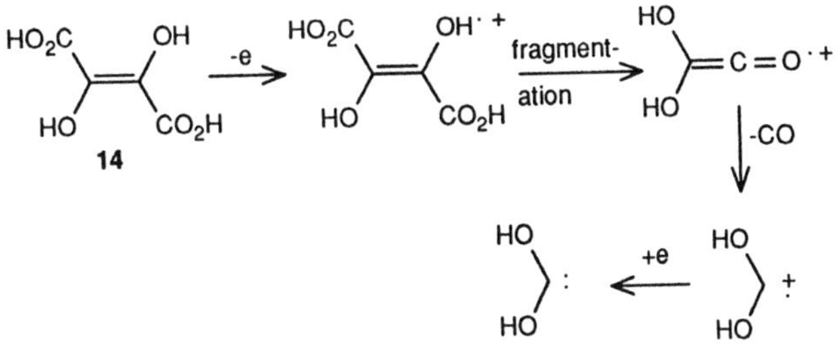

Scheme 9.

$$(RO)_2CHOAc \xrightarrow{\text{325-525 °C}} (RO)_2C: \longrightarrow RR + CO_2 \qquad (10)$$

$$(R = Me, Et)$$

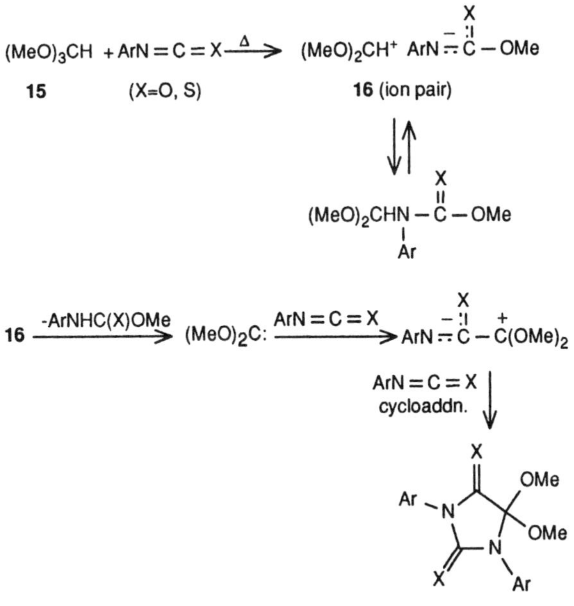

$$(MeO)_3CH + ArN=C=X \xrightarrow{\Delta} (MeO)_2CH^+ \; ArN\mathrel{\overset{\cdot\cdot}{=}}\overset{\displaystyle X}{C}-OMe$$

15   (X=O, S)   **16** (ion pair)

$$(MeO)_2CHN\overset{\displaystyle X}{-}\overset{\|}{C}-OMe$$
$$\underset{Ar}{|}$$

$$\mathbf{16} \xrightarrow{-\text{ArNHC(X)OMe}} (MeO)_2C: \xrightarrow{ArN=C=X} ArN\mathrel{\overset{\cdot\cdot}{=}}\overset{\displaystyle X}{C}-\overset{+}{C}(OMe)_2$$

$$ArN=C=X \Big\downarrow \text{ cycloaddn.}$$

Scheme 10.

3.   From norbornadienone ketals or other ketals, by thermolysis[62-67] (Eqs. 11–13).

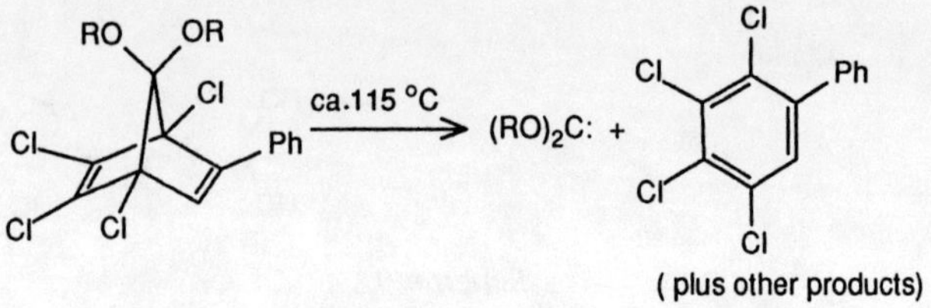

$$(RO)_2C: + \text{(plus other products)} \qquad (11)$$

$$\xrightarrow{200\ °C} \longrightarrow (MeO)_2C: + \qquad (12)$$

$$\xrightleftharpoons{} \xrightarrow{350\ °C} (MeO)_2C: \longrightarrow MeCO_2Me \qquad (13)$$

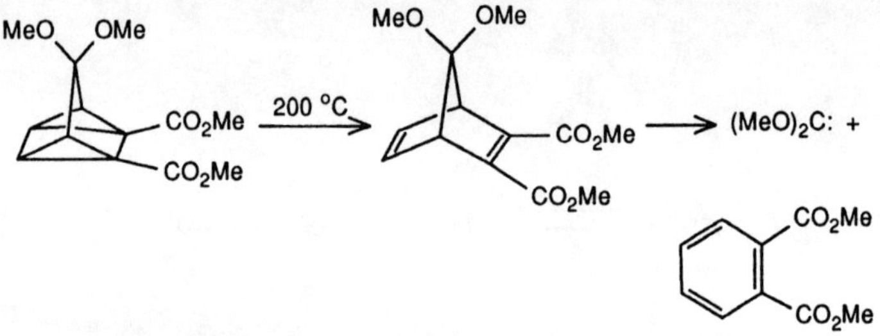

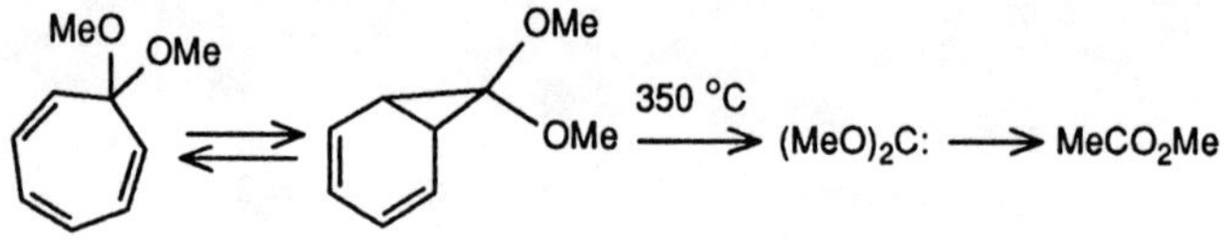

4.   From thiocarbonates (Corey–Winter[54,55] analog) (Eq. 14).

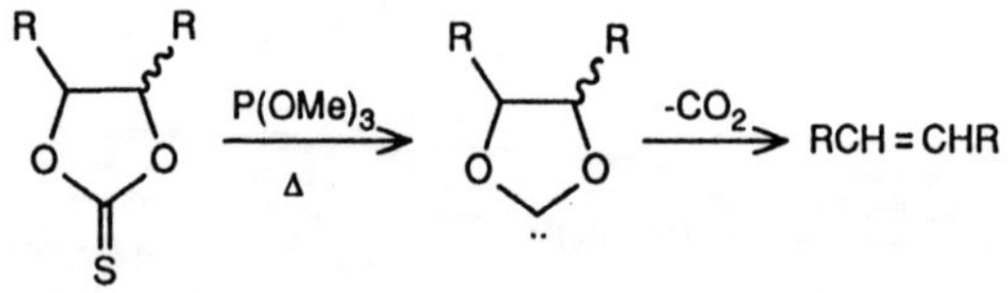

$$\xrightarrow[\Delta]{P(OMe)_3} \xrightarrow{-CO_2} RCH = CHR \qquad (14)$$

5.   From diazirines by photolysis or thermolysis[8,68-70] (Scheme 11).

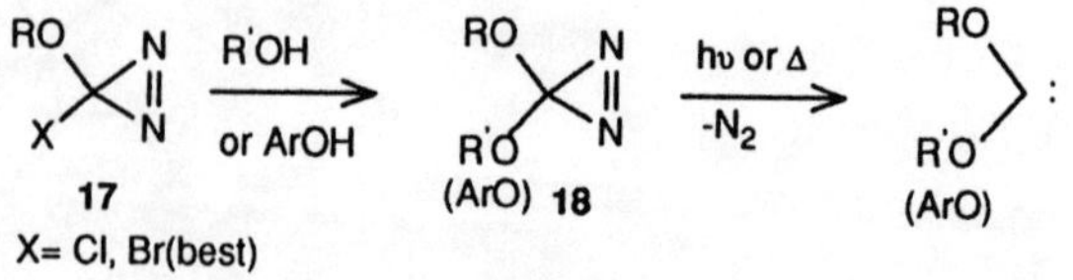

$$\mathbf{17} \xrightarrow[\text{or ArOH}]{R'OH} \mathbf{18}\ (ArO) \xrightarrow[-N_2]{h\upsilon\ or\ \Delta} (ArO)$$

X= Cl, Br(best)

Scheme 11.

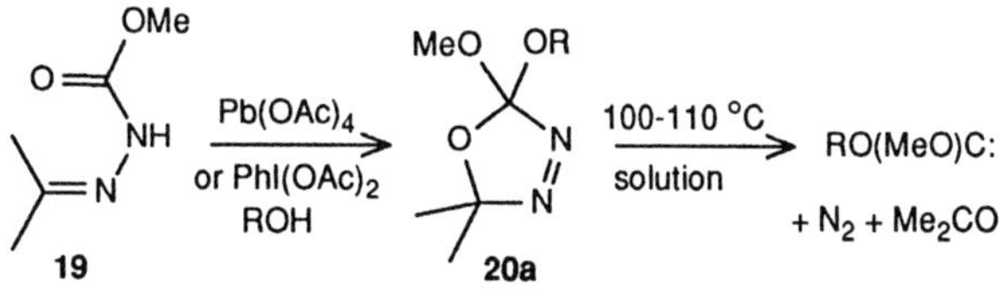

Scheme 12.

6. From 2,2-dioxyoxadiazolines (**20a, b**) by thermolysis[47,71–82] (Schemes 12, 13).

The diazirine method has been developed by the Moss group, with 3-alkoxy-3-halo-diazirines (**17**) as starting materials.[8a,68] Substitution reactions, with alcohol/alkoxide or with phenol/phenoxide provide 3,3-dialkoxy- and 3-alkoxy-3-aryloxydiazirines **18**, respectively. The diazirines lose N_2, either photochemically with ca. 350 nm light or thermally at about room temperature to afford the corresponding dioxycarbenes (Scheme 11). The diazirine approach has been very valuable, particularly for spectroscopic observations of the carbenes produced photochemically[8,69,70] and for mechanistic studies. As a rule, diazirines are not available as the

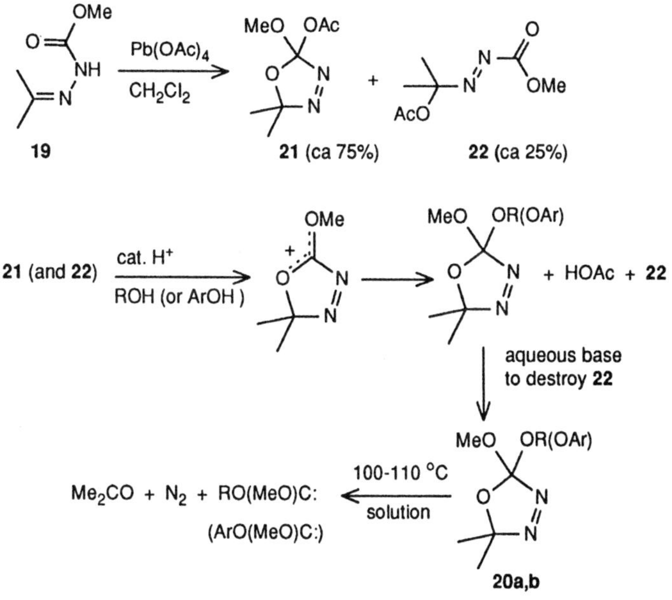

Scheme 13.

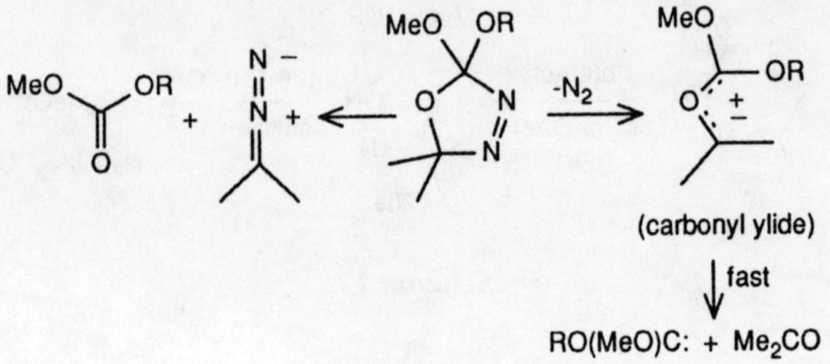

(chemical scheme: carbonate + diazopropane ⟷ dioxyoxadiazoline → carbonyl ylide, with $-N_2$)

(carbonyl ylide)

fast

$$RO(MeO)C: + Me_2CO$$

Scheme 14.

neat compounds, because they can explode, nor are they stable in solution at room temperature.

The dioxyoxadiazoline route compliments the dioxydiazirine route, affording some of the same carbenes and others that may be less easily accessed by the diazirine approach. The oxadiazolines have the advantage of shelf stability and easy handling, compared to diazirines. However, the former are not photochemical sources of dioxycarbenes. Approaches to the oxadiazolines include that of Scheme 12,[71–73,81] usable only with oxidation-resistant alcohols (not with phenols) and Scheme 13,[74–77] which is more general. The latter route involves preparation of a mixture of **21** and **22**, which is used directly to prepare a corresponding mixture of **20a** (or **20b**) and **22**. Compounds **20**, as orthoesters, are resistant to base, whereas **22** and **21** are not (Scheme 13). Thermolysis of the oxadiazolines **20** in solution at 100–110 °C leads to formation of dioxycarbene, dinitrogen, and acetone (Schemes 12, 13). That pathway is not clean in all cases and competition from fragmentation to carbonate and diazopropane occurs in up to about 20% yield in some instances (Section IV.C). The two competing fragmentations can be viewed as alternative 1,3-dipolar cycloreversions (Scheme 14), the one to a short-lived carbonyl ylide and the other to long-lived diazopropane. However, a concerted pathway to the carbene has been proposed for a particular case.[83]

D. Amino(thio)carbenes

Many amino(thio)carbenes have been generated, spurred in part by interest in thiamin,[84] the C-2 conjugate base of which **23** can be expressed as a carbene/ylide hybrid. Most of the other carbenes in this class resemble thiamin in that they are five-membered ring systems.[25,31,85–88] Systems to model thiamin itself and to assess the importance of its structural features, in relation to its function, have been designed.[89–91] Recently,[92] a dimeric thiazolato system **24** with distinct carbenic properties (from X-ray analysis) was reported.

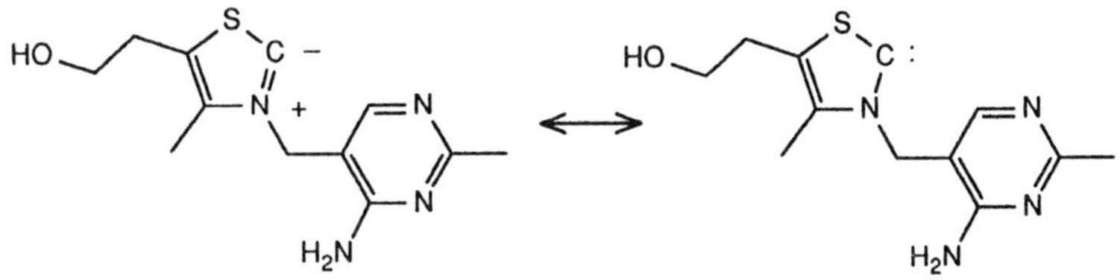

23 (conjugate base of thiamin)

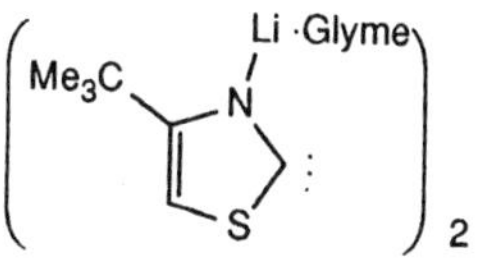

24

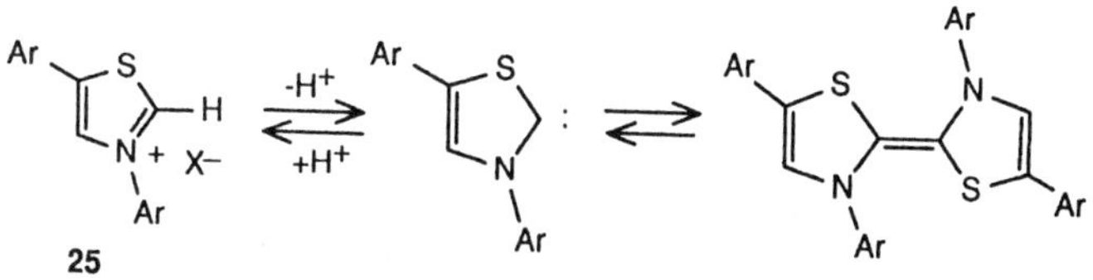

Scheme 15.

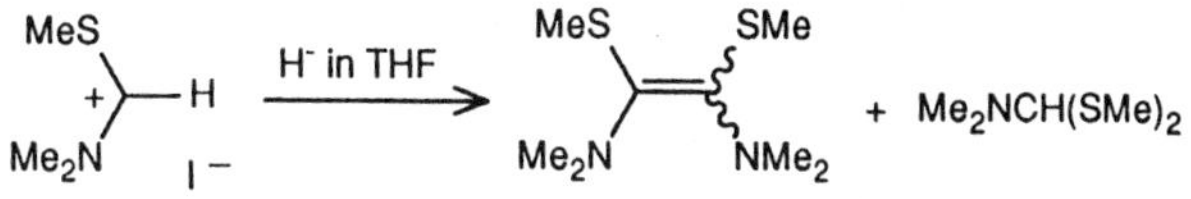

Scheme 16.

Deprotonation from C-2 of thiazoles or thiazolium salts is a general route to amino(thio)carbenes. 3,5-Diaryl-1,3,4-thiazolium salts **25** lose the proton from C-2 in solution and equilibrium is established between those salts and corresponding carbene dimers[88] (Scheme 15). Acyclic aminothiocarbenes have been generated also. Treatment of **26** with NaH in THF afforded a mixture of carbene dimers **27** and thioacetal **28**[93] (Scheme 16). Although the latter may be formed by a number of ionic pathways, carbene insertion into adventitious MeSH is an option.

E. Oxy(thio)carbenes

Until recently, carbenes in this class were quite rare. The formation of carbonyl sulfide and ethene from a Corey–Winter type of reaction suggested the intermediacy of such a carbene (Scheme 17).

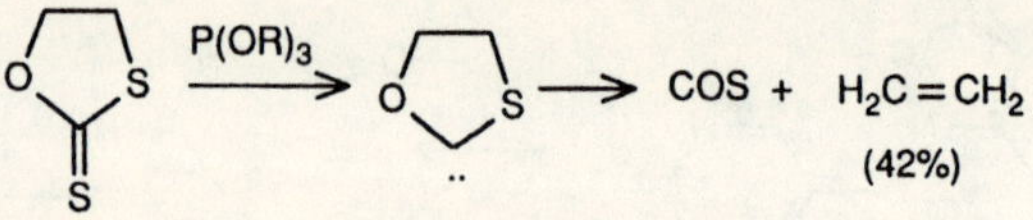

Scheme 17.

The oxadiazoline route (Section IV.C) works well for the generation of oxythio-carbenes. Crude 2-acetoxy-2-methoxy-5,5-dimethyl-Δ^3-1,3,4-oxadiazoline (**21**) could be converted to 2-methoxy-2-organothio analogues **29** by treatment with thiols in the presence of catalytic acid[94] (Scheme 18). Those products could be isolated in pure form in 44–80% yields. Their thermolysis in solution (C_6H_6, 60 °C) afforded carbene dimers **30** in 55–61% yields. The structure of the thiobenzyl E dimer was established by X-ray diffraction. Oxythiocarbenes appear to be obligatory intermediates on the path to the dimers.

The effect of a thio group on the rate constants for thermolysis of compounds **29**, relative to those for 2,2-dialkoxy analogues (Section IV. C) is worthy of mention. All four decomposed with $k \approx 2.1 \times 10^5 s^{-1}$ at 60 °C,[94] whereas dioxy analogues are slightly slower at 100 °C. While the significance of that result is not entirely clear, we speculate that the transition state for thermolysis is ylide-like, and that the sulfur atom exerts its well known anion-stabilizing effect (Scheme 19). Whether or not there is a preference for the thio substituent to rotate into the *exo* or the *endo* position (shown) is not known. A ground state energy increase, caused by substitution of SR for OR at C-2, could also contribute to the higher reactivity of **29**.[95]

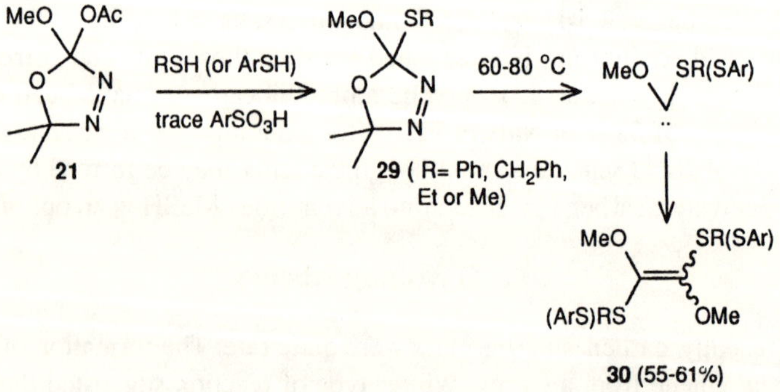

Scheme 18.

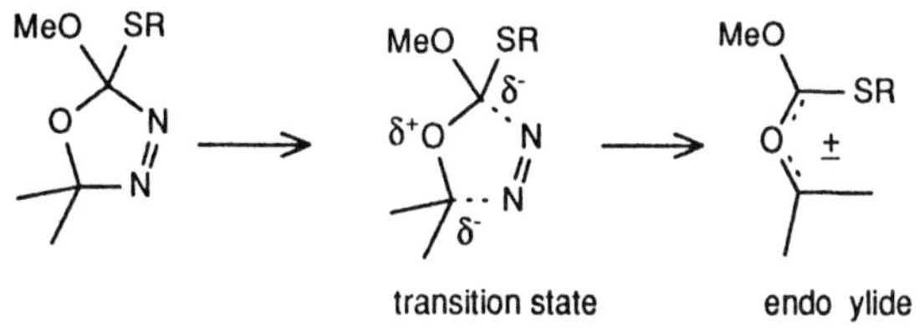

transition state   **endo ylide**

Scheme 19.

F. Dithiocarbenes

Acyclic dithiocarbenes are still rare compared to cyclic analogues. Methods for generating them as transients have been reviewed in detail[96] and only some general approaches are included here (Scheme 20).

A new method is based on the oxadiazoline approach. 2-Acetoxy-2-ethylthio-5,5-dimethyl-Δ^3-1,3,4-oxadiazoline (**31**) was prepared by oxidative cyclization (Scheme 21).[97] Treatment of the acetoxy compound **31** with ethanethiol in dichloromethane containing catalytic acid did not afford the desired 2,2-di(ethylthio)oxadiazoline (**32**) but tri(ethylthio) methane. The exchange step, when

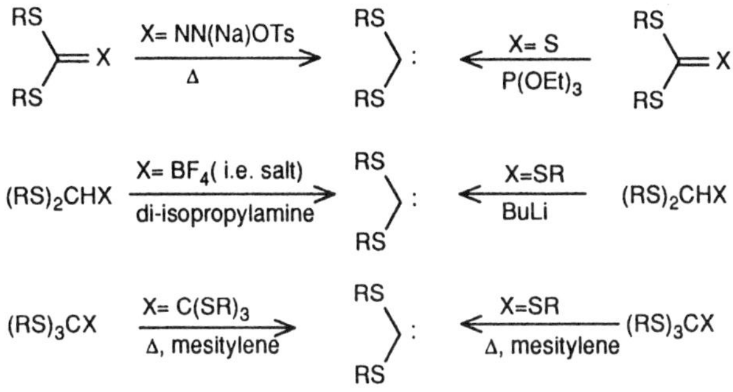

Scheme 20.

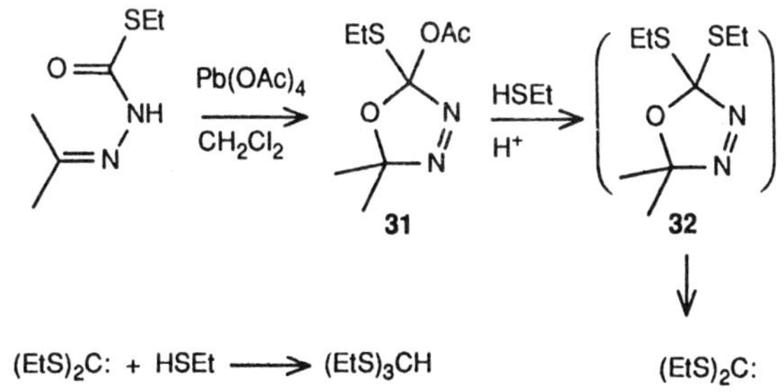

31   **32**

$(EtS)_2C: + HSEt \longrightarrow (EtS)_3CH$ $(EtS)_2C:$

Scheme 21.

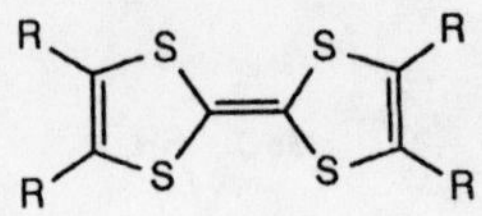

33

repeated in a refrigerator (ca. 10 °C) led to the same result after washing with
ice-cold bicarbonate, drying, and evaporation of solvent and most of the excess
EtSH in the cold under vacuum. These results can be understood in terms of rapid
decomposition of 2,2-bis(ethylthio)oxadiazoline intermediate **32**   to
bis(ethylthio)carbene and subsequent carbene insertion into EtSH. This approach
holds some promise for low temperature generation of dithiocarbenes and for direct
observation of some putative intermediates from attack of nucleophilic carbenes
on electrophilic substrates. Good experimental procedures have not yet been
worked out, however.

Cyclic dithiocarbenes are the building blocks (formally) of tetrathiafulvalenes
33, which are of high current interest as representatives of "organic metals."[98,99]
Those fulvalenes are not discussed in any detail here because free carbenes are
probably not often involved in their preparation. The recent demonstration by Chen
and Jordan[100] that thiazolium ions lead to carbene-type dimers by polar mecha-

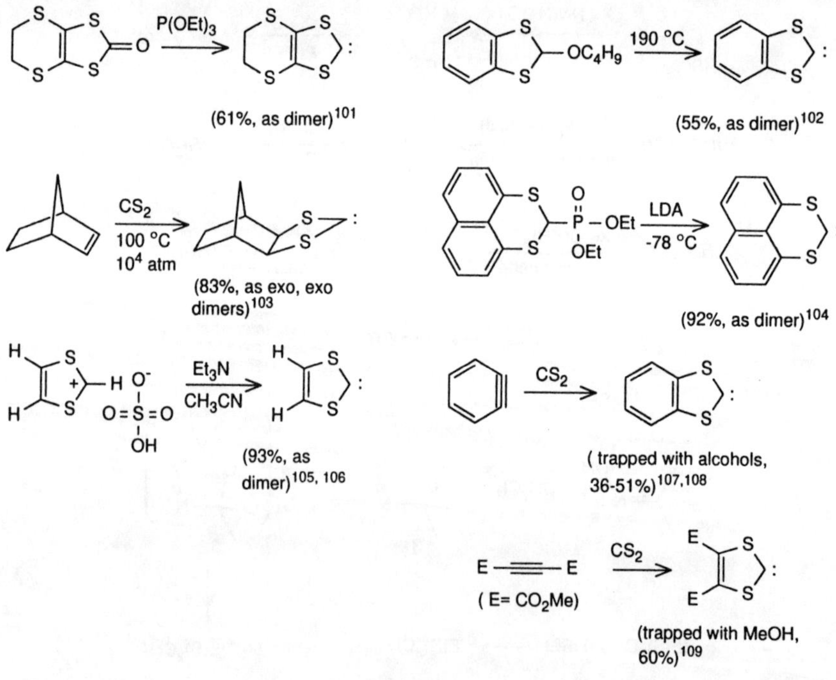

Scheme 22.

nisms, rather than by carbenic coupling, suggests that other apparent carbene dimerizations should be interpreted cautiously.

Approaches to cyclic dithiocarbene intermediates have been reviewed recently[96] and only a sample of methods appears as Scheme 22.[101–109]

IV. INTRAMOLECULAR REACTIONS OF DONOR-SUBSTITUTED CARBENES

A. Intramolecular Reactions of Diaminocarbenes

Diaminocarbene itself has a significant lifetime in the dilute gas phase,[19] indicating its kinetic stability. This result is in agreement with *ab initio* calculations on that species which placed the barrier to 1,2-H migration at more than 45 kcal/mol.[15] Diaminocarbene has not been generated in solution but substituted diaminocarbenes that were generated in solution do not appear to undergo fragmentations, skeletal rearrangements, or intramolecular insertions into CH or other bonds.[33,34,38,39] Gas phase rearrangements, by 1,2-C migration at high temperatures, are known, as in some cases illustrated in Scheme 23.[22] [2,3]-Sigmatropic rearrangements of diaminocarbenes with an N-allyl substituent have not been reported.

B. Intramolecular Reactions of Amino(oxy)carbenes

The parent, aminohydroxycarbene, has been observed in the low pressure gas phase of a mass spectrometer (Section III.B) and so has the N-methyl analogue.[45a] Both are kinetically stable on the microsecond time scale and it is not known which 1,2-H migration is kinetically preferred. The thermodynamically preferred product is undoubtedly formamide rather than formimidic acid.[45a]

Oxazolidin-2-ylidene was also kinetically stable in the mass spectrometer[47] when it was generated by neutralization of the corresponding radical cation. The same

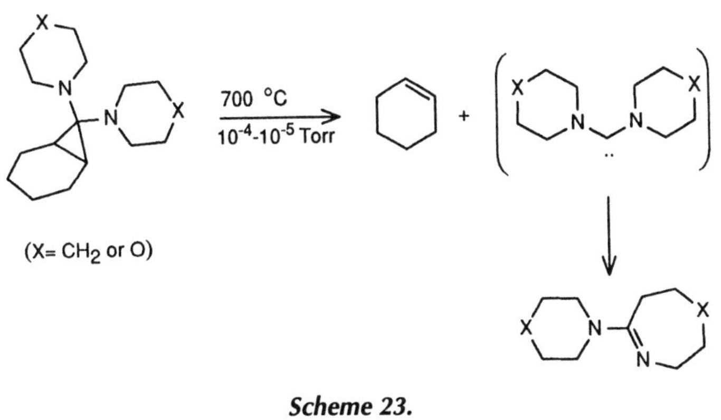

Scheme 23.

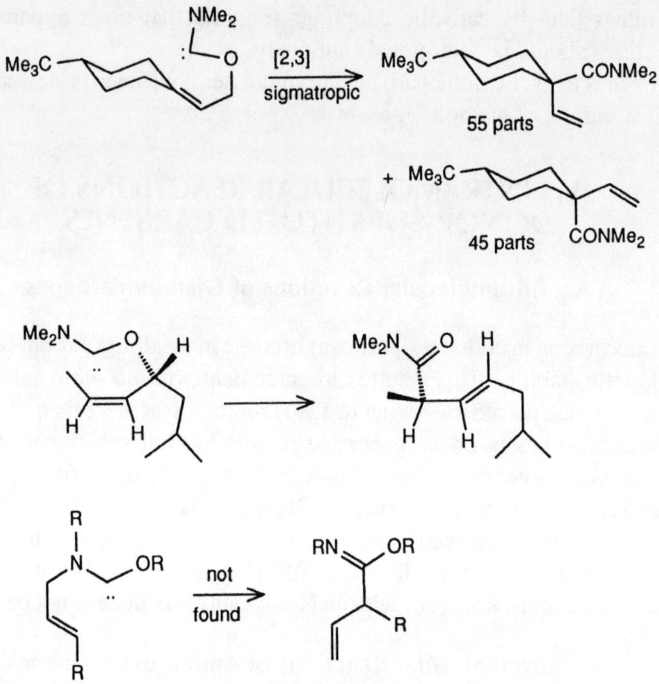

Scheme 24.

carbene, from thermolysis of oxadiazoline **3**, rearranged fully by 1,2-hydrogen migration (see Scheme 5).

An important intramolecular reaction of allyloxy(amino)carbenes is [2,3] sigmatropic rearrangement to homologous amides. That reaction has been studied in connection with its preparative and stereochemical features[49,50,110,111] (Scheme 24). Rearrangements of N-allyl analogues were not found in the literature.

C. Intramolecular Reactions of Dioxycarbenes

The simplest dioxycarbene, dihydroxycarbene, has been studied computationally[7,56,57,59,60,112,113] to predict some inherent properties of the triplet and the singlet and transition structures for reactions of the singlet with alkenes.[60] Those computations predicted that the carbene would be stable in the gas phase.

Dihydroxycarbene has since been studied experimentally in the gas phase with mass spectrometric techniques.[58] In agreement with predictions, it does not rearrange to formic acid on the microsecond time scale. The experimental result is also in agreement with the finding[114] that a single oxy substituent, as in an alkoxymethylcarbene, is sufficient to reduce the rate constant for 1,2-hydrogen migration in solution to $10^4\ s^{-1}$ or less.

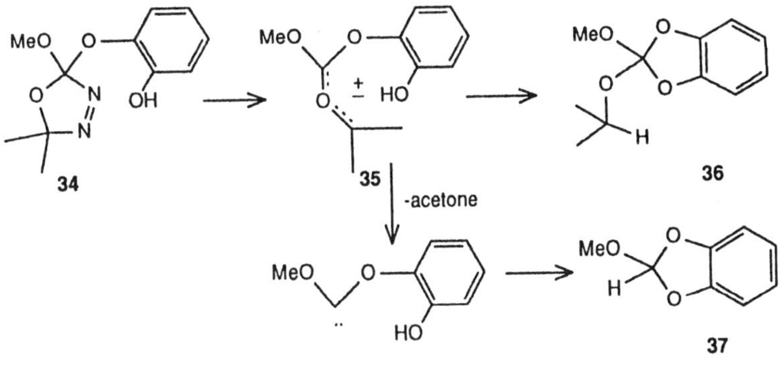

Scheme 25.

Intramolecular insertion into an OH bond is of interest because it may bear on the question of the existence of a carbonyl ylide intermediate in oxadiazoline decomposition. Thermolysis of **34** in benzene afforded **36** and **37** (Scheme 25).[115] A simple rational involves proton transfer to a carbonyl ylide intermediate **35**, either intramolecularly or by starting material, and subsequent ring closure to **36**. Fragmentation of the ylide to carbene, followed by intramolecular OH insertion, accounts for **37**. It does not follow that ylide intermediates are mandatory or that they may exist in other oxadiazoline thermolyses.[83]

[2,3]-Sigmatropic rearrangements of allyloxyalkoxycarbenes have not been reported, but migrations of alkyl groups to the carbene carbon of a dialkoxycarbene are well known. The gas-phase thermolysis of dialkoxymethyl acetates (method (3), Section III.C) afforded the esters from rearrangements of the putative carbene intermediates. Other gas-phase rearrangements of dialkoxycarbenes have been demonstrated by means of very low vapor pressure pyrolysis mass spectrometry.[77] That dialkoxycarbenes are actual intermediates could be demonstrated by generating the corresponding radical cations, neutralizing them to the carbenes in another section of the spectrometer, and subsequently re-ionizing them (NRMS experiment). It was found that ethoxymethoxycarbene rearranges by migration of either alkyl group, and so do *iso*propoxymethoxycarbene and *t*-butoxymethoxycarbene. Migration of higher alkyl groups, in preference to methyl, was observed.[77] Methoxy-2,2,2-trifluoroethoxycarbene, however, rearranged exclusively by migration of the 2,2,2-trifluoroethyl group[77] (Scheme 26). This ordering of migratory aptitudes was explained in terms of anion-like character for the migrating group, analogous to the hydride-like character of H at transition states for 1,2-migration of H-atoms. The preference for migration of branched alkyl groups, over methyl, requires effectively that carbanion stability increases with the degree of α-branching. That is not the order observed in solution but it does fit with gas-phase acidities.[116]

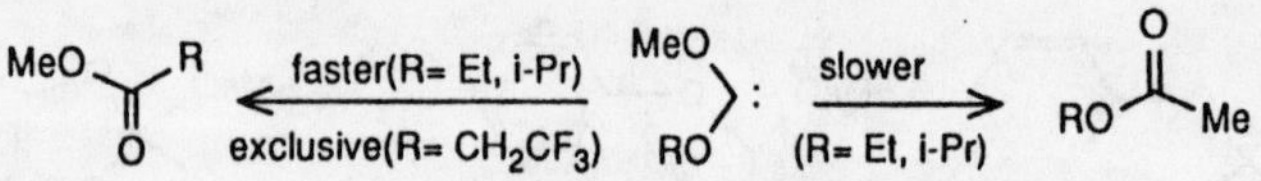

MeO–C(=O)–R ⇌ faster(R= Et, i-Pr) / exclusive(R= CH₂CF₃) MeO–C–OR : slower (R= Et, i-Pr) → RO–C(=O)–Me

Scheme 26.

Methoxy-2,2,2-trifluoroethoxycarbene, generated in solution, does not appear to rearrange.[76,117] The same behavior (slow rearrangement) characterizes the solution chemistry of other dialkoxycarbenes and of alkoxyaryloxycarbenes, which afford dimers when generated in the absence of a carbene trap.[68,72,81]

Benzyloxymethoxycarbenes **38** rearrange in solution by benzyl group migration from oxygen to carbon, affording aryl acetic acid methyl esters **39**.[115] The particular set of **38** (Scheme 27) came from oxadiazoline precursors which do not fragment cleanly in the case of a 2-benzyloxy substituent. In spite of the incomplete materials balance, it is clear that rearrangement products **39** account for at least a third of the original benzyloxy substrate. Whether or not *all* of the carbene intermediates rearrange before they can do anything else is not yet known, but carbene dimers were not found, suggesting that rearrangement is rather fast, for it is not reasonable to assume that carbene coupling could be unusually slow in the case of benzyloxymethoxycarbenes. The virtually substituent-independent yields of the products of carbene rearrangement are surprising and may mean that the rearrangement is quantitative and that other processes, competing with carbene formation rather than

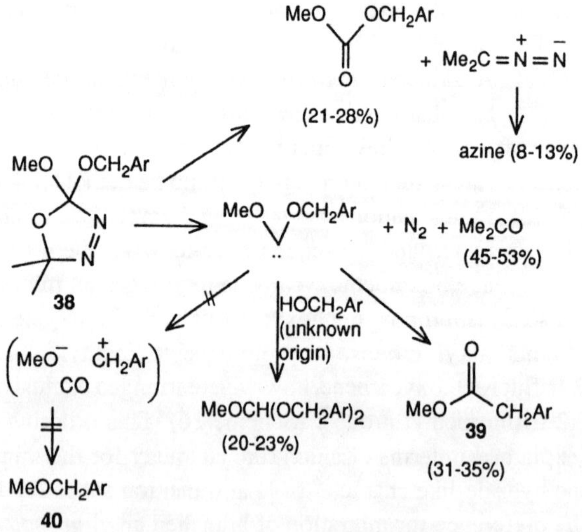

$$MeO-C(=O)-OCH_2Ar \quad (21\text{-}28\%) \quad + \quad Me_2C=\overset{+}{N}=\overset{-}{N} \quad \rightarrow \quad azine\ (8\text{-}13\%)$$

$$\mathbf{38} \quad \rightarrow \quad MeO-\ddot{C}-OCH_2Ar \quad + \quad N_2 + Me_2CO \quad (45\text{-}53\%)$$

$$\left(MeO^- \quad \overset{+}{C}H_2Ar \atop CO \right) \quad \not\Rightarrow \quad MeOCH_2Ar \quad \mathbf{40}$$

$$\xrightarrow{HOCH_2Ar\ (unknown\ origin)} MeOCH(OCH_2Ar)_2 \quad (20\text{-}23\%)$$

$$MeO-C(=O)-CH_2Ar \quad \mathbf{39} \quad (31\text{-}35\%)$$

Scheme 27.

with their rearrangement, are responsible for the low yields (Scheme 27). A carbene fragmentation, in the sense observed with benzyloxychlorocarbene,[118] would lead to benzyl cation, methoxide, and CO (Scheme 27). That reaction also does not fit the observed insensitivity of yields to the nature of the *p*-substituent, nor were the expected benzyl methyl ethers (**40**) found.

Intramolecular cycloaddition of a dialkoxycarbene carbon to a tethered triple bond is probably an important step of a sequence of reactions that convert oxadiazoline **41** to the tricyclic cyclobutanone ketal **48** (Scheme 28).[73,79,119] Unactivated double or triple bonds are poor interceptors of nucleophilic carbenes, but the

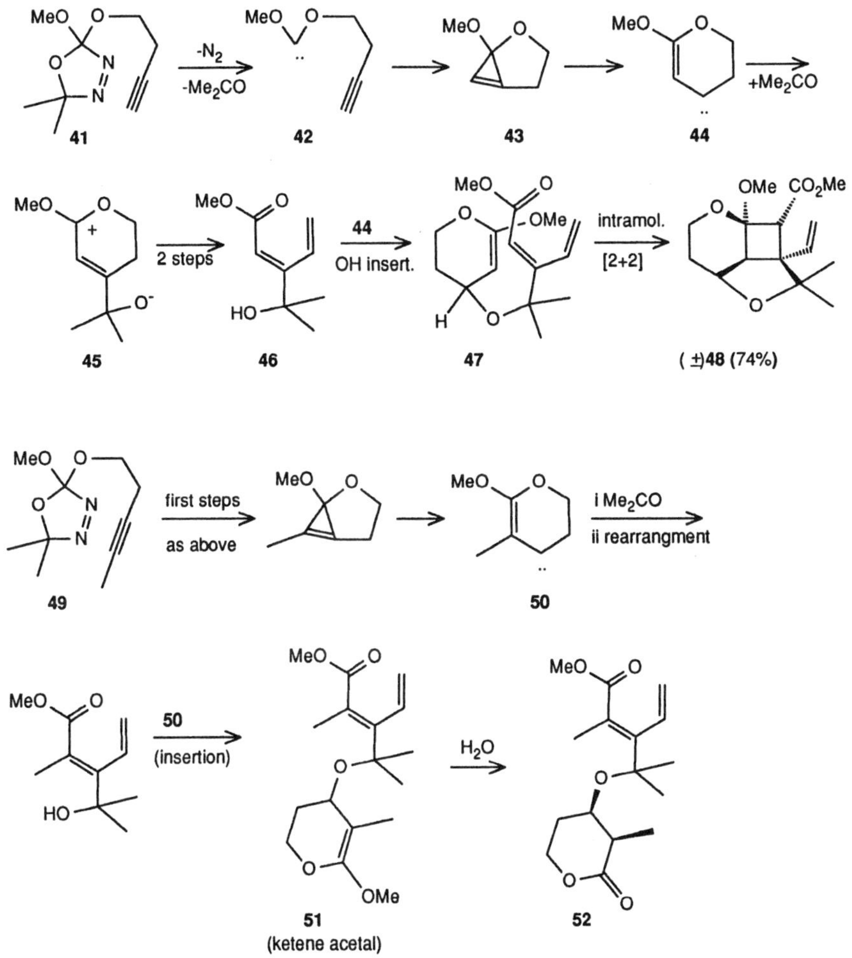

Scheme 28.

intramolecular process is efficient (overall yield of **48** = 74%). Opening of the cyclopropenone ketal **43** affords the vinylogous dialkoxycarbene **44**. Boger and co-workers,[120–127] as well as Jones and Nakamura and their groups,[128–130] have amply demonstrated the thermal ring opening of cyclopropenone ketals, as well as their nucleophilicity/basicity. Those properties make **44** a likely candidate for nucleophilic attack at the carbonyl carbon of acetone, as shown, to form **45**, which converts to **46** by proton transfer and subsequent electrocyclic ring opening. The need for some steps in the original scheme[73] has been questioned by Smith.[83] Oxadiazoline **49** with a nonterminal alkyne unit leads to the same chemistry except that **51** is too hindered for the intramolecular [2+2]-cycloaddition and survives to hydrolyze to **52**[79] (Scheme 28).

An ester-substituted triple bond leads to a different result. Although cycloaddition of a carbene intermediate to the triple bond is likely, by analogy, there isn't a need for a cyclopropene intermediate because the products are those from a formal 5-*exo* closure and could result directly from **53** (Scheme 29). There are many precedents for cycloadditions of π-delocalized vinyl carbenes, such as **53**, to benzylidene malononitrile.[121,124,126,127] Intramolecular attack of a dialkoxycarbene site at the triple bonds of **55** and **59** have also been observed.[79] Formation of both **60** and **61**

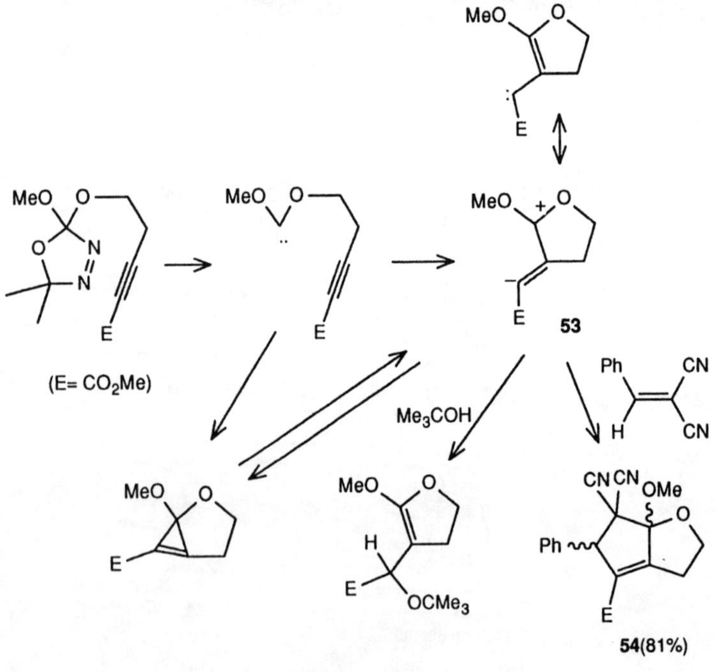

Scheme 29.

was inferred from the products of their capture with tertiary butyl alcohol, **63** and **62**, respectively. Those ketene acetals could not be isolated, with **63** hydrolyzing to **64** (5 parts) and **62** to coumarin (not shown, 1 part) (Scheme 30).

Fragmentation of cyclic, five-membered dioxycarbenes **65** to alkene and CO_2 is part of the Corey–Winter reaction for conversion of vicinal diols to alkenes[54,55] (Scheme 31). The barrier to fragmentation of the parent, to ethene and CO_2, was recently computed by Sauers[131] to be only 7.1 kcal/mol, suggesting that diverting such carbenes into other reaction channels would be difficult. It is interesting to note that the alkoxyamino analogues (Section IV.B) could be trapped in competition

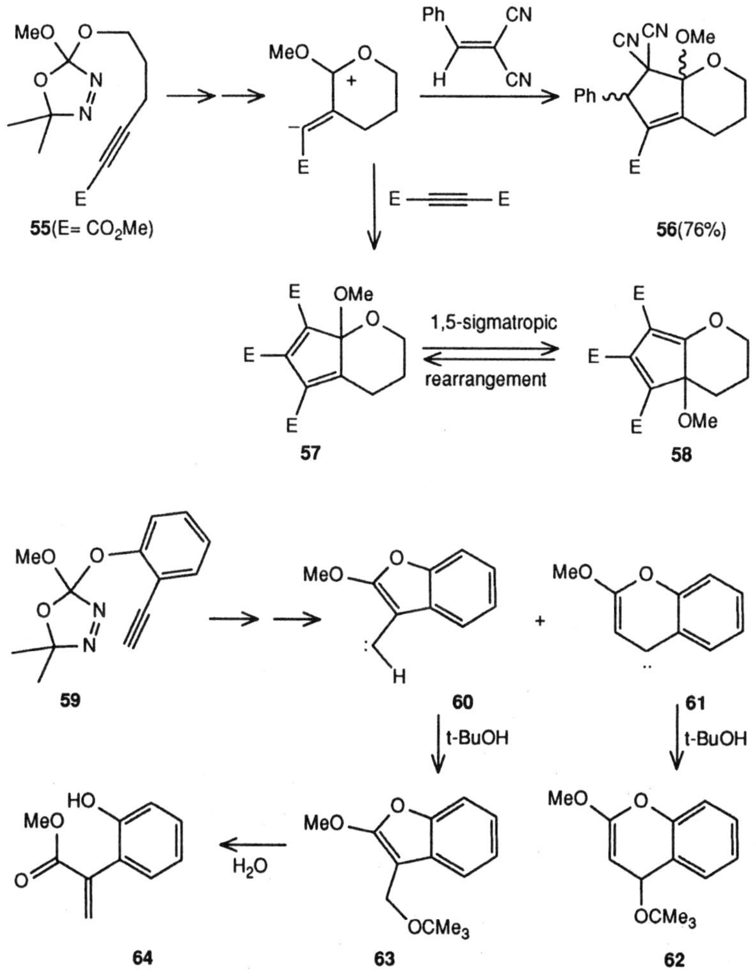

Scheme 30.

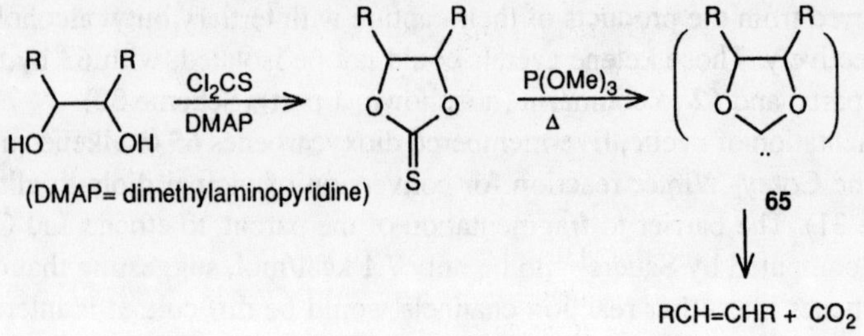

Scheme 31.

with their fragmentation, indicating an enhanced barrier, even when the nitrogen was of the amide type.[53] The barrier for the case where the nitrogen is amino has recently been computed as 17.6 kcal/mol.[132]

D. Intramolecular Reactions of Amino(thio)carbenes

Reactions in this category might include fragmentations analogous to those in the Corey–Winter reaction, to afford an alkene and an isothiocyanate, or sigmatropic rearrangements of carbenes bearing an N-allyl group. Examples of the former were not found and the known sigmatropic rearrangements emanating from the latter have been shown to be rearrangements of the carbene dimers rather than of the carbenes themselves.[133] Rearrangements by 1,2-migration of H from N to the carbenic C-atom have apparently not been reported.

E. Intramolecular Reactions of Oxy(thio)carbenes

The parent, HOCSH, is unknown and computational work to predict the product of 1,2-hydrogen migration has not appeared. Only a few oxy(thio)carbenes have been reported and their gas phase chemistry has not been examined. It is likely to be interesting, given that the thermodynamic product of alkyl migration is expected to be the carbonyl compound whereas the kinetic product could be the thiocarbonyl analogue. Recently generated alkoxy(alkylthio)-, alkoxy(phenylthio)-, and alkoxy(benzylthio)carbenes did not afford products from rearrangement in solution.[94] [2,3]-Sigmatropic rearrangements of oxythiocarbenes bearing an allylic substituent at oxygen or at sulfur have apparently not been reported. The only reported intramolecular reaction of oxy(thio)carbenes appears to be the Corey–Winter type of fragmentation, yielding carbonyl sulfide and ethene[134a] (Scheme 17).

F. Intramolecular Reactions of Dithiocarbenes

Intramolecular reactions of dithiocarbenes are rare. The parent (HSCSH) is not known nor are there examples of monoalkyl analogues that might be expected to undergo a 1,2-migration of a hydrogen atom. Carbenes bearing two thioalkyl groups are known, but 1,2-migration of an alkyl group has not been reported. 2,6-Dithiacyclohexylidene affords cyclopropane and carbon disulfide when generated at 400 °C.[134b]

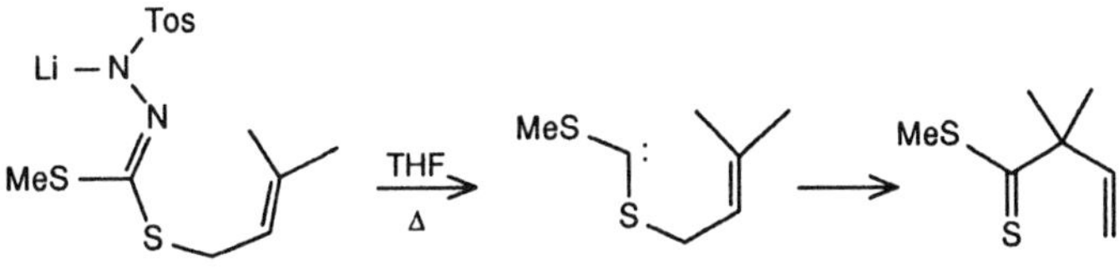

$$(15)$$

[2,3]-Sigmatropic rearrangement of S-methyl-S-(3,3-dimethylallyl)carbene (**66**, Eq. 15) was reported by Baldwin and Walker in 1972[135] and a few additional papers have appeared.[111,136]

V. INTERMOLECULAR REACTIONS OF DONOR-SUBSTITUTED CARBENES

A. Intermolecular Reactions of Diaminocarbenes

Diaminocarbenes do not react with simple alkenes, but transition states for addition of the parent (H_2NCNH_2) to ethylene, hydroxyethylene, and acrolein have been computed.[17] Diaminocarbenes do react with heterocumulenes as shown a long time ago by Hoffmann's group.[25] They were able to show that 2:1 adducts such as **68** are formed (Scheme 32), probably through a dipolar intermediate such as **67**.

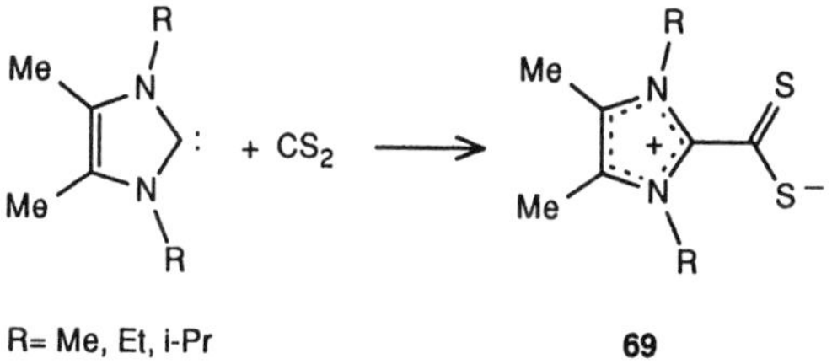

$$(16)$$

R= Me, Et, i-Pr

69

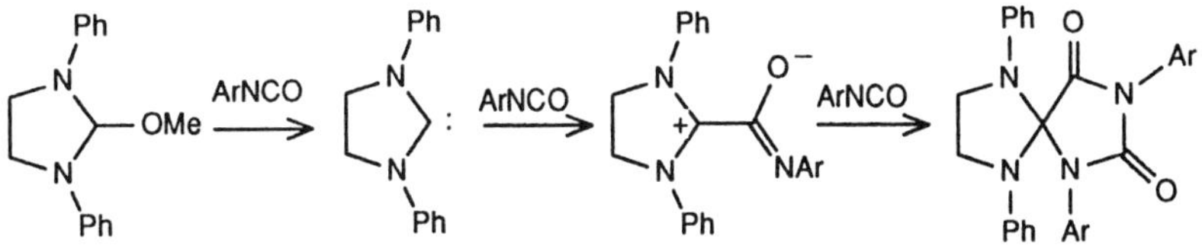

67

68

Scheme 32.

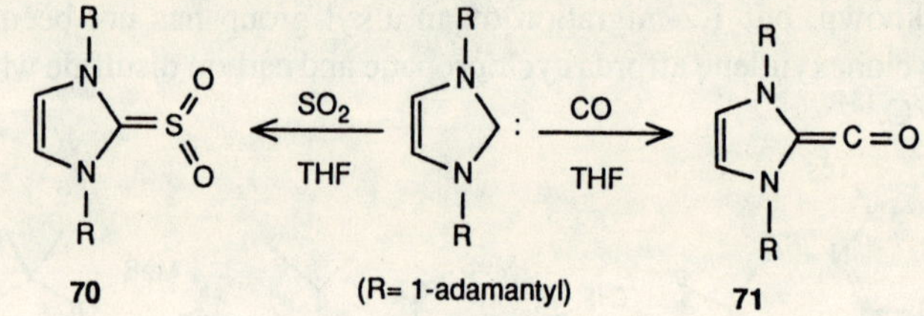

Scheme 33.

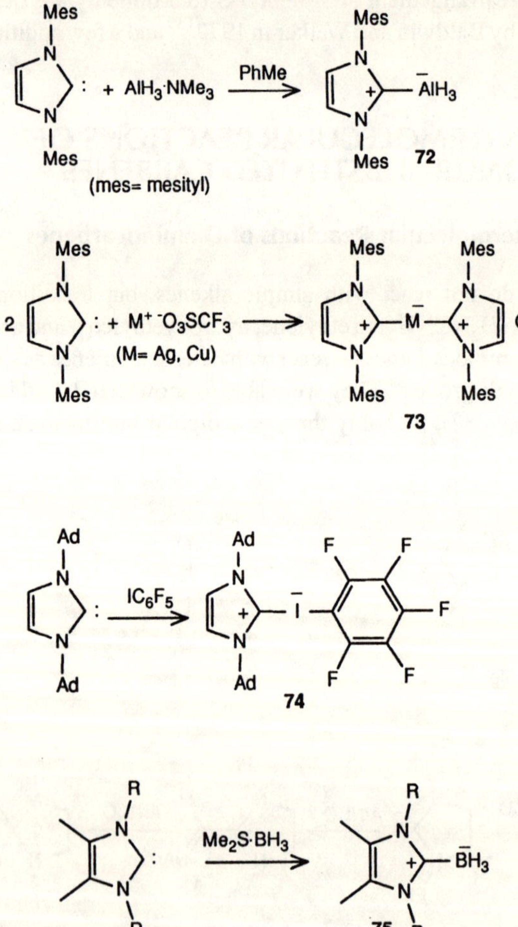

Scheme 34.

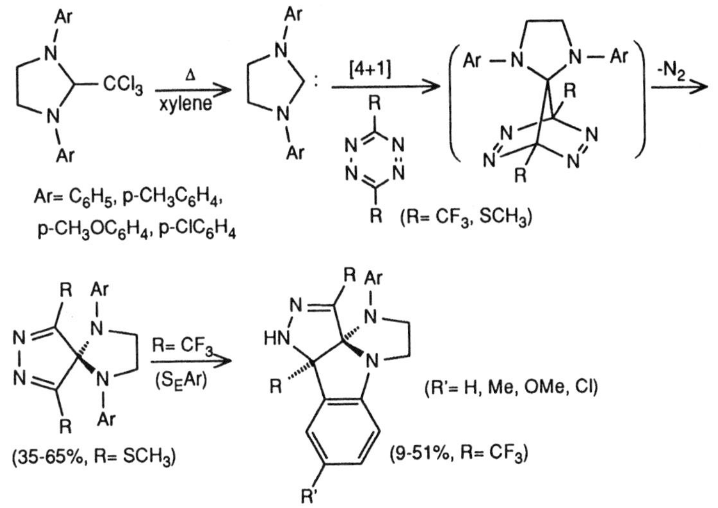

Ar= C_6H_5, p-$CH_3C_6H_4$,
p-$CH_3OC_6H_4$, p-ClC_6H_4

R (R= CF_3, SCH_3)

(35-65%, R= SCH_3)

R= CF_3
(S_EAr)

(R'= H, Me, OMe, Cl)

(9-51%, R= CF_3)

Scheme 35.

By feeding both an isocyanate and an isothiocyanate, they obtained adducts containing one unit of each and information about the relative reactivities of the heterocumulenes.

The postulated dipolar intermediate is made very plausible when viewed in the light of the reaction of other diaminocarbenes with heterocumulene CS_2 (Eq. 16).[32,137] Here the carbene carbon's coordination number increases to only three; the dipolar adducts **69** are stable products.[137] Analogous adducts have been obtained from reaction of diaminocarbene dimers with CS_2.[138] Sulfur dioxide and carbon monoxide react with 1,3-di(1-adamantyl)imidazol-2-ylidene to afford sulfene **70** and ketene **71**, respectively[139,140] (Scheme 33). Analogous recent examples of reactions of diaminocarbenes to form stable dipolar products include formation of an alane species,[141] **72** and of homoleptic complexes with copper (I) and silver (I),[142] **73**, as well as a complex with C_6F_5I,[141] **74** (Scheme 34). Stable carbene–borane adducts (**75**) from imidazol-2-ylidene and $Me_2S \cdot BH_3$ have been reported[39] also. Free 1,3-dimethylimidazolin-2-ylidene (see generalized structure in Section II) reacts with metal halides and metal oxides to form complexes of titanium, zirconium, hafnium, vanadium, niobium, tantalum, and rhenium.[143] Thus, isolable diaminocarbenes do react, not surprisingly, with appropriate electrophilic substrates.

A multistep process, which involves initial [4+1] cycloaddition of a thermally generated diaminocarbene to a tetrazine, followed by loss of N_2 and aromatic substitution[29] in case of 3,6-bis(trifluoromethyl)tetrazine, is illustrated in Scheme

35. The intramolecular aromatic substitution (S_ER) proceeded with high dias-
tereoselectivity.

B. Intermolecular Reactions of Amino(oxy)carbenes

Reactions of alkoxyaminocarbenes with heterocumulenes[51] (Scheme 36) are
analogous to those of diaminocarbenes (Section V.A). A striking feature of the
trapping of 2-oxazolidinylidenes is the fact that it competes well with fragmentation
of the carbenes.[53] As expected, amino(oxy)carbenes react with phenol and with
alcohols by OH insertion[48,53] (Scheme 36). Similar reactions of the homologous
tetrahydro-1,3-oxazin-2-ylidenes were reported.[53]

Nucleophilic carbenes also attack triple bonds in intermolecular reactions. Thermolysis of an oxadiazoline at 90 °C in benzene containing methyl propiolate gave
an interesting product of a rearrangement outlined in Scheme 37.[53c] Reaction with
dimethyl acetylenedicarboxylate (DMAD) took a similar course but with a minor
side reaction (4% yield) that is important in reactions of dimethoxycarbene with
DMAD (Section V.C).

Seitz and coworkers have trapped dimethylamino(ethoxy)carbene with 1,2,4,5-
tetrazines[144,145] to afford the corresponding isopyrazoles by sequential cycloaddition and cycloreversion (Scheme 38).

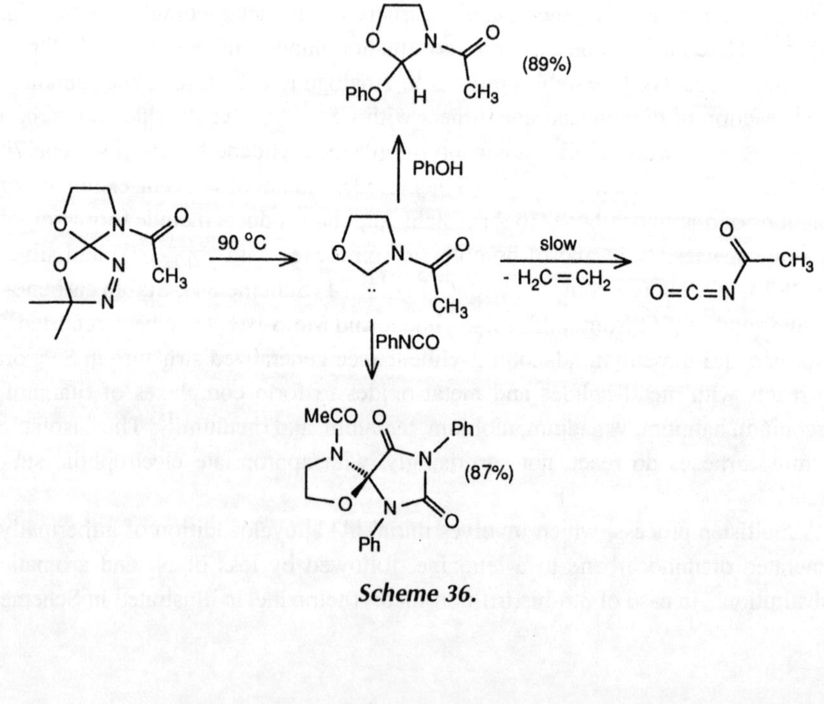

Scheme 36.

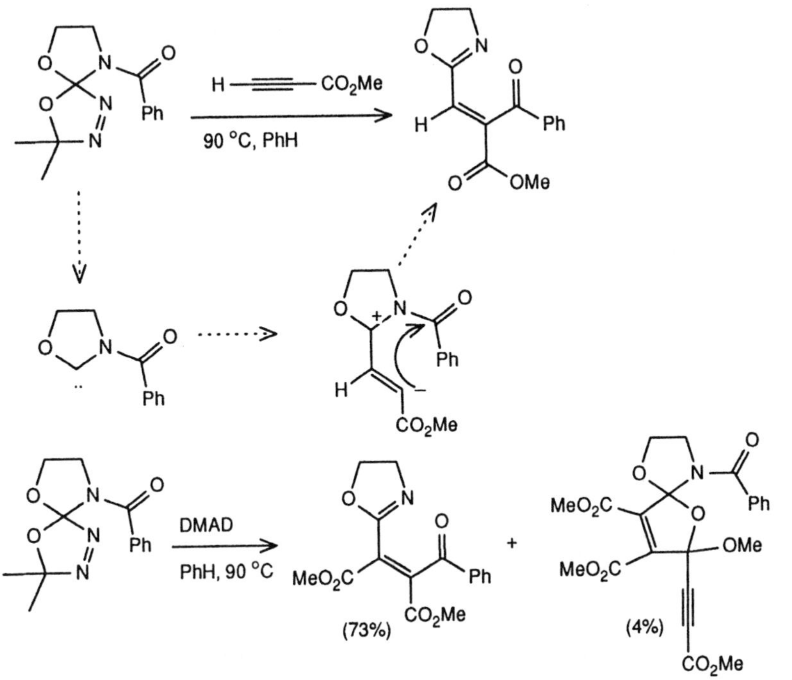

Scheme 37.

C. Intermolecular Reactions of Dioxycarbenes

Tetraalkoxyethenes are well known products from reactions that appear to involve dialkoxycarbene intermediates.[8] Carbene coupling is a particularly likely mechanism in those cases that involve photolysis or thermolysis of neutral substrates in solvents that are poor at solvating ionic intermediates. Among those substrates are the long-known norbornadienone ketals, dialkoxydiazirines, and 2,2-dialkoxy-Δ^3-1,3,4-oxadiazolines (Section III.C). In agreement with a coupling

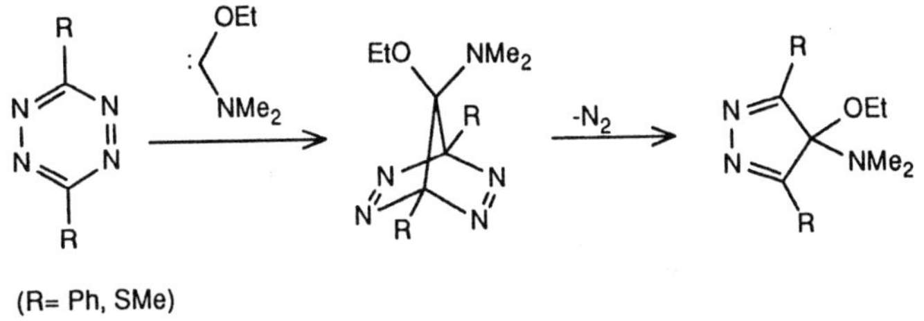

(R= Ph, SMe)

Scheme 38.

Table 2.   *E:Z* Ratios for Carbene Dimers from Oxadiazolines
(100 °C, C_6D_6)

Carbene	E-Dimer (%)	Z-Dimer (%)	E:Z
$CH_3CH_2OCOCH_3$	35	35	1.0
$CH_3CH_2CH_2OCOCH_3$	26	26	1.0
$(CH_3)_2CHOCOCH_3$	19	14	1.4
$(CH_3)_3COCOCH_3$	57	14	4.1
$C_6H_5OCOCH_3$	21	25	0.84
p-$CH_3C_6H_4OCOCH_3$	28	32	0.87
p-$CH_3OC_6H_4COCH_3$	28	29	0.96
p-$ClC_6H_4OCOCH_3$	33	23	1.4
p-$CF_3C_6H_4OCOCH_3$	33	24	1.4
p-$NCC_6H_4OCOCH_3$	27	19	1.4
p-$O_2NC_6H_4OCOCH_3$	39	13	3.0

mechanism are the recent reports of low *E:Z* discrimination in the formation of such dimers from carbene precursors bearing different oxy substituents[68,71,72,146] (Table 2).

Important intermolecular reactions of dialkoxycarbenes are those with heterocumulenes, particularly isocyanates and isothiocyanates. The older literature in this area has been thoroughly reviewed.[1] For those recently generated dialkoxycarbenes that have been trapped with such heterocumulenes, analogous behavior was found. That is, the products are 2:1 adducts like those in Schemes 36 and 32. Recently, a 1,2-bisketene was used to trap dimethoxycarbene[82] (Scheme 39).

With α,β-unsaturated carbonyl compounds, a variety of chemical reactions have been found. *N*-Phenylmaleimide is cyclopropanated at 100 °C in benzene with

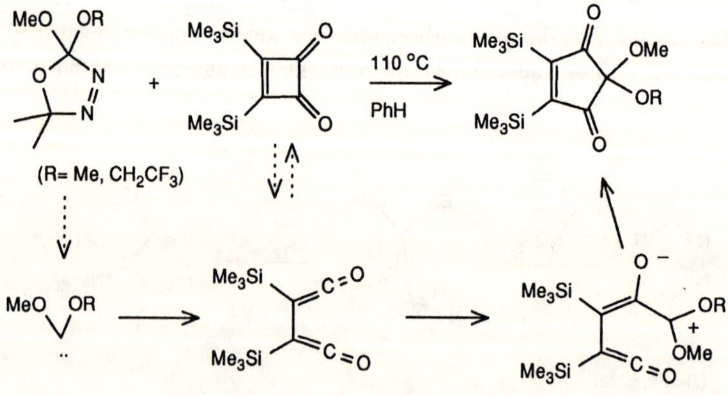

Scheme 39.

dimethoxy oxadiazoline[147a] as a carbene source. On the other hand, methoxy(2-trimethylsilylethoxy)carbene affords a surprisingly rearranged product[78] (Scheme 40). The mechanism of the reaction is not secure, but a reasonable sequence of steps has been proposed involving special properties of intermediates other than the silyloxycarbene itself.[78] A similar rearrangement occurs when that carbene reacts with C_{60}.[80]

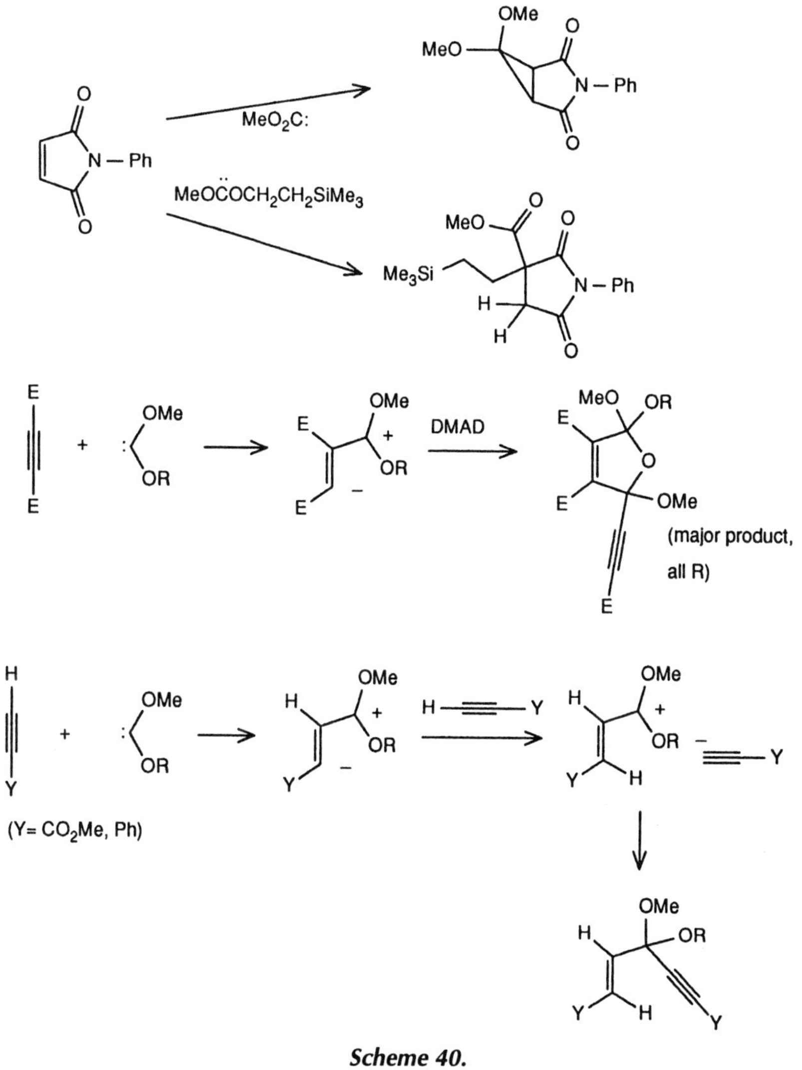

Scheme 40.

Triple bonds such as those in DMAD,[148] in propiolates, and in phenylacetylene[148] react with dialkoxycarbenes. Methyl propiolate and phenylacetylene afford products of apparent initial attack at an sp carbon, followed by proton transfer and ion pair collapse. The dipolar intermediates of Scheme 40 can equally well be represented as vinylogous dialkoxycarbenes. However, the dipolarity of the zwitterion appears to be strongly expressed, for it abstracts an alkynyl proton whereas the precursor carbene does not (Scheme 40, last entry). Vinylogous dialkoxycarbenes generated in other ways[124,126,127,149] show analogous dipolar characteristics, presumably because positive charge developed at transition states for their attack at electrophilic centers is better delocalized. However, some of their chemistry is probably initiated through electron transfer.[127]

Aryloxymethoxycarbenes[115] react with DMAD to afford products in which the aryl group has migrated from oxygen to carbon. The reaction is similar to the aryl group transfers reported by Scherowsky's group[85] and to the benzoyl group transfer observed recently from reaction of an *N*-benzoylamino(oxy)carbene with DMAD (Section V.B).

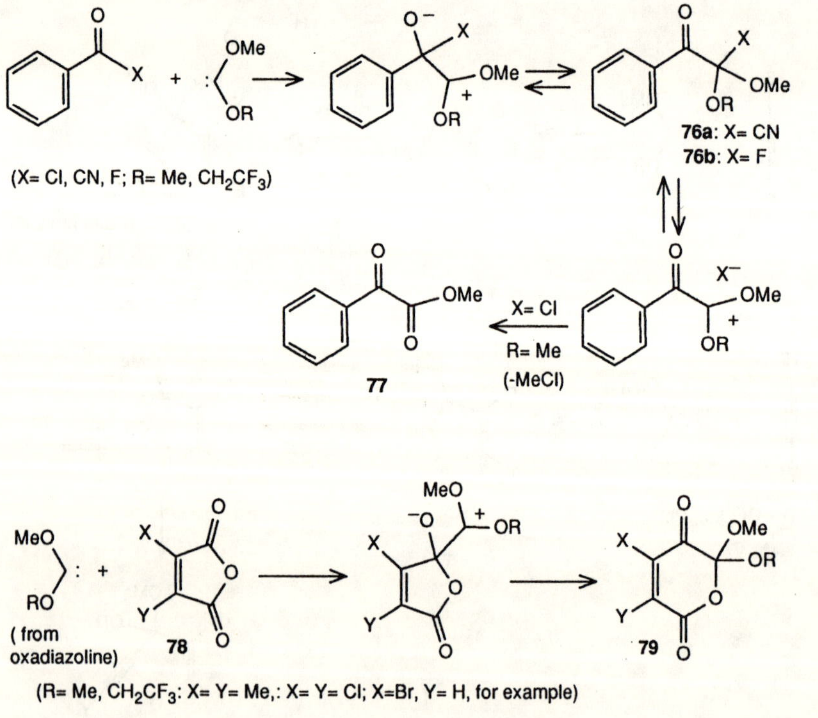

Scheme 41.

Carbonyl groups of aroyl cyanides,[150] aroyl fluorides,[150] anhydrides,[76] and some ketones[150] are interceptors for dialkoxycarbenes, as shown in Scheme 41. All of these reactions can be understood in terms of nucleophilic attack at carbonyl carbon and rearrangement of a dipolar intermediate. The latter is presumably equilibrated with the corresponding oxirane, in general. The products from overall insertion into benzoyl cyanide **76a** and fluoride **76b** are stable enough to be isolated, but that from benzoyl chloride[148] did not survive but led to chloromethane and ketoester **77** (Scheme 41). Products from insertion into maleic anhydrides **78** can be isolated but the purification of **79** is difficult.[76]

Ketones have not been extensively studied as traps for nucleophilic carbenes. 9-Fluorenone reacts with dimethoxycarbene from oxadiazoline thermolysis to afford methyl-9-methoxy-9-fluorenecarboxylate.[151] By means of half-labeled carbene (CH_3OCOCD_3) in one experiment and ^{18}O-labeled fluorenone in another, it was possible to show that the rearrangement to **80** is intramolecular and that methoxy, rather than methyl, is the migrating group. In this case, an oxirane intermediate is required. The overall process (Scheme 42) is the one-step equivalent of cyanohydrin preparation, methylation, hydrolysis, and methylation again, but its generality has not been established.

Dialkoxycarbenes do not attack unactivated double bonds, but activation can be provided in the form of strain. Thus, bicyclopropylidene reacts with a number of dialkoxycarbenes to afford acetals of dispiro[2.0.2.1]heptan-7-one (**81**) (Scheme 43).[81] The strained and substituent-activated double bond of **82** was, not surprisingly, reactive toward dimethoxycarbene. An interesting array of carbene-derived products could be rationalized in terms of addition to form diastereomeric dipolar adducts, **83** and **84**, which subsequently underwent reactions such as ring expan-

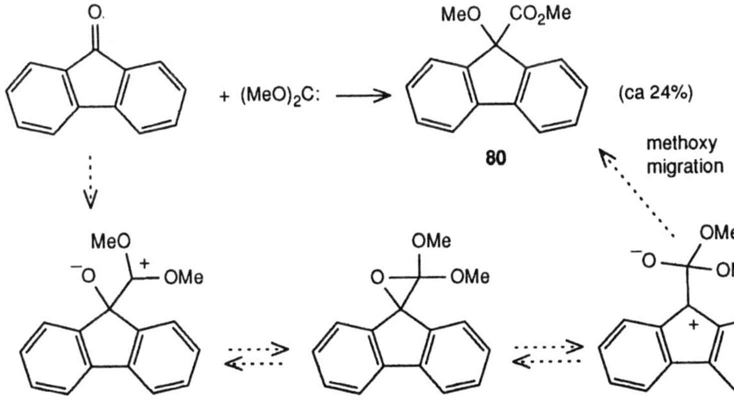

+ (MeO)₂C: → (ca 24%)

80 — methoxy migration

MeO CO₂Me; MeO; OMe; O; OMe

Scheme 42.

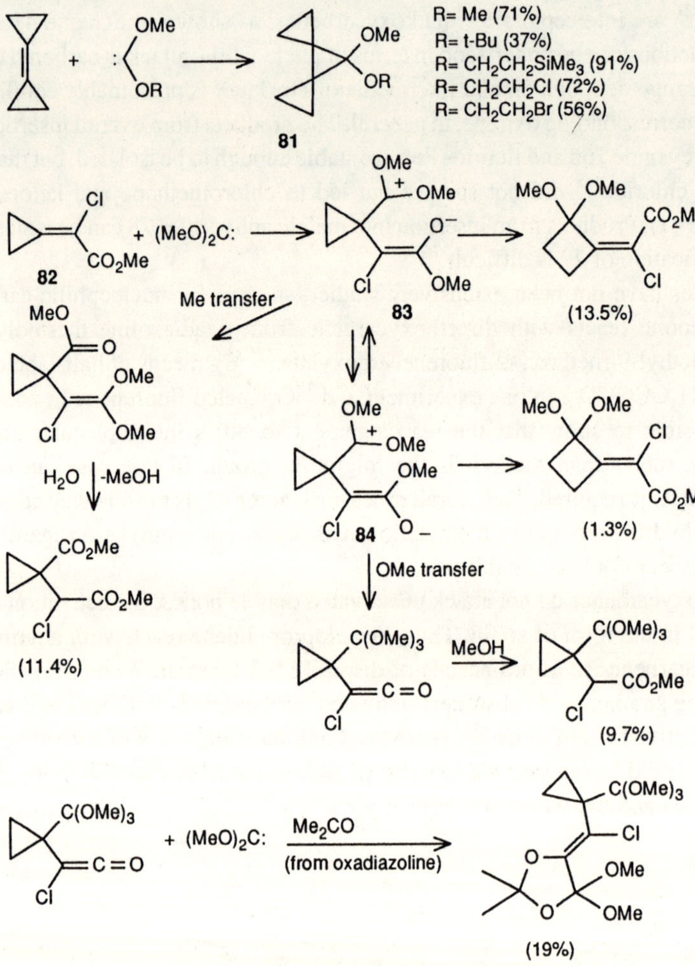

Scheme 43.

sion, methyl transfer, and methoxy transfer to lead to the eventual products[81] (Scheme 43).

Seitz and coworkers were able to trap dimethoxycarbene and diethoxycarbene in [4+1] cycloadditions to tetrazines,[144,145] generating isopyrazoles analogous to those from dimethylamino(ethoxy)carbene (Scheme 38). Vinylogous dimethoxycarbene **87**, from thermal ring opening of cyclopropanone ketal **86**, adds in the [3+2] sense to triazines **85** (Scheme 44).[152]

The electrophilic π system of C_{60}[80] is reactive toward nucleophilic carbenes.[74,130,153–155] Most dialkoxycarbenes and vinylogous[155] dialkoxycarbenes cy-

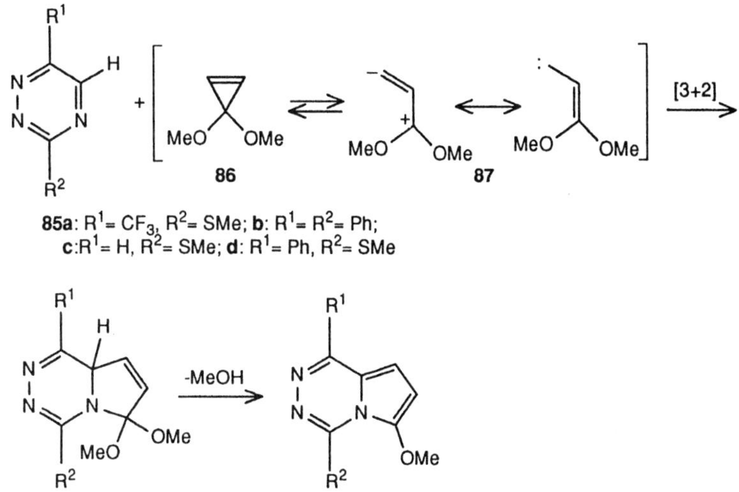
R1
H
N
N
N
N
+
MeO
OMe
86
MeO
OMe
87
MeO
OMe
[3+2]
R2
85a: R1= CF3, R2= SMe; b: R1= R2= Ph;
c:R1= H, R2= SMe; d: R1= Ph, R2= SMe
-MeOH
OMe

Scheme 44.

clopropanate C_{60}, as shown with the example in Scheme 45. In the case of methoxy-2-trimethylsilylethoxycarbene, the expected methanofullerene was not obtained but both 1,2- and 1,4-dihyrofullerenes were isolated.[80]

Aromatic substitutions by dimethoxycarbenes have been observed also.[156] For example, Sanger's reagent traps the carbene from 2,2-dimethoxyoxadiazoline to afford the fragile product **88**, observable by ^{19}F NMR spectroscopy. It could not be isolated, being hydrolyzed to the ester **89**. A reasonable mechanism (Scheme 46) invokes the classical tetrahedral intermediate common to nucleophilic aromatic substitutions. Hexafluorobenzene reacted analogously to afford the unstable pentafluorobenzoyl fluoride dimethyl acetal **90**, which hydrolyzed to **91**.[156]

The straightforward approach to unsymmetric dioxycarbenes that the oxadiazoline route offers (Section III.C) made it possible to access orthoformates with

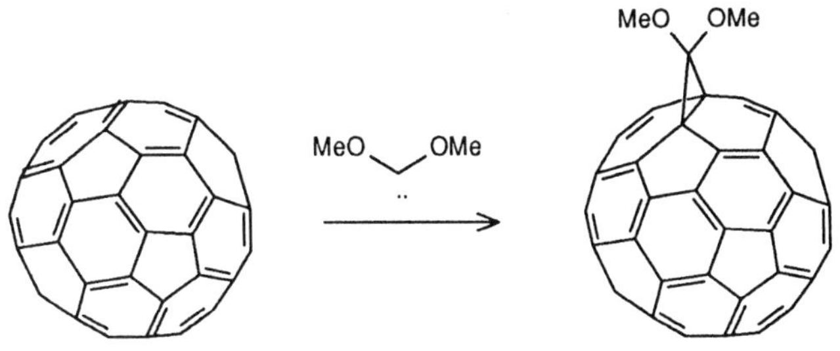
MeO
OMe
MeO
OMe

Scheme 45.

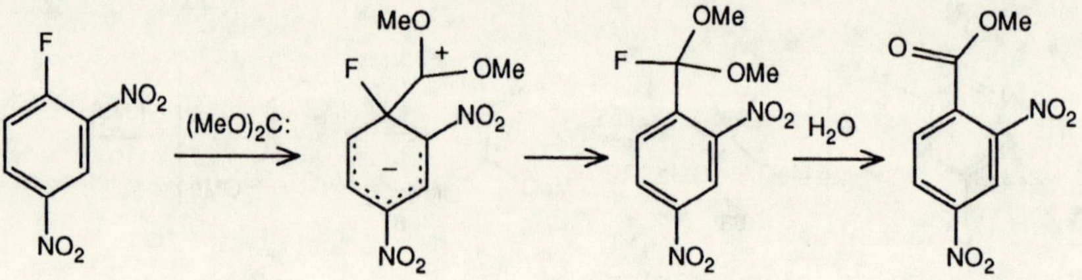

88   **89**

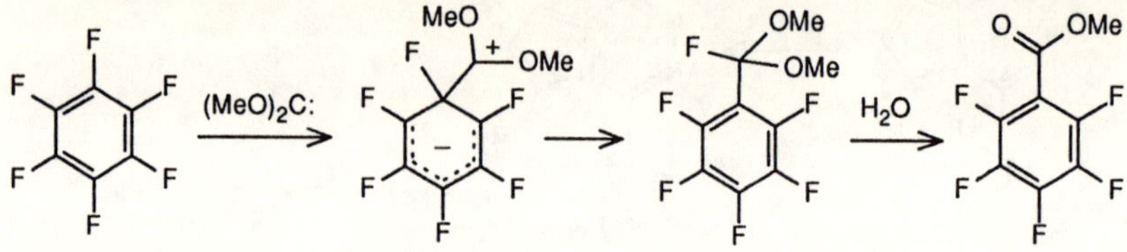

90   **91**

Scheme 46.

stereogenic orthoformyl carbon atoms, simply by trapping the carbenes with an alcohol[157] or phenol different from either oxysubstituent.[115] Some examples are shown below (Scheme 47).

Analogous C-H insertions are rare. β-Dicarbonyl compounds react with dimethoxycarbene (oxadiazoline route) to afford products of apparent C-H insertion such as **92**.[147b] β-Dicarbonyl compounds have substantial enol contents, and the mechanism of formation of **92** probably involves carbene protonation by the enol to form an ion pair (Scheme 48). Collapse of the ion pair may afford either **92** or **93**. Although the latter have not been observed, they are probably unstable and lead to **92** by dissociation back to the ion pair.

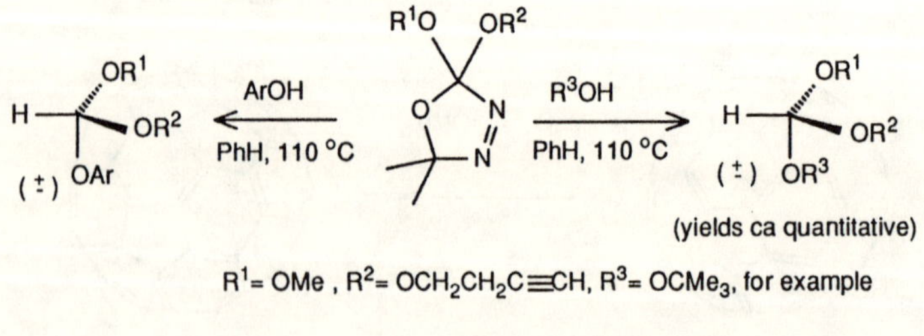

(yields ca quantitative)

$R^1 = OMe$, $R^2 = OCH_2CH_2C \equiv CH$, $R^3 = OCMe_3$, for example

Scheme 47.

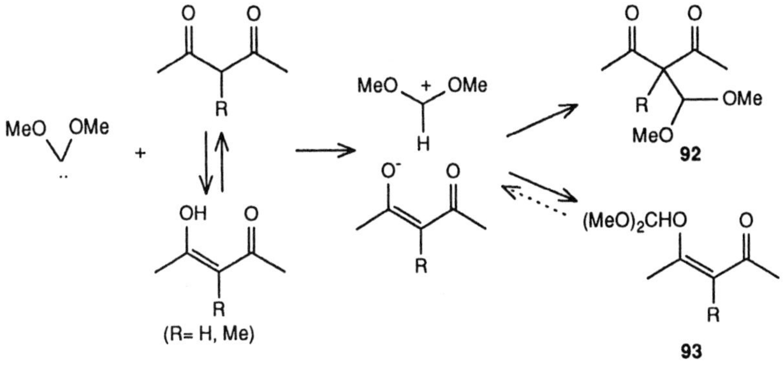

Scheme 48.

D. Intermolecular Reactions of Amino(thio)carbenes

A very interesting reaction of an aminothiocarbene with diethyl acetylenedicarboxylate, reported in 1977[85] is in Scheme 49. Thermolysis of the precursors **95**
alone afforded carbene dimers **94**, suggesting that α-elimination of water (or
alcohol) had occurred to afford the aminothiocarbene (not shown). In the presence

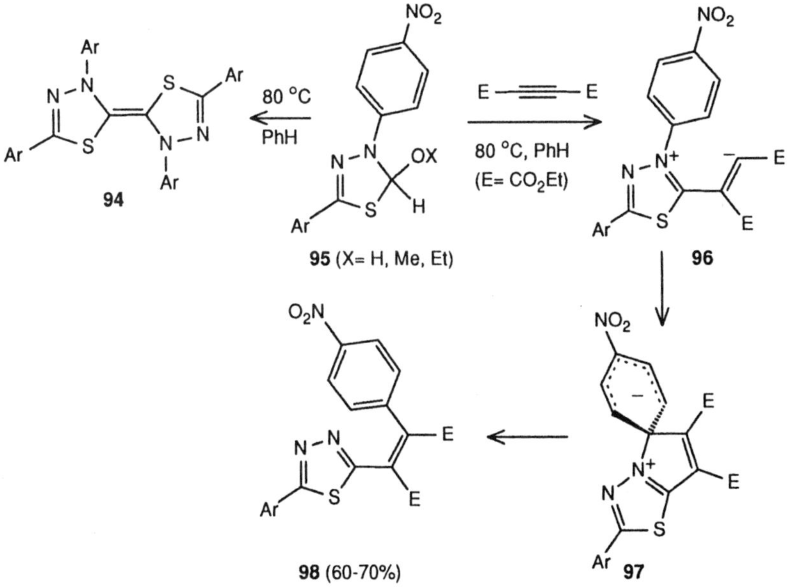

Scheme 49.

of the alkyne, the carbene was trapped to afford dipolar intermediate **96** which converted to a new dipolar intermediate **97**, leading to the final product **98**. Very similar group transfers have since been discovered with amino(oxy)carbenes and with dialkoxycarbenes (Sections V.B and C).

Addition of a nucleophilic carbene in the Michael sense may be reversible. α-Dimethylamino-α-methylthiocarbene was reported to isomerize dimethyl maleate to dimethyl fumarate.[93]

Hoffmann's group reported trapping of *N*-methylbenzothiazolidinylidene with phenyl isothiocyanate and with aryl isocyanates.[25] The mechanisms and products are analogous to those already discussed in connection with dialkoxycarbenes (Section V.C).

In at least one case, where an aminothiocarbene **99** was generated in an acid- and base-free medium, it failed to dimerize.[158] The same authors reported conversion of such carbenes to the corresponding thiones **100** with S_8, to products of cyclopropanation **101** with TCNE, and to azo compounds **102** by reaction with arene diazonium salts, for example (Scheme 50). In a case where thiazolylidene dimers are formed from thiazolium salts, there is strong evidence from crossover experiments with labeled compounds that the dimers do not arise from carbene coupling

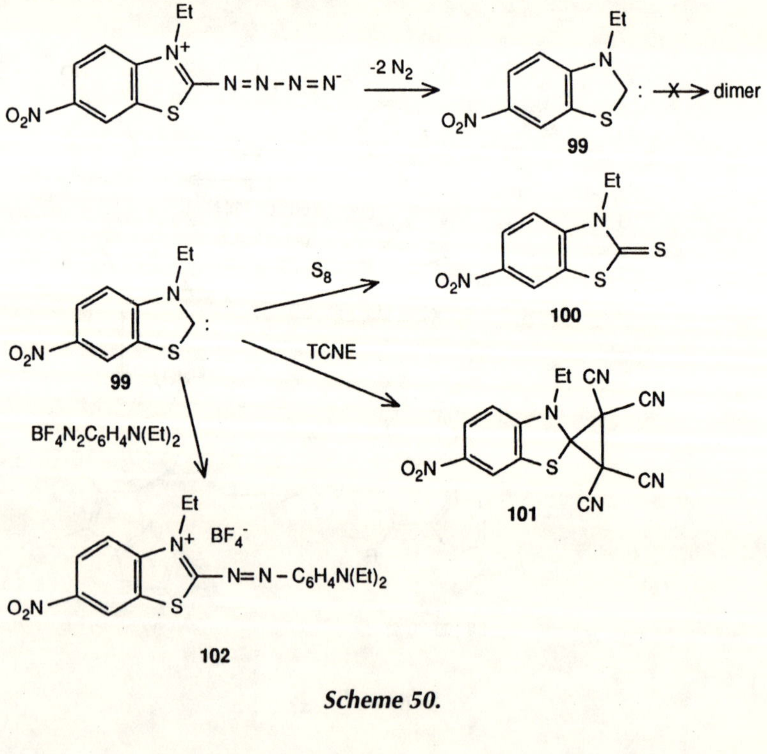

Scheme 50.

but rather from nucleophilic addition of the carbene/ylide to the thiazolium cation.[100] A kinetic study of the benzoin condensation[159] catalyzed by thiazolium salts led to the conclusion that corresponding carbene dimers, rather than the carbenes themselves, are the active catalytic species.[160,161] The authors[160] suggested that some thiamin-dependent enzymes may also operate through thiamin-derived bis-(thiazolin-2-ylidenes). A recent theoretical paper on catalysis of the benzoin condensation[162] by bis(thiazolin-2-ylidenes) suggested a biradical process as an additional possibility.

E. Intermolecular Reactions of Oxythiocarbenes

There is a little information about intermolecular reactions of oxythiocarbenes. Thermolysis of oxadiazolines **29** in benzene at 60–80 °C afforded the carbene dimers (**30**) in 55–61% yields[94] (Scheme 18). Mechanistic alternatives to carbene coupling are hard to imagine. Dichloromaleic anhydride gave the products of acyl-to-oxygen bond insertion **103** (75–89%)[94] that were less sensitive than the corresponding product from dimethoxycarbene (Section V.C). X-ray diffraction confirmed the structure **103** (R = CH$_3$) (Scheme 51). Thermolysis of **29a** or **29b** in presence of phenylisocyanate gave major products different from those afforded by dimethoxycarbene. Whereas dimethoxycarbene from an oxadiazoline precursor affords the same hydantoin **104** that was reported by Hoffmann's group,[163] the oxythiocarbenes led to oxindoles **105** (Scheme 52), as minor products and to 2:2 adducts **106** as major products. Although the reasons for the different behaviors of (MeO)$_2$C: and MeS(MeO)C: toward phenyl isocyanate are not at all clear, the products from the oxythiocarbenes clearly require aromatic substitution in an intermediate, whereas the analogous intermediate from dimethoxycarbene does a cycloaddition instead (Scheme 52). The major products (diastereomers **106**) could arise from the same initial steps, but a mechanism has not been established.[94] The strikingly different behavior of the two types of carbenes is interesting and promising, given that both are really masked carbonyl groups.

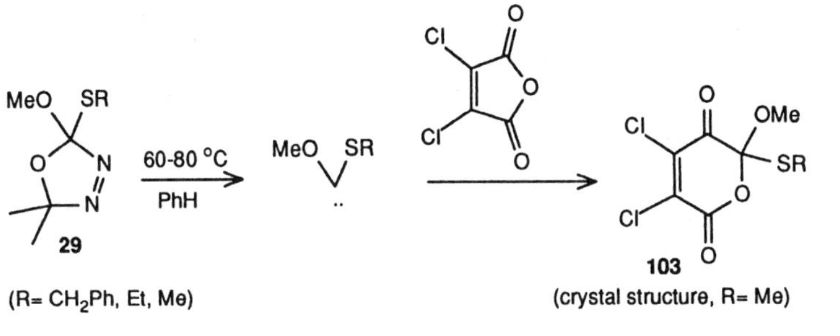

Scheme 51.

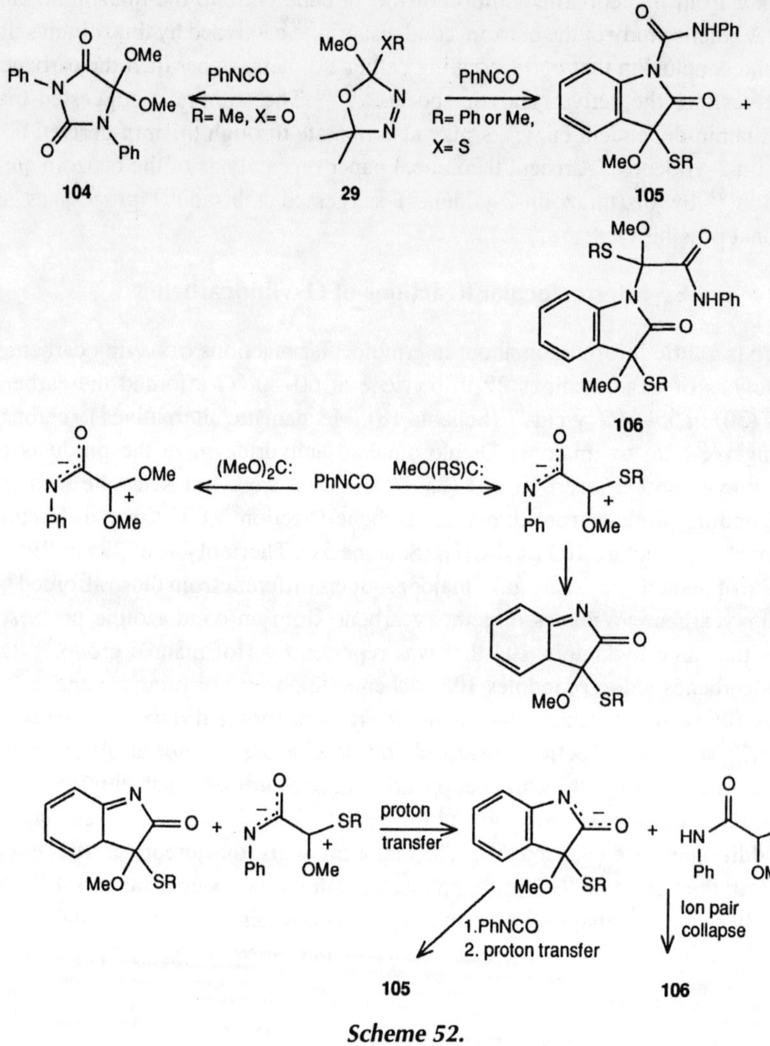

Scheme 52.

Toward DMAD, methoxy(thiomethyl)carbene behaved like dimethoxycarbene, affording diastereomeric 1:2 adducts **107** (Scheme 53).[94] Another product (6%) was dihydrofuran **108**, which may not be carbene derived but rather a product from a carbonyl ylide precursor.

Like other nucleophilic carbenes, oxy(thio)carbenes insert into OH bonds of alcohols and phenols. In general, those reactions probably involve proton transfer to the carbene carbon to form an ion pair[157] that collapses to a product that is

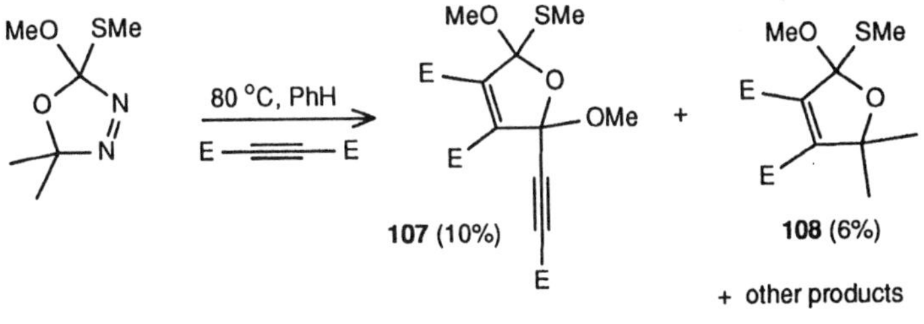

Scheme 53.

orthoformate-like. With an appropriate choice of alcohol or phenol, the products have a stereogenic center at the former carbene carbon[115] (Eq. 17).

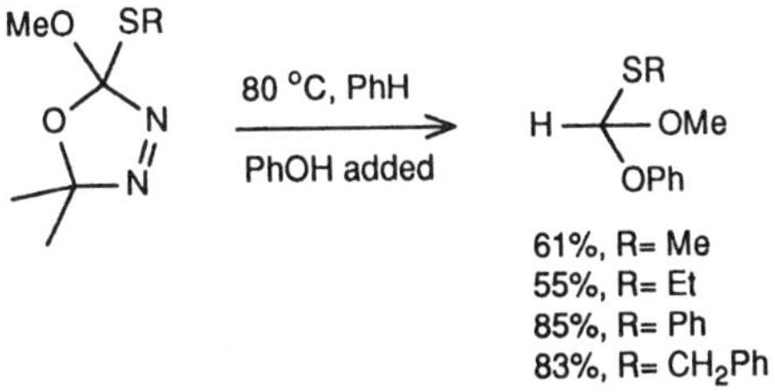

$$(17)$$

61%, R= Me
55%, R= Et
85%, R= Ph
83%, R= CH$_2$Ph

F. Intermolecular Reactions of Dithiocarbenes

A very large number of intermolecular reactions of dithiocarbenes, many of them apparent carbene couplings, have been reviewed thoroughly[1] and only a few are included here.

1,3-Benzodithiol-2-ylidenes have been trapped with sulfur and with selenium (Eq. 18).[164] Carbene complexes of chromium (0) were obtained by trapping bis(phenylthio)carbene with corresponding metal pentacarbonyls[165] (Eq. 19).

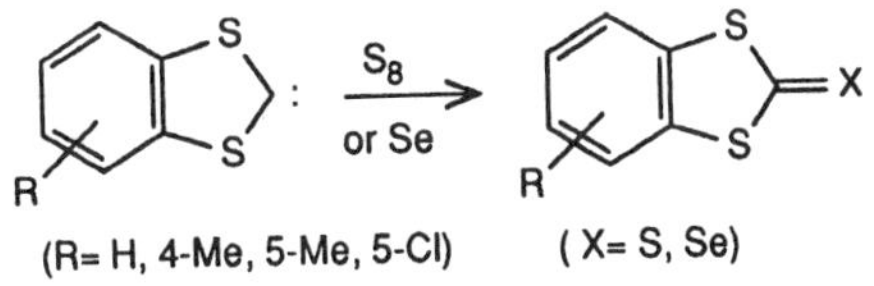

$$(18)$$

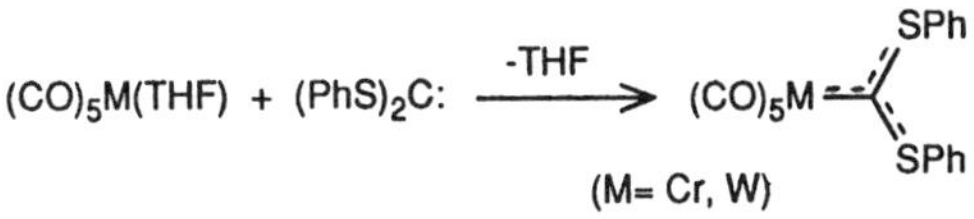

(R= H, 4-Me, 5-Me, 5-Cl)   (X= S, Se)

$$(CO)_5M(THF) + (PhS)_2C: \xrightarrow{-THF} (CO)_5M = C \begin{smallmatrix} SPh \\ SPh \end{smallmatrix} \qquad (19)$$

(M= Cr, W)

Many insertions into OH bonds are known. For example, reaction of DMAD with CS$_2$ in presence of methanol affords **109** in 60% yield,[109] while desulfurization of

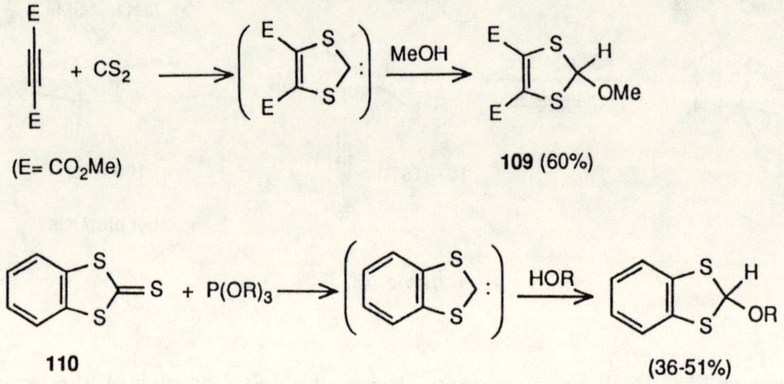

(E= CO$_2$Me)   **109** (60%)

110   (36-51%)

Scheme 54.

110 in solution containing various alcohols afforded the insertion products in 36 to 51% yields (Scheme 54). A possible insertion of (EtS)$_2$C: into the HS bond of EtSH is in Section I.F.

IV. CONCLUSION

The chemistry of metal-free carbenes bearing two heteroatom (N,O,S) donor substituents is still relatively undeveloped. It is clear that their reactivity can be tuned by choosing the heteroatoms and the substituents at those atoms. Those features, plus the fact that initially formed adducts with alkyne substrates are themselves carbenes (vinylogous carbenes/dipoles), suggests that many tandem reactions may be possible. Moreover, the development of new precursors of carbenes discussed in this chapter will undoubtedly provide opportunities for direct spectroscopic observation of more members of the family, and for determination of kinetic parameters.

ACKNOWLEDGMENTS

It is a pleasure to thank my co-workers, whose names appear in the list of references, for their enthusiastic work toward the further development of the chemistry of nucleophilic carbenes. I am grateful to David L. Pole and to John P. Pezacki, for proof-reading the manuscript.

NOTE

*The more generic nomenclature, e.g. dialkoxycarbene, rather than the more specific nomenclature, e.g. dioxacarbene, is used here. 'Dialkoxycarbene' is not ambiguous whereas 'dioxacarbene' really requires locant numbers for the positions of the oxygen atoms. For example, CH$_3$CH$_2$O(H)C: and CH$_3$OCH$_2$(H)C: are both oxabutylidenes.

REFERENCES

1. Regitz, M. Ed.; *Carbene(oide) Carbine. Houben-Weyl*; Thieme: Stuttgart, 1989.
2. Jackson, J.E.; Platz, M.S. *Advances in Carbene Chemistry*, Brinker, U.H., Ed.; JAI: Stamford, CT, 1994; pp 89–160.
3. Boehme, C.; Frenking, G. *J. Am. Chem. Soc.* **1996**, *118*, 2039–2046.
4. Sauers, R.R. *Tetrahedron Lett.* **1996**, *37*, 149–152.
5. (a) Heinemann, C.; Mueller, T.; Apeloig, Y.; Schwarz, H. *J. Am. Chem. Soc.* **1996**, *118*, 2023–2038; (b) Olsson, M.H.M.; Borowski, P.; Roos, B.O. *Theor. Chim. Acta* **1996**, *93*, 17–33; (c) Raabe, G.; Breuer, K.; Enders, D.; Teles, J.H. *Z. Naturforsch., A.* **1996**, *51a*, 95–101.
6. Fleming, I. *Frontier Orbitals and Organic Chemical Reactions*; Wiley: Chichester, 1976.
7. (a) Feller, D.; Borden, W.T.; Davidson, E.R. *J. Chem. Phys.* **1979**, *71*, 4987–4996; (b) Rondan, N.G.; Houk, K.N.; Moss, R.A. *J. Am. Chem. Soc.* **1980**, *102*, 1770–1776; (c) Mueller, P.H.; Rondan, N.G.; Houk, K.N.; Harrison, J.F.; Hooper, D.; Willen, B.H.; Liebman, J.F. *J. Am. Chem. Soc.* **1981**, *103*, 5049–5052.
8. (a) Moss, R.A.; Włostowski, M.; Shen, S.; Krogh-Jespersen, K.; Matro, A. *J. Am. Chem. Soc.* **1988**, *110*, 4443–4444; (b) Muchall, H.M.; Werstiuk, N.H.; Choudury, B.; Ma, J.; Warkentin, J.; Pezacki, J.P. *Can. J. Chem.* in press.
9. Khodabandeh, S.; Carter, E.A. *J. Phys. Chem.* **1993**, *97*, 4360–4364.
10. (a) Miller, D.M.; Schreiner, P.R.; Schaefer, H.F., III. *J. Am. Chem. Soc.* **1995**, *117*, 4137–4143; (b) Richards, C.A., Jr.; Kim, S.J.; Yamaguchi, Y.; Schaefer, H.F., III. *J. Am. Chem. Soc.* **1995**, *117*, 10104–10107.
11. Zoghbi, M.; Warkentin, J. *J. Org. Chem.* **1991**, *56*, 3214–3215.
12. Zoghbi, M.; Warkentin, J. *Can. J. Chem.* **1992**, *70*, 2967–2971.
13. Zoghbi, M.; Warkentin, J. *Can. J. Chem.* **1993**, *71*, 912–918.
14. Zoghbi, M.; Horne, S.E.; Warkentin, J. *J. Org. Chem.* **1994**, *59*, 4090–4095.
15. Heinemann, C.; Thiel, W. *Chem. Phys. Letters* **1994**, *217*, 11–16.
16. Cioslowsky, J. *Int. J. Quantum Chem. Quant. Chem. Symp.* **1993**, *27*, 309–319.
17. Moreno, M.; Lluch, J.M.; Oliva, A.; Bertran, J. *J. Chem. Soc. Perkin Trans. 2* **1986**, 183–185.
18. Farras, J.; Olivella, S.; Sole, A.; Vilarrasa, J. *J. Comp. Chem.* **1986**, *7*, 428–442.
19. McGibbon, G.A.; Kingsmill, C.A.; Terlouw, J.K. *Chem. Phys. Letters* **1994**, *222*, 129–134.
20. Clemens, D.H.; Bell, A.J.; O'Brien, J.L. *Tetrahedron Lett.* **1965**, 3257–3258.
21. Lemal, D.M.; Lovald, R.A.; Kawano, K.I. *J. Am. Chem. Soc.* **1964**, *86*, 2518–2519.
22. Vilsmeier, E.; Kristen, G.; Tetzlaff, C. *J. Org. Chem.* **1988**, *53*, 1806–1808.
23. Simchen, G. *Imminium Salts in Organic Chemistry*; Boehme, H.; Viehe, H.G., Eds.; Wiley: New York, 1979; pp 393–525.
24. Simchen, G. *Houben-Weyl, Methoden der Organischen Chemie*; Band E-5, Falbe, J., Ed.; Thieme: Stuttgart, 1985; pp 3–192.
25. Hoffmann, R.W.; Hagenbruch, B.; Smith, D.M. *Chem. Ber.* **1977**, *110*, 23–36.
26. Wanzlick, H.-W.; Schikora, E. *Chem. Ber.* **1961**, *94*, 2389–2393.
27. Wanzlick, H.-W.; Esser, F.; Kleiner, H.-J. *Chem. Ber.* **1963**, *96*, 1208–1213.
28. Wanzlick, H.-W.; Kleiner, H.-J. *Angew. Chem.* **1961**, *73*, 493.
29. Frenzen, G.; Kümmell, A.; Meyer-Dulheuer, C.; Seitz, G. *Chem. Ber.* **1994**, *127*, 1803–1806.
30. Olofson, R.A.; Thompson, W.R.; Michelman, J.S. *J. Am. Chem. Soc.* **1964**, *86*, 1865–1866.
31. Staab, H.A.; Irngartinger, H.; Mannschreck, A.; Wu, M.-T. *Liebigs Ann.* **1966**, *695*, 55–64.
32. Krasuski, W.; Nikolaus, D.; Regitz, M. *Liebigs Ann. Chem.* **1982**, 1451–1465.
33. Arduengo, A.J., III; Harlow, R.L.; Kline, M. *J. Am. Chem. Soc.* **1991**, *113*, 361–363.
34. Arduengo, A.J., III; Goerlich, J.R.; Marshall, W.J. *J. Am. Chem. Soc.* **1995**, *117*, 11027–11028.
35. Arduengo, A.J., III; Dias, H.V.R.; Harlow, R.L.; Kline, M. *J. Am. Chem. Soc.* **1992**, *114*, 5530–5534.

36. Arduengo, A.J., III; Bock, H.; Chen, H.; Denk, M.; Dixon, D.A.; Green, J.C.; Herrmann, W.A.; Jones, N.L.; Wagner, M.; West, R. *J. Am. Chem. Soc.* **1994**, *116*, 6641–6649.

37. Kuhn, N.; Kratz, T. *Synthesis* **1993**, 561–562.

38. (a) Enders, D.; Breuer, K.; Raabe, G.; Runsink, J.; Teles, J.H.; Melder, J.-P.; Ebel, K.; Brode, S. *Angew. Chem. Int. Ed. Engl.* **1995**, *34*, 1021–1023; (b) Enders, D.; Breuer, K.; Runsink, J.; Teles, J.H. *Liebigs Ann.* **1996**, 2019–2028.

39. Kuhn, N.; Henkel, G.; Kratz, T.; Kreutzberg, J.; Boese, R.; Maulitz, A.H. *Chem. Ber.* **1993**, *126*, 2041–2045.

40. Dixon, D.A.; Arduengo, A.J., III. *J. Phys. Chem.* **1991**, *95*, 4180–4182.

41. Arduengo, A.J., III; Dixon, D.A.; Kumashiro, K.K.; Lee, C.; Power, W.P.; Zilm, K.W. *J. Am. Chem. Soc.* **1994**, *116*, 6361–6367.

42. Schaftenaar, G.; Postma, R.; Ruttink, P.J.A.; Burgers, P.C.; McGibbon, G.A.; Terlouw, J.K. *Int. J. Mass Spectrom. Ion Processes* **1990**, *100*, 521–544.

43. McGibbon, G.A.; Terlouw, J.K.; Burgers, P.C. *Rapid Commun. Mass Spectrom.* **1993**, *7*, 138–141.

44. Hop, C.E.C.A.; Chen, H.; Ruttink, P.J.A.; Holmes, J.L. *Org. Mass Spec.* **1991**, *26*, 476, 480.

45. (a) McGibbon, G.A.; Burgers, P.C.; Terlouw, J.K. *Int. J. Mass Spectrom. Ion Processes* **1994**, *136*, 191–208, and references therein; (b) Ruttink, P.J.; Burgers, P.C.; Terlouw, J.K. *Int. J. Mass Spectrom. Ion Processes* **1995**, *145*, 35–43.

46. Jutzi, P.; Schröder, F.-W. *Angew. Chem. Int. Ed. Engl.* **1971**, *10*, 339.

47. Wong, T.; Warkentin, J.; Terlouw, J.K. *Int. J. Mass Spectrom. Ion Processes* **1992**, *115*, 33–52.

48. Brown, J.M.; Place, B.D. *J. Chem. Soc. Chem. Commun.* **1971**, 533–534.

49. Buchi, G.; Cushman, M.; Wuest, H. *J. Am. Chem. Soc.* **1974**, *96*, 5563–5565.

50. Chan, K.-K.; Saucy, G. *J. Org. Chem.* **1977**, *42*, 3828–3832.

51. Reiffen, M.; Hoffmann, R.W. *Chem. Ber.* **1977**, *110*, 37–48.

52. Wanzlick, H.-W.; Steinmaus, H. *Chem. Ber.* **1968**, *101*, 244–251.

53. (a) Couture, P.; Terlouw, J.K.; Warkentin, J. *J. Am. Chem. Soc.* **1996**, *118*, 4214–4215; (b) Couture, P.; Warkentin, J. *Can. J. Chem.* **1997**, *75*, 1264–1280; (c) Couture, P.; Warkentin, J. *Can. J. Chem.* **1997**, *75*, 1281–1294.

54. Corey, E.J.; Winter, R.A.E. *J. Am. Chem. Soc.* **1963**, *85*, 2677–2678.

55. Corey, E.J.; Carey, F.A.; Winter, R.A.E. *J. Am. Chem. Soc.* **1965**, *87*, 934–935.

56. Goddard, J.D.; Yamaguchi, Y.; Schaefer, H.F., III. *J. Chem. Phys.* **1992**, *96*, 1158–1166.

57. Räsänen, M.; Raaska, T.; Kunttu, H.; Murto, J. *J. Mol. Struct. (Theochem)* **1990**, *208*, 79–90.

58. (a) Burgers, P.C.; McGibbon, G.A.; Terlouw, J.K. *Chem. Phys. Letters* **1994**, *224*, 539–543; (b) Wiedmann, F.A.; Cai, J.; Wesdemiotis, C. *Rapid Commun. Mass Spectrom.* **1994**, *8*, 804–807.

59. Bock, C.W.; Redington, R.L. *J. Chem. Phys.* **1986**, *85*, 5391–5400.

60. Moreno, M.; Lluch, J.M.; Oliva, A.; Bertran, J. *J. Chem. Soc. Perkin Trans. 2* **1985**, 131–135.

61. Oele, P.C.; Louw, R. *Tetrahedron Lett.* **1972**, *48*, 4941–4944.

62. Lemal, D.M.; Lovald, R.W.; Harrington, R.W. *Tetrahedron Lett.* **1965**, 2779–2785.

63. Lemal, D.M.; Gosselink, E.P.; McGregor, S.D. *J. Am. Chem. Soc.* **1966**, *88*, 582–600.

64. Lemal, D.M.; Gosselink, E.P.; Ault, A. *Tetrahedron Lett.* **1964**, 579–585.

65. Hoffmann, R.W.; Hauser, H. *Tetrahedron Lett.* **1964**, 197–201.

66. McDonald, R.M.; Krueger, R.A. *J. Org. Chem.* **1966**, *31*, 488–494.

67. Hoffmann, R.W. *Acc. Chem. Res.* **1985**, *18*, 248–253.

68. (a) Moss, R.A.; Włostowski, M.; Terpinski, J.; Kmiecik-Lawrynowicz, G.; Krogh-Jespersen, K. *J. Am. Chem. Soc.* **1987**, *109*, 3811–3812; (b) Ge, C.-S.; Jefferson, E.A.; Moss, R.A. *Tetrahedron Lett.* **1993**, *34*, 7549–7552.

69. Eisenthal, K.B.; Moss, R.A.; Turro, N.J. *Science* **1984**, *225*, 1439–1445.

70. Du, X.-M.; Fan, H.; Goodman, J.L.; Kesselmayer, M.A.; Krogh-Jespersen, K.; LaVilla, J.A.; Moss, R.A.; Shen, S.; Sheridan, R.S. *J. Am. Chem. Soc.* **1990**, *112*, 1920–1926.

71. El-Saidi, M.; Kassam, K.; Pole, D.L.; Tadey, T.; Warkentin, J. *J. Am. Chem. Soc.* **1992**, *114*, 8751–8752.

72. Kassam, K.; Pole, D.L.; El-Saidi, M.; Warkentin, J. *J. Am. Chem. Soc.* **1994**, *116*, 1161–1162.

73. Kassam, K.; Warkentin, J. *J. Org. Chem.* **1994**, *59*, 5071–5075.

74. Win, W.W.; Kao, M.; Eiermann, M.; McNamara, J.J.; Wudl, F.; Pole, D.L.; Kassam, K.; Warkentin, J. *J. Org. Chem.* **1994**, *59*, 5871–5876.

75. Sharma, P.; Warkentin, J. *Tetrahedron Lett.* **1995**, *36*, 7591–7594.

76. Pole, D.L.; Warkentin, J. *Liebigs Ann.* **1995**, 1907–1914.

77. Suh, D.; Pole, D.L.; Warkentin, J.; Terlouw, J.K. *Can. J. Chem.* **1996**, *74*, 544–558.

78. Pole, D.L.; Sharma, P.K.; Warkentin, J. *Can. J. Chem.* **1996**, *74*, 1335–1340.

79. Kassam, K.; Warkentin, J. *Can. J. Chem.* **1997**, *75*, 120–128.

80. González, R.; Wudl, F.; Pole, D.L.; Sharma, P.K.; Warkentin, J. *J. Org. Chem.* **1996**, *61*, 5837–5839.

81. de Meijere, A.; Kozhushkov, S.I.; Yufit, D.S.; Boese, R.; Haumann, T.; Pole, D.L.; Sharma, P.K.; Warkentin, J. *Liebigs Ann.* **1996**, 601–612.

82. Colomvakos, J.D.; Egle, I.; Ma, J.; Pole, D.L.; Tidwell, T.T.; Warkentin, J. *J. Org. Chem.* **1996**, *61*, 9522–9527.

83. Smith, W.B. *J. Org. Chem.* **1995**, *60*, 7456–7460.

84. Kluger, R. *Chem. Rev.* **1987**, *87*, 863–876.

85. Scherowsky, G.; Dünnbier, K.; Höfle, G. *Tetrahedron Lett.* **1977**, 2095–2098.

86. Olofson, R.A.; Landesberg, J.M. *J. Am. Chem. Soc.* **1966**, *88*, 4263–4265.

87. Olofson, R.A.; Landesberg, J.M.; Houk, K.N.; Michelman, J.S. *J. Am. Chem. Soc.* **1966**, *88*, 4265–4266.

88. Scherowsky, G. *Chem. Ber.* **1974**, *107*, 1092–1101.

89. Leeper, F.J.; Smith, D.H.C.; Doyle, N.J.; Raithby, P.R. *J. Chem. Soc. Perkin Trans. 2* **1995**, 777–784.

90. Leeper, F.J.; Smith, D.H.C. *J. Chem. Soc. Chem. Commun.* **1990**, 961–963.

91. Leeper, F.J.; Smith, D.H.C. *J. Chem. Soc. Perkin Trans. 1* **1995**, 861–867.

92. Boche, G.; Hilf, C.; Harms, K.; Marsch, M.; Lohrenz, J.C.W. *Angew. Chem. Int. Ed. Engl.* **1995**, *34*, 487–489.

93. Nakai, T.; Okawara, M. *J. Chem. Soc. Chem. Comm.* **1970**, 907–908.

94. Er, H.-T.; Pole, D.L.; Warkentin, J. *Can. J. Chem.* **1996**, *74*, 1480–1489.

95. Hine, J.; Klueppel, A.W. *J. Am. Chem. Soc.* **1974**, *96*, 2924–2929.

96. Schank, H. *Houben-Weyl, Methoden der Organischen Chemie*; Regitz, M., Ed.; Thieme: Stuttgart, 1989; pp 1697–1705.

97. Er, H.-T.; Warkentin, J. 1995, unpublished.

98. Aitken, R.A. *Progress in Heterocyclic Chemistry*; Suschitzky, H.; Skriven, E.F.V. ,Eds.; Pergamon: Oxford, 1992; pp 136–149.

99. Yamashita, Y.; Tomura, M.; Tanaka, S. *J. Chem. Soc. Perkin Trans. 1* **1996**, 3358.

100. Chen, Y.-T.; Jordan, F. *J. Org. Chem.* **1991**, *56*, 5029–5038.

101. Varma, K.S.; Bury, A.; Harris, N.J.; Underhill, A.E. *Synthesis* **1987**, 837–838.

102. Nakayama, J. *Synthesis* **1975**, 168.

103. Plieninger, H.; Heuck, C.C.; Bühler, R. *Tetrahedron* **1972**, *28*, 73–81.

104. Yui, K.; Aso, Y.; Otsubo, T.; Ogura, F. *Chem. Lett.* **1986**, 781–784.

105. Tatemitsu, H.; Nishikawa, E.; Sakata, Y.; Misumi, S. *Synth. Met.* **1987**, *19*, 565–568.

106. Coffen, D.L.; Chambers, J.Q.; Williams, D.R.; Garrett, P.E.; Canfield, N.D. *J. Am. Chem. Soc.* **1971**, *93*, 2258–2268.

107. Nakayama, J. *Synthesis* **1975**, 38–39.

108. Nakayama, J. *J. Chem. Soc. Perkin Trans. 1* **1975**, 525–530.

109. Hartzler, H.D. *J. Am. Chem. Soc.* **1973**, *95*, 4379–4389.

110. Hoffmann, R.W. *Angew. Chem. Int. Ed. Engl.* **1979**, *18*, 563–640.

111. Nakai, T.; Mikami, K. *Chemistry Lett.* **1979**, *12*, 1081–1084.

112. Francisco, J.S. *J. Chem. Phys.* **1992**, *96*, 1167–1175.

113. Reddington, R.L.; Bock, C.W.; Aboab, B. *J. Mol. Struct. (Theochem)* **1990**, *224*, 89–112.

114. Moss, R.A.; Liu, W.; Ge, C.-S. *J. Phys. Org. Chem.* **1993**, *6*, 376–379.

115. El-Saidi, M.; Warkentin, J. 1996, unpublished.

116. Lowry, T.H.; Richardson, K.S. *Mechanism and Theory in Organic Chemistry*; Harper and Row: New York, 1987; pp 293–301, and references therein.

117. Ge, C.-S.; Jefferson, E.A.; Moss, R.A. *Tetrahedron Lett.* **1993**, *34*, 7549–7552.

118. Moss, R.A.; Wilk, B.K.; Hadel, L.M. *Tetrahedron Lett.* **1987**, *28*, 1969–1972.

119. Kassam, K.; Venneri, P.; Warkentin, J. *Can. J. Chem.* **1997**, *75*, 1256–1263.

120. Boger, D.L.; Brotherton, C.E.; Georg, G.I. *Tetrahedron Lett.* **1984**, *25*, 5615–5618.

121. Boger, D.L.; Brotherton, C.E. *J. Am. Chem. Soc.* **1984**, *106*, 805–807.

122. Boger, D.L.; Brotherton, C.E. *J. Org. Chem.* **1985**, *50*, 3425–3427.

123. Boger, D.L.; Brotherton, C.E. *J. Am. Chem. Soc.* **1986**, *108*, 6713–6719.

124. Boger, D.L.; Brotherton, C.E. *J. Am. Chem. Soc.* **1986**, *108*, 6695–6708.

125. Boger, D.L.; Brotherton, C.E.; Georg, G.I. *Org. Synth.* **1987**, *65*, 32–40.

126. Boger, D.L.; Wysocki, J.J. *J. Org. Chem.* **1988**, *53*, 3408–3421.

127. Boger, D.L.; Brotherton-Pleiss, C.E. *Advances in Cycloaddition*; Curran, D.P., Ed.; JAI Press: Stamford, CT, 1990; pp 147–216.

128. Tokuyama, H.; Isaka, M.; Nakamura, E. *J. Am. Chem. Soc.* **1992**, *114*, 5523–5530.

129. Chambers, G.R.; Jones, M., Jr. *J. Am. Chem. Soc.* **1980**, *102*, 4516–4518.

130. Nakamura, E. *J. Synth. Org. Chem. Jpn.* **1994**, *52*, 935–945.

131. Sauers, R.R. *Tetrahedron Lett.* **1994**, *35*, 7213–7216.

132. Sauers, R.R. personal communication, 1996.

133. Baldwin, J.E.; Walker, J.A. *J. Am. Chem. Soc.* **1974**, *96*, 596–597.

134. (a) Jones, F.N.; Andreades, S. *J. Org. Chem.* **1969**, *34*, 3011–3014; (b) Hoffmann, R.W.; Barth, W. *J. Chem. Soc. Perkin Trans. 2* **1983**, 1687–1692.

135. Baldwin, J.E.; Walker, J.A. *J. Chem. Soc. Chem. Commun.* **1972**, 354–355.

136. Nakai, T.; Mikami, K. *Chem. Lett.* **1978**, *11*, 1243–1244.

137. Kuhn, N.; Bohnen, H.; Henkel, G. *Z. Naturforsch., B.* **1994**, *49*, 1473–1480.

138. Nakayama, J.; Akiyama, I. *J. Chem. Soc. Chem. Commun.* **1992**, 1522.

139. Skrypnik, Y.G.; Lyashchuk, S.N. *Sulfur Lett.* **1994**, *17*, 287–294.

140. Lyashchuk, S.N.; Skrypnik, Y.G. *Tetrahedron Lett.* **1994**, *35*, 5271–5274.

141. Arduengo, A.J., III; Dias, H.V.R.; Calabrese, J.C.; Davidson, F. *J. Am. Chem. Soc.* **1992**, *114*, 9724–9725.

142. Arduengo, A.J., III; Dias, H.V.R.; Calabrese, J.C.; Davidson, F. *Organometallics* **1993**, *12*, 3405–3409.

143. Herrmann, W.A.; Öfele, K.; Elison, M.; Kühn, F.E.; Roesky, P.W. *J. Organometallic Chem.* **1994**, *480*, C7–C9.

144. Kümmell, A.; Seitz, G. *Tetrahedron Lett.* **1991**, *32*, 2743–2746.

145. Gerninghaus, C.; Kümmell, A.; Seitz, G. *Chem. Ber.* **1993**, *126*, 733–738.

146. Couture, P.; El-Saidi, M.; Warkentin, J. *Can. J. Chem.* **1997**, *75*, 320–332.

147. (a) Pole, D.L.; Warkentin, J. 1996; unpublished; (b) Pole, D.L.; Couture, P.; Warkentin, J. *J. Chem. Soc. Perkin Trans. 2* **1997**, 1565–1570.

148. Hoffmann, R.W.; Lilienblum, W.; Dittrich, B. *Chem. Ber.* **1974**, *107*, 3395–3407.

149. Tokuyama, H.; Yamada, T.; Nakamura, E. *Synlett* **1993**, 589–591.

150. Couture, P.; Pole, D.L.; Warkentin, J. 1996, unpublished.

151. Couture, P.; Pole, D.L.; Warkentin, J. *J. Org. Chem.* **1997**, *62*, 4065–4067.

152. Frenzen, G.; Rischke, M.; Seitz, G. *Chem. Ber.* **1993**, *126*, 2317–2323.

153. Diederich, F.; Isaacs, L.; Philp, D. *Chem. Soc. Rev.* **1994**, 243–255.

154. Isaacs, L.; Diederich, F. *Helv. Chim. Acta* **1993**, *76*, 2454–2464.

155. Tokuyama, H.; Nakamura, E.; Nakamura, M. *Tetrahedron Lett.* **1993**, *34*, 7429–7432.

156. Ross, J.P.; Couture, P.; Warkentin, J. *Can. J. Chem.* **1997**, *75*, 1331–1335.

157. Kirmse, W. *Advances in Carbene Chemistry*; Brinker, U.H., Ed.; JAI Press: Stamford, CT, 1994; pp 1–57.
158. Balli, H.; Grüner, H.; Maul, R.; Schepp, H. *Helv. Chim. Acta* **1981**, *64*, 648–655.
159. Breslow, R.; Kool, E. *Tetrahedron Lett.* **1988**, *29*, 1635–1638.
160. Lopez-Calahorra, F.; Rubires, R. *Tetrahedron* **1995**, *51*, 9713–9728.
161. Lappert, M.F.; Maskell, R.K. *J. Chem. Soc. Chem. Commun.* **1982**, 580–581.
162. Marti, J.; Lopez-Calahorra, F.; Bofill, J.M. *J. Mol. Struct. (Theochem)* **1995**, *339*, 179–194.
163. Hoffmann, R.W.; Steinbach, K.; Dittrich, B. *Chem. Ber.* **1973**, *106*, 2174–2184.
164. Nakayama, J.; Sugiura, H.; Hoshino, M. *Tetrahedron Lett.* **1983**, *24*, 2585–2588.
165. Raubenheimer, H.G.; Swanepoel, H.E. *J. Organometallic Chem.* **1977**, *141*, C21–C22.

INDEX